NATO ASI Series

Advanced Science Institutes Series

A series presenting the results of activities sponsored by the NATO Science Committee, which aims at the dissemination of advanced scientific and technological knowledge, with a view to strengthening links between scientific communities.

The Series is published by an international board of publishers in conjunction with the NATO Scientific Affairs Division

A Life Sciences B Physics	Plenum Publishing Corporation London and New York
C Mathematical and Physical Sciences D Behavioural and Social Sciences E Applied Sciences	Kluwer Academic Publishers Dordrecht, Boston and London
F Computer and Systems Sciences G Ecological Sciences H Cell Biology	Springer-Verlag Berlin Heidelberg New York London Paris Tokyo

Series H: Cell Biology Vol. 26

The ASI Series Books Published as a Result of Activities of the Special Programme on CELL TO CELL SIGNALS IN PLANTS AND ANIMALS

This book contains the proceedings of a NATO Advanced Research Workshop held within the activities of the NATO Special Programme on Cell to Cell Signals in Plants and Animals, running from 1984 to 1989 under the auspices of the NATO Science Committee.

The books published as a result of the activities of the Special Programme are:

Vol. 1: **Biology and Molecular Biology of Plant-Pathogen Interactions.** Edited by J. A. Bailey. 1986.

Vol. 2: **Glial-Neuronal Communication in Development and Regeneration.** Edited by H. H. Althaus and W. Seifert. 1987.

Vol. 3: **Nicotinic Acetylcholine Receptor: Structure and Function.** Edited by A. Maelicke. 1986.

Vol. 4: **Recognition in Microbe-Plant Symbiotic and Pathogenic Interactions.** Edited by B. Lugtenberg. 1986.

Vol. 5: **Mesenchymal-Epithelial Interactions in Neural Development.** Edited by J. R. Wolff, J. Sievers, and M. Berry. 1987.

Vol. 6: **Molecular Mechanisms of Desensitization to Signal Molecules.** Edited by T. M. Konijn, P. J. M. Van Haastert, H. Van der Starre, H. Van der Wel, and M. D. Houslay. 1987.

Vol. 7: **Gangliosides and Modulation of Neuronal Functions.** Edited by H. Rahmann. 1987.

Vol. 9: **Modification of Cell to Cell Signals During Normal and Pathological Aging.** Edited by S. Govoni and F. Battaini. 1987.

Vol. 10: **Plant Hormone Receptors.** Edited by D. Klämbt. 1987.

Vol. 11: **Host-Parasite Cellular and Molecular Interactions in Protozoal Infections.** Edited by K.-P. Chang and D. Snary. 1987.

Vol. 12: **The Cell Surface in Signal Transduction.** Edited by E. Wagner, H. Greppin, and B. Millet. 1987.

Vol. 19: **Modulation of Synaptic Transmission and Plasticity in Nervous Systems.** Edited by G. Hertting and H.-C. Spatz. 1988.

Vol. 20: **Amino Acid Availability and Brain Function in Health and Disease.** Edited by G. Huether. 1988.

Vol. 21: **Cellular and Molecular Basis of Synaptic Transmission.** Edited by H. Zimmermann. 1988.

Vol. 23: **The Semiotics of Cellular Communication in the Immune System.** Edited by E. E. Sercarz, F. Celada, N. A. Mitchison, and T. Tada. 1988.

Vol. 24: **Bacteria, Complement and the Phagocytic Cell.** Edited by F. C. Cabello and C. Pruzzo. 1988.

Vol. 25: **Nicotinic Acetylcholine Receptors in the Nervous System.** Edited by F. Clementi, C. Gotti, and E. Sher. 1988.

Vol. 26: **Cell to Cell Signals in Mammalian Development.** Edited by S. W. de Laat, J. G. Bluemink, and C. L. Mummery. 1989.

Cell to Cell Signals in Mammalian Development

Edited by

S.W. de Laat
J.G. Bluemink
C.L. Mummery

Hubrecht Laboratory
The Netherlands Institute for Developmental Biology
Uppsalalaan 8, 3584 CT Utrecht, The Netherlands

Springer-Verlag
Berlin Heidelberg New York London Paris Tokyo
Published in cooperation with NATO Scientific Affairs Division

Proceedings of the NATO Advanced Research Workshop on Cell-to-Cell Signalling in Mammalian Development held in Les Arcs, France, from February 21–26, 1988

Library of Congress Cataloging-in-Publication Data. NATO Advanced Research Workshop on Cell to Cell Signals in Mammalian Development (1988 : Les Arcs, Savoie, France) Cell to Cell signals in mammalian development / edited by S. W. de Laat, J. G. Bluemink, C. L. Mummery. p. cm.—(NATO ASI series. Series H, Cell biology ; vol. 26) "Proceedings of the NATO Advanced Research Workshop on Cell to Cell Signals in Mammalian Development held in Les Arcs, France, from February 21–26, 1988"—T.p. verso. "Published in cooperation with NATO Scientific Affairs Division."

ISBN-13: 978-3-642-73144-0 e-ISBN-13: 978-3-642-73142-6
DOI: 10.1007/978-3-642-73142-6

1. Mammals—Cytology—Congresses. 2. Mammals—Development—Congresses. 3. Cell interaction—Congresses. I. Laat, S. W. de (Siegfried W.), 1945- . II. Bluemink, J. G. (Johannes Gerardus), 1935- III. Mummery, C. L. (Christine L.), 1953- . IV. North Atlantic Treaty Organization. Scientific Affairs Division. V. Title. VI. Series. QL739.15.N37 1988 599'.0333—dc 19 88-36813

Softcover reprint of the hardcover 1st edition 1989

2131/3140-543210 – Printed on acid-free paper

Preface

Developmental biology is the area of the Life Sciences that deals with the regulatory principles underlying the generation of a new organism. Virtually all multicellular organisms originate from a single cell, the fertilized egg cell or zygote, through a complex and integrated process of cell multiplication, cell differentiation and organisation of cells into tissues and organs in the right positions and proportions within an embryo.

Until quite recently, our knowledge of the molecular and cellular mechanisms that guide mammalian development was minimal. The rapid advances of molecular biology and cell biology have radically changed this situation however. Major contributions have come from (i) the recognition that regulatory genes identified in invertebrate morphogenesis have their evolutionarily conserved homologues in vertebrates; these include certain proto-oncogenes and homeotic genes which can now be readily characterized; (ii) the identification and molecular characterization of genes and gene products involved in cell-to-cell signalling such as polypeptide growth factors and extracellular matrices which may in their turn control morphogenesis, cell proliferation and differentiation; and (iii) the use of cultured equivalents of early embryonic cells, such as embryo-derived stem cell lines, and the application of transgenic methodology.

The NATO Advanced Research Workshop on "Cell-to-Cell Signalling in Mammalian Development" held in Les Arcs, France, from February 21 to 26, 1988, provided an opportunity to review and discuss intensively the most recent advances in this area of research. The organizers had in mind to create a platform for discussion among colleagues who view the problems of mammalian development and, more specifically, the molecular nature and developmental significance of cell-to-cell signals, from different perspectives. Presentations and discussions focused on (i) hoxgenes, their products and their potential role in mammalian morphogenesis; (ii) growth factors, their receptors and signal transduction, and their role in guiding cell multiplication, differentation and transformation; and (iii) cell adhesion and extracellular matrix molecules, as short-range cell-to-cell signals in mammalian development.

This book presents the contents of the main lectures and a selection of relevant poster contributions presented during this workshop. It is intended to make the proceedings of this meeting accessible to a larger audience. We hope that it will stimulate further exchange of views and news in this fascinating area of research.

The organizers of this Workshop wish to acknowledge the NATO Scientific Affairs' Division for their generous support, and the International Society of Developmental Biologists, the European Developmental Biology Organisation and the Hubrecht Foundation for their contributions. We also wish to thank Ms. Sabina Rusch for her excellent secretarial assistance, and Drs. Peter Gruss, Joseph Schlessinger and Jean Paul Thierry for co-organizing the workshop.

September 1988

Siegfried W. de Laat
John J. Bluemink
Christine L. Mummery
Utrecht, The Netherlands

CONTENTS

I. HOMEODOMAIN CONTAINING GENES: STRUCTURE AND ROLE IN DEVELOPMENT

II. EXTRACELLULAR MATRICES AND CELL ADHESION MOLECULES

III. GROWTH FACTORS, RECEPTORS AND ONCOGENES: STRUCTURE AND FUNCTION

IV. GROWTH FACTORS, RECEPTORS AND ONCOGENES: ROLE IN DIFFERENTIATION AND DEVELOPMENT

The Murine Homeobox Protein Hox 1.1 Is Differentially Expressed in Embryonal Mesoderm and Ectoderm

Frank Schulze, Michael Kessel and Peter Gruss

Dept. Molecular Cell Biology
Max Planck Institute of Biophysical Chemistry
Am Fassberg
3400 Göttingen
West Germany

Introduction

Structural conservation is usually indicative of functional conservation. If this were also true of developmental control genes then the isolation of sequences homologous to developmental control genes of Drosophila could lead to the identification of genes with a related function in the mammalian genome. Indeed, based on this hypothesis, several candidates for mammalian developmental control genes have been identified. Undoubtedly most effort has gone into delineating the possible role of murine homeobox genes. At least 18 members of the Hox multigene family have been found, most of which are located on mouse chromosomes 6 and 11 (see Martin et al. 1987 for references and nomenclature). Another class of putative control gene is characterized by a common finger motif, a structure first suggested for the *Xenopus* transcription factor TFIIIA (Miller et al. 1985; for review see Klug and Rhodes 1987). Murine finger containing genes have been identified and individual members of this family have been shown to be specifically active in developing and adult neurons (Chowdhury et al. 1987; K. Chowdhury et al., submitted). Finally, a candidate for a third class of murine developmental control gene (Pax) bears a high degree of homology to *Drosophila paired box* (U. Deutsch et al., submitted). This conserved element is present in *Drosophila* segmentation genes such as *paired, gooseberry proximal* and *gooseberry distal* (Bopp et al. 1986; Côté et al. 1987).

The best studied example of conserved sequences with a possible role in murine developmental control is however the homeobox (see Gehring and Hiromi 1986 and Scott and Carroll 1987 for review). Most if not all of the mouse Hox genes are expressed during embryogenesis in a region specific manner (Dony and Gruss 1987; Gaunt 1987; Gaunt et al. 1986; Utset et al. 1987; Toth et al. 1987; Holland and Hogan 1988; Breier et al. 1988). These studies were conducted using ^{35}S-labelled RNA as a probe in *in situ* experiments and thus only in exceptional cases do they allow resolution down to the single cell level. Furthermore,

NATO ASI Series, Vol. H26
Cell to Cell Signals in Mammalian Development
Edited by S. W. de Laat et al.

as recently shown for *Drosophila caudal* and *hunchback* (Mlodzik and Gehring 1985; Tautz 1988), the level of RNA expressed does not always reflect the amount of protein translated. Therefore, immunohistochemistry using antibodies directed against a Hox protein should be employed in order to reach an unambigous answer.

To this end we studied the expression of Hox 1.1 in developing mouse embryos. The Hox 1.1 protein sequence, as deduced from a cDNA sequence, consists of 229 amino acids with a molecular weight of 25,740 Dalton (Kessel et al. 1987). Antibodies against synthetic peptides were shown to bind to a nuclear protein which is associated with chromatin during meiosis (Kessel et al. 1987; Schulze et al. 1988). We employed these antibodies in immunohistochemical analyses of mouse embryos at different stages and detected the antigen in the neural tube, spinal ganglia and also in sclerotomic structures of 12.5-day-old embryos. However, expression was restricted to sclerotomes of the thoracic region. Moreover, the ectodermal activity was not in register with the mesoderm expression and thus the Hox1.1 expression must be spatially and temporally tightly regulated.

Results

Experimental Strategy

The first experimental evidence demonstrating differential expression of murine homeobox RNA was provided for Hox 1.1 (m6) in differentiating F9 cells (Colberg-Poley et al. 1985a, b). Monoclonal antibodies directed against the Hox1.1 protein were subsequently used to show that, in the F9 cell system, the Hox 1.1 protein was expressed with similar kinetics (Kessel et al. 1987; Schulze et al. 1988). In order to decipher the possible function of murine homeobox genes during development the spatial and temporal expression pattern has to be established. Initially "Northern analyses" helped to determine the temporal expression profile (Colberg-Poley et al. 1985a, b). The spatial pattern, however, cannot be adequately determined with this method and *in situ* techniques are required in order to identify the respective signal at the single cell level. In particular, immunofluorescence analyses using the monoclonal antibodies available should allow a precise localization of the protein in the developing embryo.

Expression of the Hox 1.1 Protein in Mesoderm

Northern analyses revealed the highest amount of stable Hox 1.1 RNA in 12.5 day old embryos (Colberg-Poley et al. 1985a, b). Therefore, frozen parasagittal sections of similarly staged mouse embryos were prepared and, in order to localize the Hox 1.1 protein, monoclonal antibodies (Kessel et al. 1987; Schulze et al. 1988) were employed. As a control the sections were stained with DAPI, a dye labelling any nuclei present. A schematical representation of the data to be published (K. Mahon et al., submitted) is presented in Figure 1A. The section displayed shows somites and spinal ganglia in the cervical and thoracic regions of the embryo in black indicating the presence of the Hox 1.1 protein. These observations are in agreement with the RNA *in situ* analysis (K. Mahon et al., submitted). Interestingly sclerotomes represent the primordium for the axial skeleton. In this respect it is important to note that the Hox 1.1 gene is not expressed in all sclerotomes but only in the sclerotomes of the thoracic region. Thus, cranially located cervical sclerotomes and caudally located sclerotomes beginning with L1 do not express Hox 1.1. The data further reveal a nuclear location of the Hox 1.1 protein, a feature expected from tissue culture analyses (Kessel et al. 1987; Schulze et al. 1988) and work on *Drosophila* homeobox proteins (Carroll and Scott 1985; DiNardo et al. 1985).

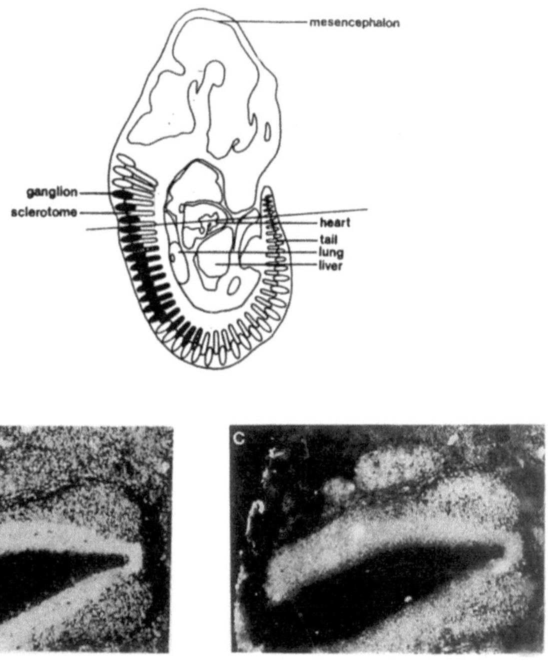

Figure 1A-C. Hox 1.1 expression in the 12-day-old mouse embryo. **A** Schematical representation of Hox 1.1 RNA and Hox 1.1 protein expression. The figure summarizes the data from RNA and protein analyses (K. Mahon et al., submitted). Hox 1.1 positive ganglia and sclerotomes are filled in in black. The bar indicates the site of the cross section shown in **B** and **C**. **B** Cross section through a 12-day mouse embryo. The nuclei are stained with DAPI. **C** Immunofluorescence analysis of the section shown in **B** using the monoclonal antibody p1.5. Note the nuclear staining of the neural tube and the spinal ganglion cells.

Expression of the Hox 1.1 Protein in Ectoderm

As demonstrated schematically in Fig. 1A spinal ganglia - ectodermal descendants of neural crest cells - express the Hox 1.1 protein. In Fig. 1C data are presented demonstrating the presence of Hox 1.1 protein in cells of the neural tube and in spinal ganglia. The area displayed represents a transverse section taken from the cervical region (see bar in Fig. 1A). DAPI stain was taken as a control in order to label every nucleus present on the section (Fig. 1B). Using indirect immunofluorescence the Hox 1.1 protein was identified with the help of the p1.5 antibodies (Schulze et al. 1988). In comparison with Fig. 1B only neural tube cells and cells of the spinal ganglion are specifically labelled. Again, not all spinal ganglia harbor the protein indiscriminately. Only a subset beginning at C4 (see Fig. 1A) and ending at L1 (K. Mahon et al., submitted) express both Hox 1.1 RNA and protein. Further analysis confirmed this and demonstrates in a parasagittal section that the neural tube expresses the protein (Fig.

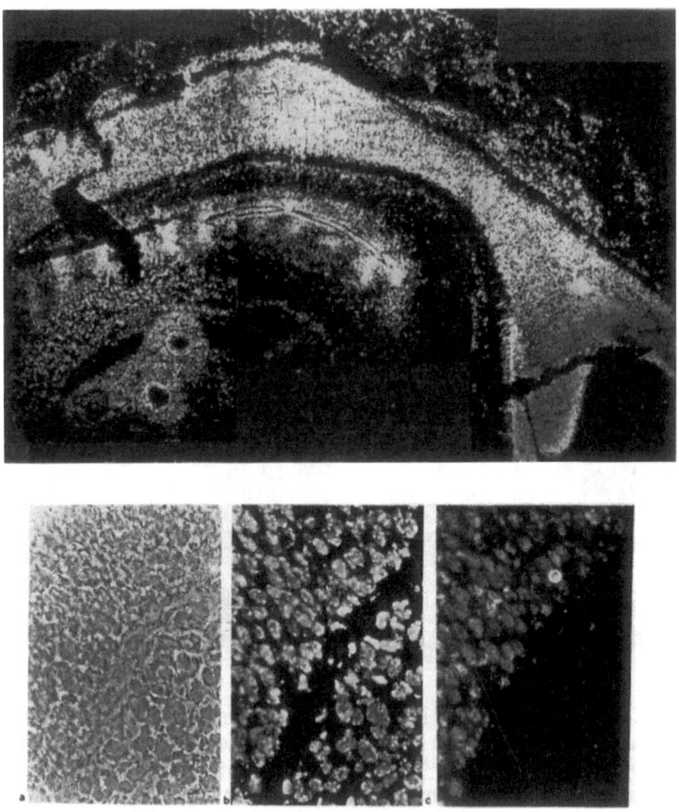

Figure 2a'-c'. Identification of the Hox 1.1 protein in neural tube cells at day 12 of development. The upper part of the figure shows an overview of a section in the region of the neural tube; all nuclei were visualized by DAPI staining. The lower part shows enlargements in phase contrast (**a'**), with DAPI stain (**b'**), and immunofluorescence (**c'**) employing the p1.5 monoclonal antibody. Note the positive region in the neural tube cells, while the paraxial mesenchyme shows the fluorescence.

2). In contrast to DAPI stained cells (Fig. 2b') only the subset representing neural tube cells reveals the Hox 1.1 protein (Fig. 2c'). Thus, ectoderm-derived cells of the neural tube and spinal ganglia express the Hox 1.1 protein.

Discussion

If the mouse homeobox genes play a role in the regulation of differentiation processes, one should expect a temporally and spatially differentiated expression of these genes in the murine embryo. Thus far, no data on the expression of homeobox proteins during murine embryogenesis exist. Only recently have initial data on the expression of Hox 1.3 RNA (Dony and Gruss 1987) and its protein (Odenwald et al. 1987) been published. The Hox 1.3 protein could be shown to be expressed in Purkinje neurons of the cerebellum, pyramidal and dental neurons of the hypocampus and motor neurons of the spinal cord. In this study we applied immunohistochemical techniques using monoclonal antibodies to embryos in order to localize the Hox 1.1 protein, a technique allowing detection of the proteins down to the one cell level even in heterogeneous tissues. Since the Hox 1.1 RNA is most strongly expressed in 12.5-day embryos, we analyzed embryos at this stage.

In 12.5-day mouse embryos 43-48 somites are found at different stages of maturation. At this point in development the original segmentation of the mouse embryo is beginning to disappear. The cells originating from different regions of the somite begin to proliferate and form the segmented axial mesoderm and are involved in the formation of different tissues such as vertebrae, ribs, and muscle (for review see Hogan et al. 1985). The Hox 1.1 protein was shown to be present in sclerotomes. Interestingly, the cervical and the first two thoracic sclerotomes show no fluorescence and likewise in the sclerotomes towards the posterior part of the mouse fluorescence becomes weaker and disappears altogether. These results were obtained in parasagittal, tangential and cross sections of the mouse and lead to the conclusion that expression of the protein is not dependent on the degree of maturation of the somites. Fluorescence could also be shown in embryonal ribs of 13-day-old embryos (data not shown), indicating that expression of the protein has been retained since the ribs develop directly from the somites.

Furthermore, the protein could be shown to be expressed in the neural tube. The paraxial mesenchyme surrounding the neural tube shows no immunofluorescence (Figs. 1, 2), demonstrating the sharply delimited expression of the protein in this area. A decrease in fluorescence towards the posterior part of the embryo can also be seen for the neural tube. The ganglia which derive from the neural crest also show strong fluorescence in the nucleus, while the first three cervical ganglia are negative for fluorescence. Towards the posterior part of the mouse the fluorescence decreases in the ganglia and disappears completely. The formation of the ganglia is controlled by the segmented paraxial mesoderm of the somites (Hogan et al. 1985). Thus, regions positive for immunofluorescence represent tissue induced by the mesoderm.

For further analysis 11-day-old embryos were dissected and examined by immunofluorescence (data not shown). Here also fluorescence could be detected in the neural tube and in the somites. These results establish that in mesodermal development the protein is expressed starting in the somites, through the sclerotomes and up to the precursors of the ribs. The expression of the protein is spatially and temporally restricted. Thus, in this respect the Hox 1.1 protein fulfills what is expected of a gene product regulating embryogenesis. The protein is mainly expressed in mesodermal tissue or in ectodermal tissue induced by mesodermal tissue. An important role is ascribed to the embryonal mesoderm in the regulation of pattern formation processes (primary and secondary induction; Carlson 1981). The results obtained in this study support the hypothesis that murine homeobox genes have a controlling function during mesoderm differentiation. Further studies on vertebrate homeobox genes will provide insight into the molecular mechanisms regulating structure and regional specification in vertebrate embryos. Both fields of experimental embryology and genetics follow the assumption that local specification occurs in mesodermal somites, the spinal cord, and in the neural crest. Moreover, there is evidence that the mesoderm influences specification in some regions of the neural ectoderm (Gurdon 1987).

Up to now the mechanisms of passing on and storing information and of communication between cells have remained unclarified. All the more interesting it is then that the homeobox containing genes are expressed in many different embryonal tissues which derive from the somitic mesoderm, lateral mesoderm, neural crest, and neural tube. Others have come to similar conclusions with *Drosophila* (Lawrence and Johnston 1984; Akam and Martinez-Arias 1985; Ghysen et al. 1985; Martinez-Arias 1986; Hooper 1986).

In this study we were able to show that the protein is not expressed evenly throughout the tissues but only in particular sections. This leads to the notion that the homeogene under consideration carries out a particular function in the development of defined regions along the anterior-posterior axis. If this assumption is correct and if one sums up all previous expression analyses, it is conceivable that each large region of the mouse embryo requires the expression of a particular set of homeotic genes for its specification. This theory is supported by the fact that localization of RNA expression of different homeobox genes demonstrates that the areas of expression overlap but are not identical. For further analysis it will be necessary to study the expression of other homeobox genes and so, piece by piece, gain a closer insight into the role of these genes and their regulation. Gene transfer experiments to put homeodomain proteins into a new surroundings or to specifically destroy endogenous homeobox genes (e.g. by homologous recombination) might finally lead to an understanding of their actual function.

References

Akam ME, Martinez-Arias MA (1985) The distribution of Ultrabithorax transcripts in Drosophila embryos. EMBO J 4:1689-1700

Bopp D, Barri M, Baumgartner S, Frigerio G, Noll M (1986) Conservation of a large protein domain in the segmentation gene paired and in functionally related genes of Drosophila. Cell 47:1033-1040

Breier G, Dressler GR, Gruss P (1988) Primary structure and developmental expression pattern of Hox 3.1, a member of the murine Hox 3 homeobox gene cluster. EMBO J, in press

Carlson BM (1981) Patten's foundations of embryology, 4th edn. McGraw-Hill, New York

Carroll SB, Scott MP (1985) Localization of the fushi tarazu protein during Drosophila embryogenesis. Cell 43:47-57

Chowdhury K, Deutsch U, Gruss P (1987) A multigene family encoding several "finger" structures is present and differentially active in mammalian genomes. Cell 48:771-778

Colberg-Poley AM, Voss SD, Chowdhury K, Stewart CL, Wagner E, Gruss P (1985a) Clustered homeo boxes are differentially expressed during murine development. Cell 43:39-45

Colberg-Poley AM, Voss SD, Chowdhury K, Gruss P (1985b) Structural analysis of mouse genes containing homeo box sequences and their expression in embryonal carcinoma cells. Nature 314:713-718

Côté S, Preiss A, Haller J, schuh R, Kienlin A, Seifert E, Jäckle H (1987) The gooseberry-zipper region of Drosophila: five genes encode different spatially restricted transcripts in the embryo. EMBO J 6:2793-2801

DiNardo S, Kuner JM, Theis J, O'Farrell PH (1985) Development of embryonic pattern in D. melanogaster as revealed by accumulation of the nuclear engrailed protein. Cell 43:59-69

Dony C, Gruss P (1987) Specific expression of the Hox 1.3 homeobox gene in murine embryonic structures originating from or induced by the mesoderm. EMBO J 6:2965-2975

Gaunt SJ (1987) Homeobox gene Hox 1.5 expression in mouse embryos: earliest detection by in situ hybridization is during gastrulation. Development 101:51-60

Gaunt SJ, Miller JR, Powell DJ, Duboule D (1986) Homeobox gene expression in mouse embryos varies with position by the primitive streak stage. Nature 324:662-664

Gehring WJ, Hiromi Y (1986) Homeotic genes and the homeobox. Ann Rev Genet 20:147-173

Ghysen A, Jan LY, Jan YN (1985) Segmental determination in Drosophila central nervous system. Cell 40:943-948

Gurdon JB (1987) Embryonic induction - molecular prospects. Development 99:285-306

Hogan B, Holland P, Schofield P (1985) How is the mouse segmented? Trends Genet 1:67-74

Holland PWH, Hogan BLM (1988) Spatially restricted patterns of expression of the homeobox containing gene Hox 2.1 during mouse embryogenesis. Development, in press

Hooper J (1986) Homeotic gene function in the muscles of Drosophila larvae. EMBO J 5:2321-2329

Kessel M, Schulze F, Fibi M, Gruss P (1987) Primary structure and nuclear localization of a murine homeodomain protein. Proc Natl Acad Sci USA 84:5306-5310

Klug A, Rhodes D (1987) "Zinc fingers": a novel protein motif for nucleic acid recognition. Trends Biochem Sci 12:464-469

Lawrence PA, Johnston P (1984) Role of the engrailed gene in the internal organs of Drosophila. EMBO J 3:2839-2844

Martin GR, Boncinelli E, Duboule D, Gruss P, Jackson IJ, Krumlauf R, Lonai P, McGinnis W, Ruddle F, Wolgemuth D (1987) Nomenclature for homeobox containing genes. Nature 325:21-22

Martinez-Arias A (1986) The Antennapedia gene is required and expressed in parasegments 4 and 5 of the Drosophila embryo. EMBO J 5:135-141

Miller J, McLachlan AD, Klug A (1985) Repetitive zinc-binding domains in the protein transcription factor IIIa from Xenopus oocytes. EMBO J 4:1609-1614

Mlodzik M, Fjose A, Gehring WJ (1985) Isolation of caudal, a Drosophila homeobox containing gene with maternal expression, whose transcripts form a concentration gradient at the pre-blastoderm stage. EMBO J 4:2961-2969

Odenwald WF, Taylor CF, Palmer-Hill FJ, Friedrich V, Tani M, Lazzarini RA (1987) Expression of a homeodomain protein in non contact inhibited cultured cells and postmitotic neurons. Genes Devel 1:482-496

Schulze F, Chowdhury K, Zimmer A, Drescher U, Gruss P (1988) The murine homeobox gene product, Hox 1.1 protein, is growth controlled and associated with chromatin. Differentiation, in press

Scott MP, Carroll SB (1987) The segmentation and homeotic gene network in early Drosophila development. Cell 51:689-698

Tautz D (1988) Regulation of the Drosophila segmentation gene hunchback by two maternal morphogenetic centers. Nature, in press

Toth LE, Slawin KL, Pintar JE, Nguyen-Huu CM (1987) Region-specific expression of mouse homeobox genes in the embryonic mesoderm and central nervous system. Proc Natl Acad Sci USA 84:6790-6794

Utset MF, Awgulewitsch A, Ruddle FH, McGinnis W (1987) Region-specific expression of two mouse homeobox genes. Science 235:1379-1382

STRUCTURE, EXPRESSION AND EVOLUTIONARY RELATIONSHIPS OF MURINE HOMEBOX GENES IN THE HOX 2 CLUSTER

N.Papalopulu, A.Graham, J.Lorimer, R.Kenny, J.McVey
and R.Krumlauf

National Institute for Medical Research
The Ridgeway, Mill Hill
London, NW7 1AA
United Kingdom

Abstract

We have characterized the murine homeobox locus, Hox 2, localized on chromosome 11. This gene complex contains at least seven homeoboxes and represents one of the two major mouse clusters, the other being Hox 1. Analysis of two members in the locus, Hox 2.1 and Hox 2.6, shows that the genes are differentially regulated in embryonal carcinoma cells and have both temporally and spatially restricted patterns of expression in mouse embryos. Analysis of RNA isolated from chick embryos reveals that Hox 2.1 related transcripts are expressed in a pattern similar to mouse embryos, suggesting that regulatory regions may be conserved for this gene in the two species. The predicted protein sequences for these Hox 2 genes contain domains, besides the homeodomain, with substantial homology to other vertebrate and *Drosophila* homeobox proteins. The data suggest that there has been strong selective pressure to conserve other domains, which may functionally interact with or compliment the homeodomain. Comparison of Hox 2.1 and Hox 2.6 to other mouse homeobox genes reveals close sequence identity to specific genes in the Hox 1 cluster, suggesting that these two loci arose by duplication and divergence of an ancestral cluster. Hox 2.6 itself was found to be related to the human gene C13 and the *Xenopus* Xhol A gene. This group of genes is also related to the *Drosophila Dfd* homeotic gene, and appears to represent a '*Dfd*' class of vertebrate homeobox gene that arose before the divergence between arthropods and vertebrates. The large degree of identity in several domains of mouse homeobox proteins makes it difficult to identify a role for any particular member without a gene-specific probe. To circumvent this problem we have inserted an additional domain into the Hox 2.1 gene that will generate a modified form of the protein. When tissue-culture cells are transfected with the altered Hox 2.1 constructs, an antibody raised against the marker epitope detects the modified Hox 2.1 protein and shows that it is located in the nucleus.

NATO ASI Series, Vol. H26
Cell to Cell Signals in Mammalian Development
Edited by S. W. de Laat et al.

Introduction

Several genes associated with the control of positional information in the Drosophila embryo, contain a conserved region that encodes a 61 amino acid domain, termed the homeobox (McGinnis et al.1984a). The presence, in a broad range of vertebrates, of sequences related to the homeobox (McGinnis et al, 1984b), has allowed the isolation of a new family of genes from many species. In the mouse, 18-20 genes have been identified so far, which contain sequences homologous to the *Drosophila* Antennapedia *(Antp)* class of homeoboxes (Colberg-Poley et al, 1985a&b; Hart et al, 1985; Hauser et al, 1985; Dubuole et al, 1986; Krumlauf et al, 1987; Odenwald et al, 1987). The murine homeobox containing proteins have not been found to display significant homology to the *Drosophila* proteins in regions other than the homeodomain, with the exception of the en-1 and en-2 genes related to the *Drosophila* Engrailed gene (Joyner et al, 1985). This has lead to speculation that the mammalian genes may represent a family which have "recruited" the homeodomain, possibly utilizing its DNA binding potential, and evolved a different role in vertebrates from that performed by the *Drosophila* analogs. The suggestion that the mammalian homeobox genes may play a role in development is derived from their patterns of gene expression. The majority of mouse and human genes examined reveal complex temporally and spatially restricted patterns of expression during embryogenesis (Colberg-Poley et al, 1985a; Awgulewitsch et al, 1986; Wolgemuth et al, 1986; Krumlauf et al, 1987; Dony and Gruss, 1987; Holland and Hogan, 1988). This expression often continues through later stages of organogenesis and fetal development, into adult life. In the embryonic central nervous system, many murine genes have similar overlapping, but not identical, regional localizations of their transcripts, which do not correlate with morphological differences (Awgulewitsch et al, 1986; Krumlauf et al, 1987; Dony and Gruss, 1987; Holland and Hogan, 1988). This feature is shared with the *Drosophila* homeobox genes (Harding et al, 1985), and it may be that these overlapping patterns reflect positional information. The complex tissue-specific expression of these genes in fetal and adult stages may be consistent with their distribution in embryogenesis, and a common functional role. However, there is no clear aspect of the patterns which can be used to relate these features, implying that the murine genes may play multiple roles during development. In the absence of direct genetic evidence concerning the functional nature of individual mammalian homoebox genes, it is important to examine common features of structure and expression shared between the mammalian and other vertebrate genes. In this way, it may be possible to focus on conserved aspects of vertebrate development that are related to the role of homeobox proteins. The results presented in this report describe the structural and evolutionary relationships of genes from the murine Hox 2 cluster with other vertebrate and *Drosophila* homeobox genes.

Results

The Hox 2 locus- Hox 2.6 and Hox 2.1 structure

Homeobox containing genes in the mouse genome have been found in some cases to be tightly clustered (Hart et al, 1985; Hauser et al, 1985; Dubuole et al, 1986). By comparison *Drosophila* homeobox, genes while also clustered, are in general much larger in size and are spread over greater chromosomal distances (Kaufman et al, 1980). We have isolated overlapping cosmid clones which define one of the major clusters of homeobox genes in the mouse, Hox 2. The Hox 2 cluster, as represented in our cosmid clones contains seven homeoboxes in an 80 kb region, located on chromosome 11 (Fig 1). This includes the two new members identified in this study, termed Hox 2.6 and Hox 2.7, in addition to the five previously observed members (Hart et al, 1987). At present Hox 2.5 and Hox 2.7 represent the most 5′ and 3′terminal members, respectively, of the cluster. Isolation of additional flanking sequences will be required to determine if there are more distal members in the locus. Sequence analysis indicates that all seven genes have the same orientation with respect to the direction of transcription, as has been observed for all linked mouse homeobox genes identified.

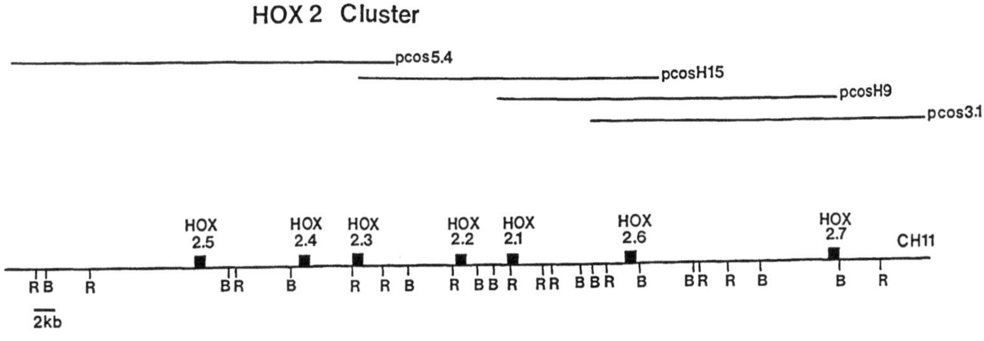

Figure 1. Structure of the Hox 2 cluster. Seven homeoboxes (dark squares) detected by hybridization and sequencing cosmid clones spanning the complex, are positioned in the diagram. Cosmid clones are shown above the map, and restriction sites are: EcoRI (R) and BamHI (B).

In order to examine the gene structure and predict potential protein sequences, we have isolated cDNA clones for two genes, Hox 2.1 (Krumlauf et al, 1987) and Hox 2.6. The clones were selected from a cDNA library prepared from RNA extracted from 8 1/2 d mouse embryos Farner et al, 1986). The sequencing of cDNAs from both genes revealed that they have long open reading frames which contain homeodomains located near the carboxy terminus, and a conserved hexapeptide region common to most homeobox proteins. Hox 2.1 is predicted to be a protein of 269 amino acids (AA), with 193 AA in the N-terminal region preceding the homeodomain and 15 AA on the carboxy terminal side. Similarly, Hox 2.6 is comprised of 250 AA, with 160 AA preceding the homeodomain and 29 AA between the domain and the carboxy terminus. Analysis of the Hox 2.1 and Hox 2.6 genomic and cDNA sequences shows several basic features held in common with other mouse and vertebrate homeobox genes. The genes have two exons, and the protein coding information is located in the 5'portion of the mRNAs generating long 3'untranslated regions (~1kb). The hexapeptide is located close to the homeodomain on the N-terminal side, but is encoded on a separate exon from the homeodomain, with the splice site located 15-40 bp upstream of the homeobox. The predicted proteins are very rich in proline and serine, and due to the distribution of prolines through out the non-homeo domain should generate a non-helical structure.

Comparison of Hox 2 with other murine genes

The predicted Hox 2.1 and Hox 2.6 proteins have sequence identity in the homeo domain (88%,54/61 AA), but no other significant identity was observed outside this region. In a homology search of the homeobox sequences available from other murine genes, the homeobox of Hox 2.1 was found to be most similar to Hox 1.3 (100%) and Hox 2.6 to Hox 1.4 (93%,57/61 AA). These results are interesting because Hox 1.3 and Hox 1.4 are two members of another major homeobox cluster located on chromosome 6 (Dubuole et al 1986). Based on the number of genes in these two clusters, their intergenic distances, direction of transcription and patterns of expression, it appears that they could be related via duplication and divergence of an ancestral cluster. Comparison of the entire Hox 2.1 and Hox 1.3 (Odenwald et al, 1987) proteins by dot-matrix alignment (Fig 2A; Garbern, Odenwald and Krumlauf, unpublished) reveals that several regions of these proteins are highly conserved. The identity in the region of the homeo domain extends 15-20 AA upstream and downstream to include the hexapeptide domain and the carboxy terminus. Overall this extended C-terminal region shows 90% homology (86/95 AA) between Hox 2.1 and Hox 1.3. Another significant block of homology includes the N-terminus of the proteins,where 23 Of the first 25 AA and a total of 44 of 58 AA are identical. Many of the differences in this region represent conservative amino acid changes. The remaining central region of the proteins displays no obvious homology at the nucleic

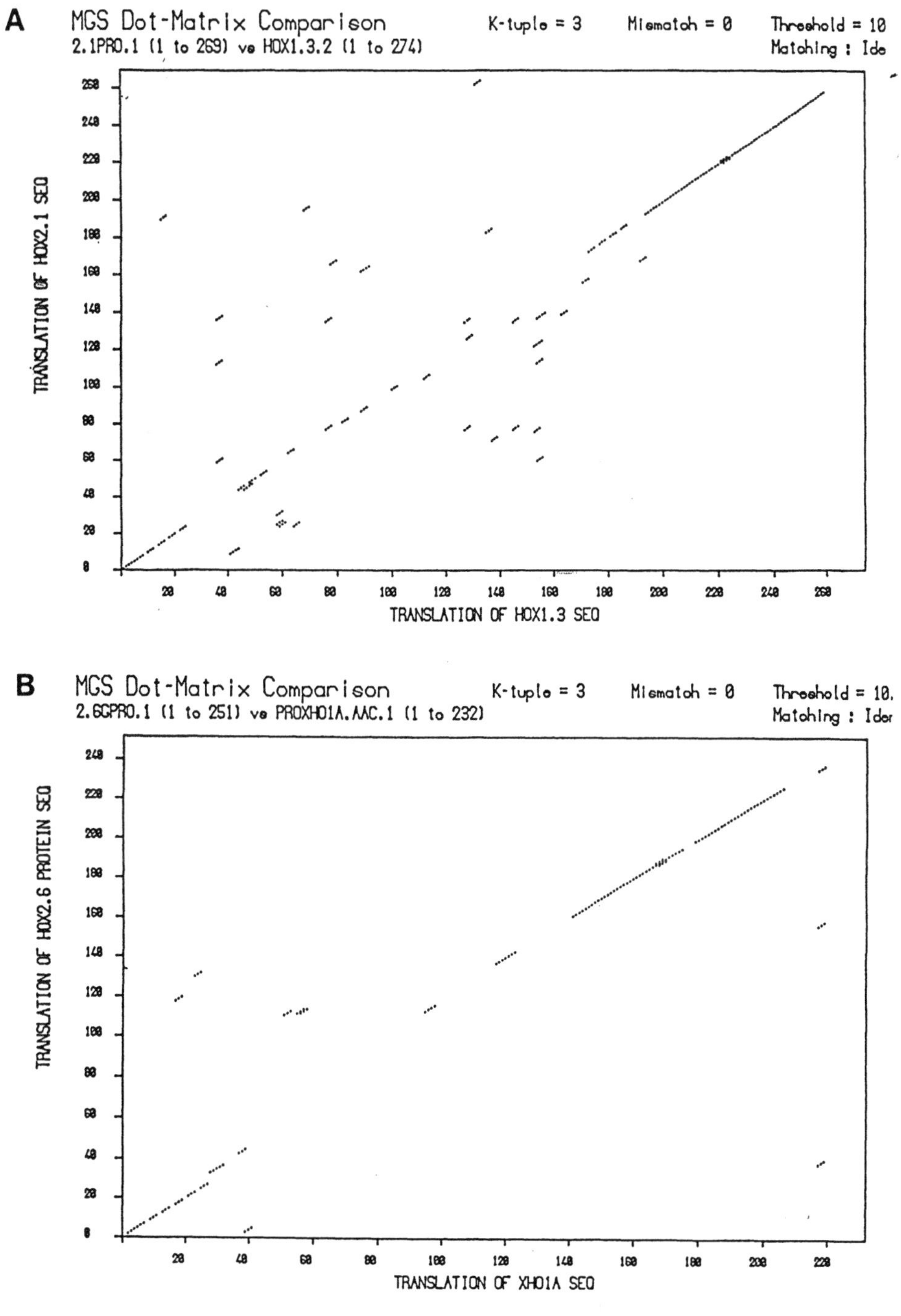

Figure 2. Dot-matrix protein sequence comparisons. (A) Comparison of the mouse Hox 2.1 and Hox 1.3 proteins. (B) Comparison of the mouse Hox 2.6 and *Xenopus* Xho1A proteins.

acid or protein level, except that they are rich in serine (25%, 28/112 AA). The pattern of a conserved N-terminal and extended homeodomain separated by a divergent central domain is shared between Hox 1.1 (Kessel et al, 1987) and Hox 2.3 (Meijlink et al 1987). Hox 2.6 is also related to Hox 1.4, based on the partial Hox 1.4 sequence available (Dubuole et al, 1986). Taken together these results support the idea that the Hox 1 and Hox 2 clusters are indeed related by a duplication event. There appears to have been strong selective pressure in duplicated members to conserve domains in addition to the homeodomain, most notably the N-terminal region. It may therefore be possible to classify different groups of non-linked homeobox genes based on conserved sequences in the N-terminal region.

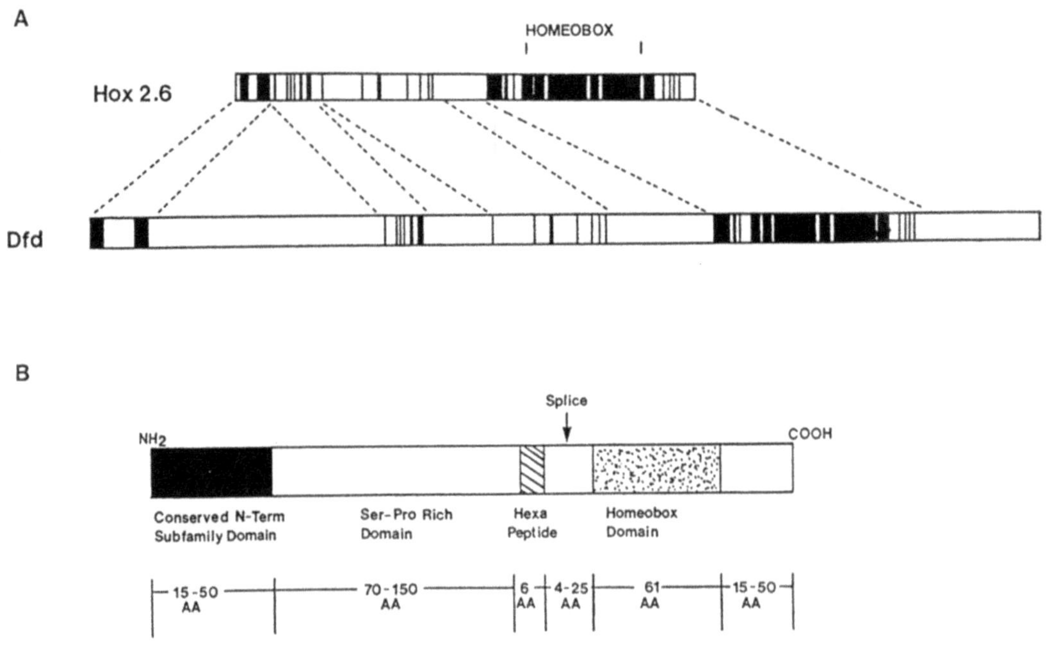

Figure 3. Homology between Hox 2.6 and *Drosophila Dfd*. (A) Comparison of the mouse Hox 2.6 protein with *Dfd*, where dark shaded areas represent amino acids that are identical in the two proteins. The dashed lines relate the areas where the homology occurs in the proteins. (B) Generalized homeobox protein structure based on comparison of several vertebrate genes.

Hox 2.6 is related to the Drosophila Dfd gene

Sequence comparison of Hox 2.6 with other vertebrate homeobox genes revealed extensive homology with the human c13 (Mavilio et al, 1986) and *Xenopus* XhoIA (Harvey et al, 1986) genes. Figure 2B shows a dot-matrix alignment between Hox 2.6 and XhoIA. The results clearly show a conserved N-terminal domain, with 26 of the first 30 AA identical, and an extended homeo domain homology. These same regions are also conserved between c13 and Hox 2.6. It is striking that both the regions of homology and the degree of conservation shared between a mouse (Hox 2.6) and *Xenopus* (XhoIA) homeobox gene, are the same as those seen in two mouse genes (Hox 2.1 and Hox 1.3). It has been noted (Regulski et al, 1987), that the *Drosophila Dfd* homeotic gene was related to the c13 and XhoIA genes. We have compared *Dfd* with Hox 2.6 (Fig 3A), and observed that regions of homology exist. The N-terminal domain of Hox 2.6, shared with c13 and XhoIA, is also present in *Dfd*, although not in a continuous run of amino acids. Thus Hox 2.6 represents another member of a vertebrate "*Dfd*" class of homeobox proteins. In view of the conservation detailed above between mouse genes, and between mouse genes and other vertebrate genes, a general structure for many of the vertebrate homeobox proteins is diagrammed in Fig. 3B. The role played by the N-terminal region, and the diverged central domain in the specificity of each protein is an important point which must be addressed in the future.

Marking Hox 2.1 Protein

Since several regions of mouse homeobox proteins appear to be conserved in other mouse proteins, it is difficult to generate a gene-specific probe. Peptide antibodies raised against the diverged central domain present one solution. However this region is often rich in serine and proline, which can result in antibodies that recognize these rich regions in other proteins. To circumvent this problem we have inserted into the central domain of Hox 2.1, a sequence coding for a peptide. A monoclonal antibody raised against this epitope, which does not recognize other proteins. This provides a means of detecting the modified protein when the gene is introduced into tissue-culture cells or transgenic mice. Figure 4 demonstrates a transfection of the modified Hox 2.1 gene into Cos cells. Hox 2.1 protein is observed in the nuclei of some cells in the population, presumably the transfected cells (Fig. 4A). The number of positive cells is consistent with the transfection efficiency. A control DAPI staining (Fig. 4B) shows negative cells in the same sample, suggesting that the method is specific for the transfected cells. This indicates that the peptide epitope is accessible in the protein, and will be useful in marking and possibly purifying the protein in gene transfer experiments.

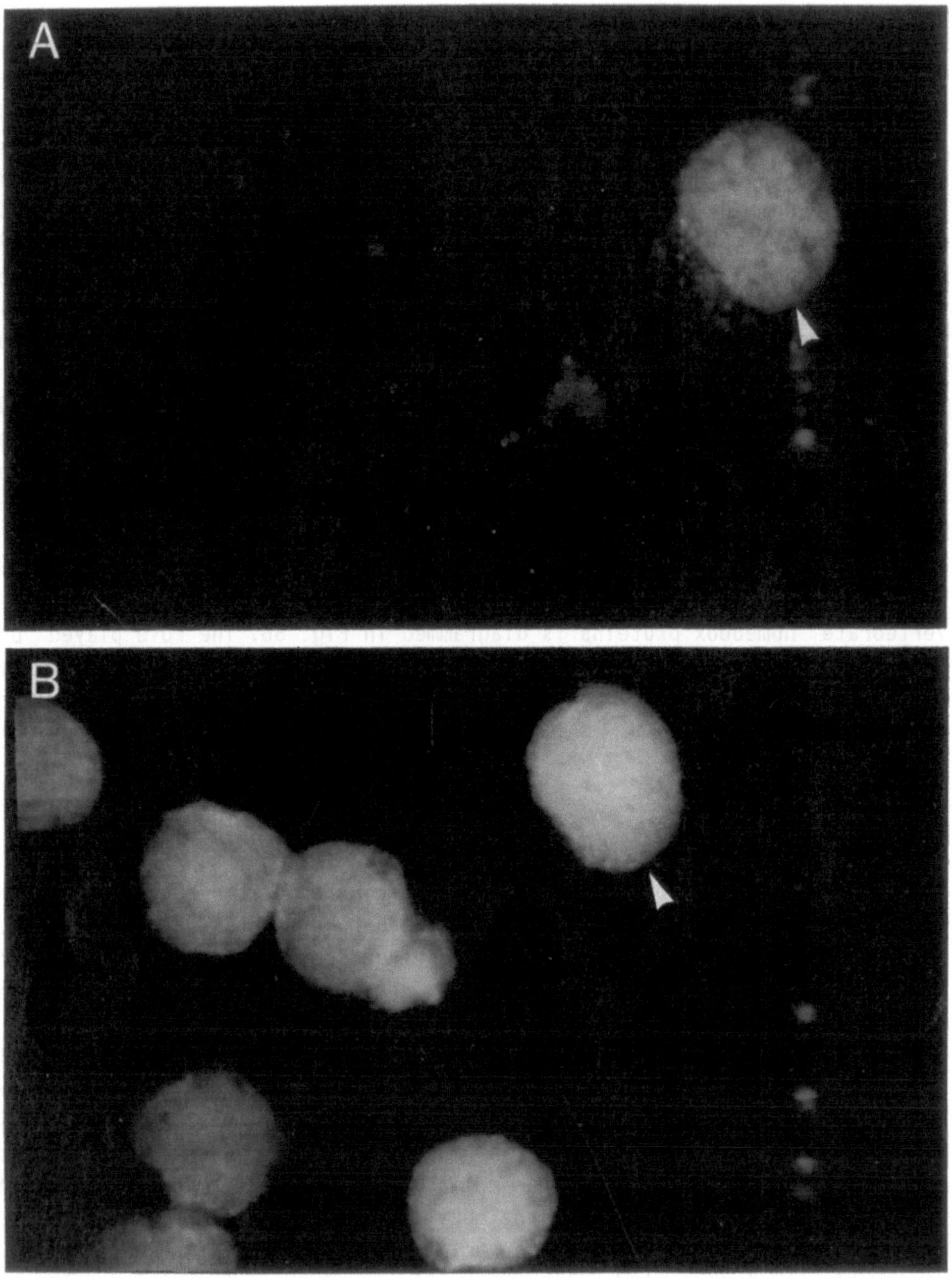

Figure 4. Marked Hox 2.1 protein is localized in the nucleus. (A) Cos cells transfected with a modified gene and stained with specific first antibody and FITC labelled second antibody. (B) Control DAPI staining of the same cells to show location of FITC negative (non-transfected) cells.

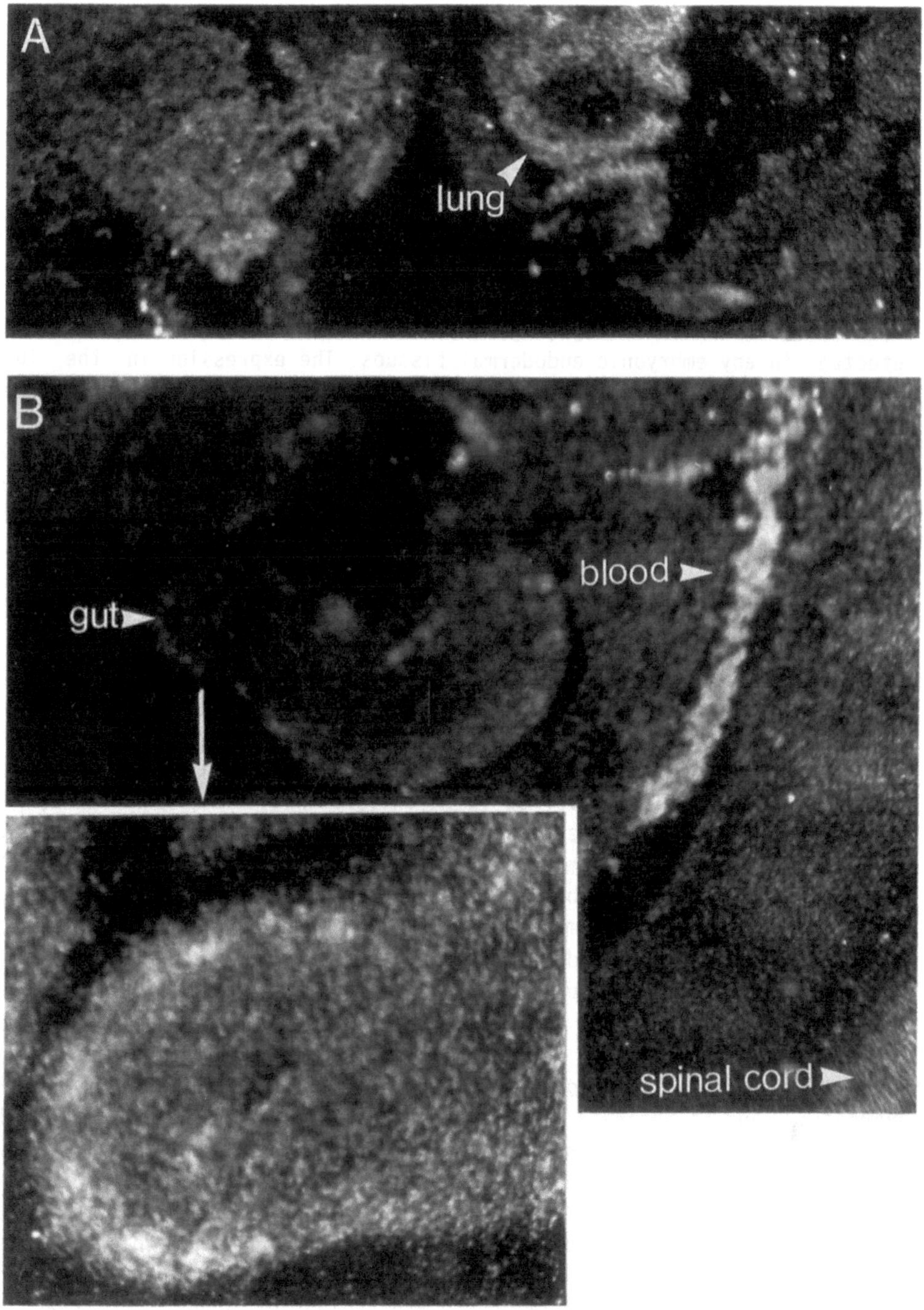

Figure 5. *In situ* hybridization of Hox 2.1 in lung and gut of 12 1/2d mouse embryos. (A) Illustrates expression in mesodermal mesenchyme and not endodermal epithelia of the lung. (B) Expression in gut is located in neural crest derived myenteric plexis.

Conserved expression of Hox 2.1

Analysis of Hox 2.1 expression by *in situ* hybridization and northern blotting has shown a complex pattern in the mouse embryo (Krumlauf et al, 1987; Holland and Hogan, 1988). Expression is seen in the ectodermal derivatives: spinal cord, hind brain and ganglia of the peripheral nervous system (Fig. 5B). There are spatial restrictions to this pattern, in that anterior regions such as mid brain and fore brain do not express these genes, while starting from the hind brain posterior regions display high levels of expression. Expression is also observed (Fig. 5) in some mesodermal derivatives: lung (Fig. 5A),stomach and kidney; but has not been detected in any embryonic endodermal tissues. The expression in the lung illustrates that it is the mesodermal mesenchyme and not the endodermal epithelia that make Hox 2.1 (Fig. 5A). Expression in the gut (Fig. 5B) is punctate and located in the neural crest derived myenteric plexis. No obvious features can be identified in the mesodermal patterns of expression of Hox 2.1, which can be explained by positional expression in the ectoderm.

To ask if these patterns of expression are unique to the mouse, we have used a Hox 2.1 probe to examine expression in chick embryos. This probe does not contain the homeobox, but spans the conserved N-terminal domain. RNA extracted from both chick and mouse tissues were analyzed by northern blotting (Fig. 6). High stringency washing conditions were utilized, and the results indicate that Hox 2.1 has related transcripts in the chick embryo. Transcripts are observed in chick carcasses, which contain spinal cord and in lung. The mouse tissues display a typical pattern of Hox 2.1 expression (Fig. 6). A number of transcripts are detected in the chick samples, similar to the number observed in mouse tissues (Fig. 6; Krumlauf et al, 1987), suggesting complex transcription occurs in the chick homeobox genes. These results in addition to preliminary *in situ* hybridization data indicate that the patterns of expression of Hox 2.1 related transcripts in chick are nearly identical to those in mouse embryos. Expression is observed in the chick spinal cord, and the level does not extend to regions anterior to the hind brain. The dorsal root ganglia expresses, and as observed for the mouse Hox 2.1 (Fig 5B), expression in the gut is punctate. This suggests that regulatory elements may also have been conserved in related genes between these species, and that the complex pattern of Hox 2.1 expression in ectodermal and mesodermal derivatives is not unique to the mouse. It will be necessary to clone the chick related genes to directly compare these sequences. However, in comparing sequences between Hox 2.6, c13 and Xho1A, there are regions of identity upstream of the protein coding region that extend 5′ of the start site of the major transcripts. These might represent conserved control regions.

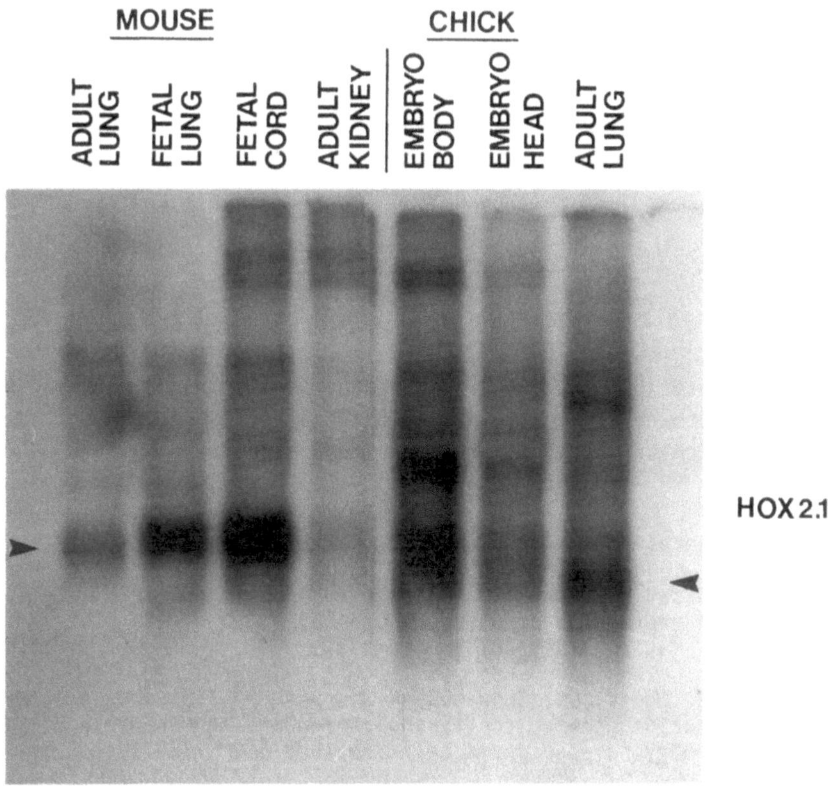

Figure 6. Expression of Hox 2.1 in mouse and chick embryos. Northern blot analysis of poly A+ RNA (5ug/lane) from various embryonic and adult tissues, probed with mouse Hox2.1 single-stranded RNA. Final washing conditions were RNAse A 2ug/ml in 2xssc for 30 min. then .1xssc, .1% SDS at 50 C. Arrows indicate major transcripts.

Conclusion

We have presented results characterizing the murine Hox 2 locus and some of its members. This locus appears to be related to the Hox 1 locus via duplication and many of the members of these clusters still display a high degree of sequence identity in specific regions of their respective proteins. It is not yet known whether there are other clusters related to these two clusters, and they therefore may form part of a much larger family of duplicated gene clusters. The similarity between different mouse proteins raises the question of whether they are functionally related or can compliment each other. The homology observed between different mouse genes also extends to related genes in other species, and suggests that strong selective pressure has been exerted to maintain these conserved domains. They therefore may play an important role in the functional specificity of the proteins possibly via interaction with the homeo domain. The conservation in the patterns of expression of Hox 2.1 between chick and mouse embryos implies that not only the proteins, but their regulatory regions are highly conserved. It will be interesting in the future to see if other classes of vertebrate genes are related to Drosophila genes such as *Dfd*.

Acknowledgements

We thank Dr.s Peter Holland and Brigid Hogan for discussion of this work; Dr. Leanne Wiedemann for comments on the manuscript; and Dr. Dennis Summerbell for chick embryos. This work was supported by the Medical Research Council.

Literature References

Awgulewitsch A, Utset MF, Hart CF, McGinnis W & Ruddle FH (1986) Spatial restriction in expression of a mouse homeobox locus within the central nervous system. Nature 320: 328-335

Colberg-Poley AM, Voss SD, Chowdhury K & Gruss P (1985a) Structural analysis of murine genes containing homeobox sequences and their expression in embryonal carcinoma cells. Nature 314: 713-718

Colberg-Poley AM, Voss SD, Chowdhury K, Stewart CL, Wagner EF, & Gruss P (1985b) Clustered homeoboxes are differentially expressed during murine development. Cell 43: 39-45

Dony C & Gruss P (1987) Specific expression of the Hox 1.3 homeobox gene in murine embryonic structures originating from or induced by the mesoderm. EMBO 6: 2965-2975

Dubuole D, Baron A, Mahl P & Galliot B (1986) A new homeobox is present in overlapping cosmid clones which define the mouse Hox 1 locus. EMBO 5: 1973-1980

Farnher K, Hogan BML & Flavell R (1987) Transcription of H-2 and Qa genes in embryonic and adult mice. EMBO 6: 1265-1274

Harding K, Wedeen C, McGinnis W & Levine M (1985) Spatially regulated expression of homeotic genes in *Drosophila*. Science 229: 1236-1242

Hart CP, Awgulewitsch A, Fainsod A, McGinnis W & Ruddle FH (1985) Homeobox gene complex on mouse chromosome 11: molecular cloning, expression in embryogenesis and homology to a human homeobox locus. Cell 43: 9-18

Hart CP, Bogarad LD, Fainsid A & Ruddle FH (1987) Polypurine/polypyrimidine sequence elements of the murine loci, Hox-1,-2, and -3. NAR 15: 5495

Harvey RP, Tabin CJ & Melton DA (1986) Embryonic expression and nuclear localization of *Xenopus* homeobox (Xhox) gene products. EMBO 5: 1237-1244

Hauser CA, Joyner AL, Klein RD, Learned TK, Martin GR & Tijan R (1985) Expression of homologous homeobox containing genes in differentiated human teratocarcinoma cells and mouse embryos. Cell 43: 19-28

Holland P & Hogan BML (1988) Spatially restricted patterns of expression of the homeobox containing gene Hox 2.1 during murine embryogenesis. Development (in press)

Joyner AL, Kornberg T, Coleman KG, Cox DR & Martin GR (1985) Expression during embryogenesis of a mouse gene with sequence homology to the *Drosophila* Engrailed gene. Cell 43: 29-37

Kaufman TC, Lewis R & Wakimoto (1980) Cytogenic analysis of chromosome 3 in *Drosophila*: the homeotic gene complex in polytene chromosome interval 84A-B. Genetics 94: 115-133

Kessel M, Schulze F, Fibi M & Gruss P (1987) Primary structure and nuclear localization of a murine homeodomain protein. PNAS 84: 5306-5310

Krumlauf R, Holland PWH, McVey JH & Hogan BML (1987) Developmental and spatial patterns of expression of the mouse homeobox gene, Hox 2.1. Development 99: 603-613

Mavilio F, Simeone A, Giampaolo A, Faiella A, Zappavigna V, Acampora D, Poiana G, Russo G, Peschle C & Boncinelli E (1986) Differential and stage-related expression in embryonic tissues of a new human homeobox gene. Nature 324: 664-667

Meijlink F, De Laaf R, Verrijzer P, Destree O, Kroezen V, Hilkens J & Deschamps J (1987) A mouse homeobox containing gene on chromosome 11: sequence and tissue-specific expression. NAR 15: 6773-6786

McGinnis W, Levine ML, Hafen E, Kuroiwa A & Gehring WJ (1984a) A conserved DNA sequence in homeotic genes of the *Drosophila* , Antennapedia and Bithorax complexes. Nature 308: 428-433

McGinnis W, Garber RL, Wirz J, Kuroiwa A & Gehring WJ (1984b) A homologous protein-coding sequence in *Drosophila* genes and its conservation in other metazoans. Cell 37: 403-408

Odenwald WF, Taylor FJ, Palmer-Hill FJ, Friedrich Jr V, Tani M & Lazzarini RA (1987) Expression of a homeodomain protein in noncontact-inhibited cultured cells and in postmitotic neurons. Genes & Devel. 1: 482-496

Regulski M, McGinnis N, Chadwick R & McGinnis W (1987) Developmental and molecular analysis of Deformed; a homeotic gene controlling *Drosophila* head development. EMBO 6: 767-777

Wolgemuth DJ, Englemyer E, Duggal RN, Gizang-Ginsberg E, Mutter GL, Ponzetto C, Viviano C & Zakeri ZF (1986) Isolation of a mouse cDNA coding for a developemntally regulated testis-specific transcript containing homeobox homology. EMBO 5: 1229-123

Regulation of Expression of the Hox2.3 Gene

Frits Meijlink, Lia de Laaf, Peter Verrijzer, Wim de Graaff and Jacqueline Deschamps
Hubrecht Laboratory
Netherlands Institute for Developmental Biology
Uppsalalaan 8
3584 CT UTRECHT
The Netherlands

INTRODUCTION

Only few vertebrate genes have been described, that can be unambiguously linked to developmental processes. For this reason, investigators of vertebrate embryology concentrate their attention on genes for which the evidence implicating them in pattern formation is only circumstantial. The homeobox containing genes (1), found throughout the animal kingdom including mice and humans, exemplify such genes. In the genetically well-characterized fruitfly Drosophila melanogaster their involvement in various stages of the segmentation of the embryo and in neuronal development (1,2) is beyond doubt, although their precise function still has to be established.

Sequence homology with Drosophila counterparts (the very possession of a homeobox) originally provided the motivation to study mouse and other vertebrate homeobox genes. Additional structural similarities and similarities in expression patterns were uncovered later. Although formal proof of a developmental role is still lacking (no vertebrate 'loss of function' homeobox gene mutant has yet been described) it is now generally accepted that vertebrate homeobox genes play a role in at least some development-related processes.

We are concentrating our studies on the mouse Hox2.3 gene, which is a member of the Hox 2 cluster on chromosome 11 (3). Eventually the analysis of mouse mutants with abnormal Hox2.3 activity should demonstrate the function of this gene. Prerequisite to the feasibility of such an approach is a phase of descriptive studies, including structural analysis of the gene itself and establishment of its

NATO ASI Series, Vol. H26
Cell to Cell Signals in Mammalian Development
Edited by S. W. de Laat et al.

expression pattern.

Here we present data concerning the expression of the Hox2.3 gene in embryos, paying special attention to comparative analyses with other homeobox genes. We also report experimental data on the influence of retinoic acid on homeobox gene expression in embryonal carcinoma and cultured primary embryo cells. Finally, we report the results of preliminary studies on protein-DNA interactions in the promoter region of the Hox2.3 gene.

RESULTS

Structure of the Hox 2.3 gene

We screened genomic and cDNA libraries using probes containing *Drosophila melanogaster* homeobox sequences under low-stringency hybridization conditions. From homeobox-containing clones that were isolated in this way, homologous probes were prepared to re-screen the same and other libraries under high-stringency conditions (4). Clones isolated included several overlapping genomic clones corresponding to different regions in the Hox 1 and Hox 2 clusters and cDNAs that were eventually by sequence analysis identified as corresponding to Hox 2.3 and Hox 1.4. (5,6). The Hox 2 cluster containing Hox2.3 had been described previously (3). We will discuss here only the Hox2.3 containing clones. They include cDNAs from two different libraries and three overlapping genomic clones containing the entire Hox2.3 gene. A combination of restriction mapping and sequence analysis resulted in a rather detailed genomic restriction map of the Hox 2.3 gene, including the flanking sequences and a portion of the Hox 2.4 gene which is located upstream from Hox2.3 (4). The site of polyadenylation and the exact location of the single intron and the exons were derived from comparison of the cDNA and genomic sequences, and by S1 nuclease protection analysis (see below). The sequence of the Hox 2.3 gene including the complete coding region and its flanking sequences has been published (4,7).

Comparison of the primary sequence of Hox2.3 with other homeobox genes

Per definition, the protein encoded by Hox 2.3 resembles other Hox-encoded proteins in the homeodomain. The highest similarity was found in comparison with the Drosophila melanogaster Antennapedia (8) and the mouse Hox 1.1 (9) gene, which differ within the homeodomain in only one amino acid. All homeodomain proteins described so far display significant similarity at their amino termini (4).

A conserved hexapeptide (consensus Ile Tyr Pro Trp Met Arg)

Ile	Tyr	Pro	Trp	Met	Arg
Leu	Phe				Lys
Val					

reported to be present in homeodomain proteins at distances varying from 4–19 amino acids is present in Hox2.3 at a distance of five amino acids. Many homeodomain proteins already described have an acidic carboxy-terminal tail. Characteristic for the Hox2.3 protein is a stretch of six glutamic acid residues. Such a peptide is also found in a limited number of other homeodomain proteins, namely the Xenopus laevis gene XlHBox2 (10), the human gene HHOcl (11), and mouse Hox1.1 (9). These same proteins display considerable similarity to Hox2.3 throughout the protein. This is demonstrated in Figure 1. We conclude from the similarities between the three genes that they are truly homologous and have presumably very similar functions. The similarity with Hox 1.1 is in agreement with the proposal of Odenwald et al. (12) and others, that a portion of the Hox2 and Hox 1 clusters arose by some duplication event from a common ancestor gene cluster. This is based on sequence similarities between Hox 2.1 and Hox1.3, between Hox 1.4 and Hox2.6 and on conservation of the intergenic sequences in the regions between Hox 2.3 and Hox2.1, and Hox 1.1 and Hox 1.3. Comparative analyses show that at the nucleotide level, the sequence similarities between Hox2.3, XlHbox2, HHOcI and Hox1.1 are very limited in the 3′ untranslated region, but very high in the 5′ untranslated region (for instance, 99% between Hox2.3 and HHOcl). Although even unrelated homeobox genes display significant similarity in the 100 nucleotides preceding the protein-coding region (13), even higher homology was observed within the homologous genes mentioned here and with Hox1.1.

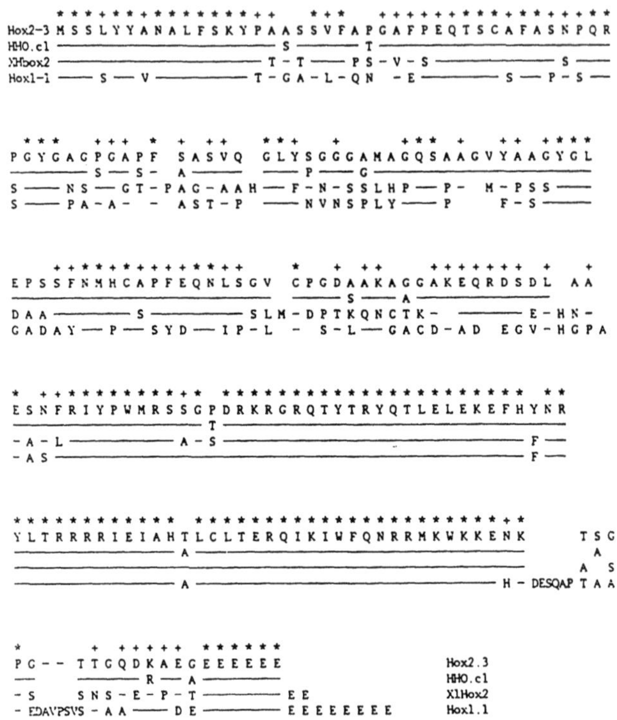

Fig. 1. Comparison of protein sequences of four homeo domain proteins.
The derived amino acid sequences of Hox2-3 (4), its human and Xenopus homologues HHO.c1 (11) and XlHbox2 (10), respectively, and mouse Hox1.1 (9) are compared. A line indicates an identical amino acid, changes are indicated by the appropriate letter. An asterisk indicates identity in all four proteins in the specific position, a + sign identity in three out of four. Note conservation of the cysteine residues.

Expression of Hox2.3 in adult animals

Hox2.3 is expressed in specific regions in the developing embryo and in several organs of the adult animal, suggesting that it can fulfil functions in several different cellular processes. Fig. 2 shows Northern blot analysis of RNAs isolated from adult organs. The highest levels of expression are detected in kidney and testis. Levels above background were also observed in brain and in ovary. An interesting possibility is, that the latter signal originates from the oocyte, since the Hox2.3 homologue in Xenopus laevis was reported to be expressed during oogenesis (14); we have, however, not yet been able to confirm this for the mouse. The first lane in Fig. 2 represents RNA from a total embryo homogenate, demonstrating the relative strength of this signal as compared to that of the adult organs. The level of expression observed in the embryo is very high, considering that at this stage the signal originates almost entirely from the spinal cord (see below). The same filters were hybridized with probes for B-actin and the neurofilament gene NF 68.

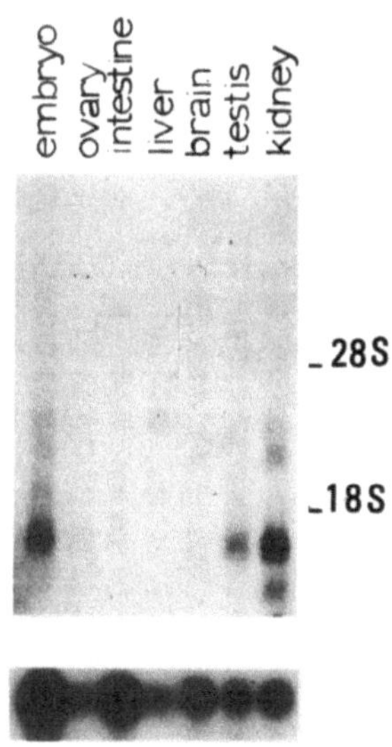

Fig. 2. Expression of Hox2.3 in adult mice compared to that in embryos. Approximately 5 μg of poly(A)$^+$ RNA from the tissues as indicated were loaded on a formaldehyde/agarose gel and analyzed by Northern hybridization. The embryo sample is a total homogenate from a 13.5 day embryo. The filter was hybridized with a Hox2.3 probe and subsequently with a β-actin probe (lower panel).

Expression in EC and ES cells: effect of retinoic acid

Expression of homeobox genes in embryonal stem cells has been studied by several research groups. Embryonal carcinoma (EC) and embryonal stem cells (EK or ES cells) serve as a model for the early stages of embryogenesis for which it is difficult to obtain sufficient material for molecular biological analysis. High

(15,16) or limited to non-existing (16–18) induction of homeobox genes during differentiation was reported. We were interested to know whether the differences in the results could have been caused by the specific protocol used in these experiments to induce cell differentiation. Therefore we compared the expression of Hox 2.3 and other homeobox genes in EC and ES cells that were induced to differentiate according to alternative protocols.

We have shown that accumulation of high levels of transcripts from Hox2.1 and Hox1.1 (19), Hox2.3 and En-1 (20) and Hox 1.4 (this paper) correlates with retinoic acid-induced EC cell differentiation rather than with differentiation *per se*. When an ES cell line and several EC lines, among which C1003, were induced to differentiate by alternative methods involving or not involving of a chemical inducer, accumulation of high levels of homeobox mRNA only occurred in the presence of the chemical inducer, in particular retinoic acid (Fig. 3). Therefore, the observed retinoic acid-mediated stimulation of homeobox mRNA accumulation applies to the five homeobox genes considered in these studies and is, therefore, not restricted to members of the family of *Antennapedia*-like homeobox-genes.

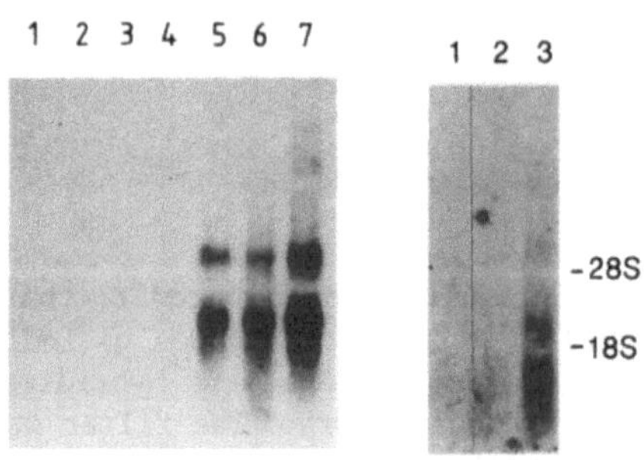

Fig. 3. Expression of the Hox2.3 gene in EC and ES cells.
Left panel. C1003 EC cells (lane 1) and in their differentiated derivatives obtained after the cells were grown for one, three and five days in serum-free medium (lanes 2, 3 and 4) or for one, two and three days in serum-free medium supplemented with 0,1 μM RA (lanes 5, 6, and 7).
Right panel. Expression in EK-cs1 stem cells (lane 1), in cells induced to differentiate by being grown in suspension as aggregates for three days and subsequently plated on tissue culture surface for two days (lane 2) and in cells treated similarly but in the presence of 1 μM RA(lane 3). 30 μg of total RNA was present in each lane.

The conclusions from these experiments is that all five homeobox genes, lying on three different chromosomes, that we tested respond similarly to the presence of retinoic acid and must therefore share common structural features making them susceptible to this induction.

Comparative analysis of expression)of five homeobox genes in mouse embryos: dissection experiments

To obtain an estimate of the location of Hox2.3 expression and to be able to directly compare it with that of various other homeobox genes, we dissected embryos of 13 day p.c. into viscera, limb buds, cranio facial tissue, and in six CNS fractions: three from the brain and three from the spinal cord. RNA was extracted from these tissues and was then subjected to standard Northern hybridization analysis using probes for Hox 2.3, Hox 1.1, Hox2.1, Hox1.4 and En-1. To define the dissection method, the diagram in Fig. 4 specifies how the brain and spinal cord fractions were obtained. The results are displayed in Fig. 4. Each filter was subsequently hybridized to a homeobox gene probe as indicated, to NF 68 as a neuronal marker and to beta actin as a control for the amount of stable mRNA present. While all five genes are mainly active in nervous tissue, we observe specific differences in expression from one gene to another. The majority of Hox2.3 transcripts are found in the three fractions of the spinal cord with weak expression in the brain and in the viscera and limb buds. The Hox1.1 gene, which is structurally strongly related to Hox2.3 (see above) has a very similar distribution along the spinal cord, whereas the other genes all display specific characteristic patterns. En-1, as was previously reported by Joyner et al. (21), is expressed more anteriorly, and Hox1.4 is also strongly expressed in the posterior brain. The pattern of expression of Hox 2.1 is as expected in view of the published data from in situ hybridization (22,23). These dissection data illustrate that the homeobox genes considered here are all being expressed in the central nervous system, but that there are very specific and characteristic differences between them.

<u>Expression in primary rat cells</u>

In an attempt to link our studies on EC cells with the data obtained from <u>in vivo</u> experiments, we studied expression of homeobox genes in primary embryonic brain cells.

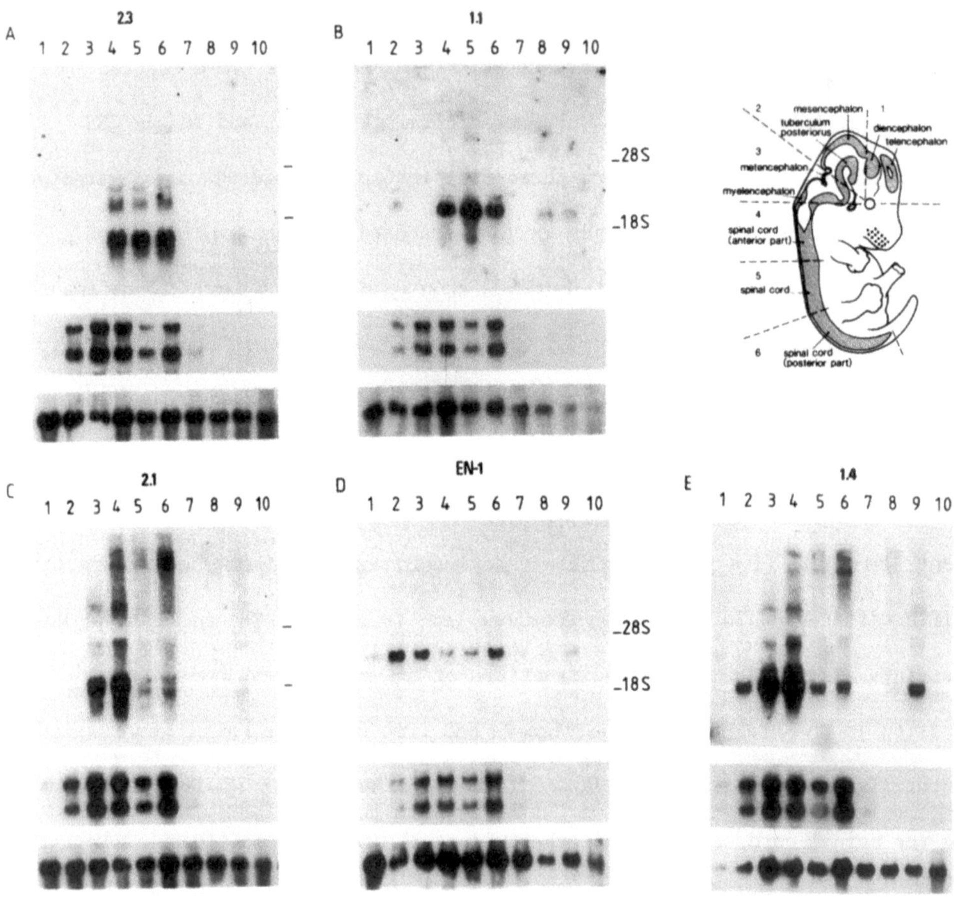

<u>Fig. 4</u>. Hox2-3 and Mo-Enl expression in rat primary mid brain cells.
RNA was isolated from rat primary mid brain cells grown for two days in serum-containing medium (lane 1) or grown for two days in the same medium supplemented with 1 μM RA (lane 2). 30 μg of total RNA was layered in each lane. The same RNA filter was successively hybridized to Hox2.3 (panel A), a Mo-enl probe (18) (panel B), NF68 (underneath panel A) and a Beta actin probe (underneath panel B).

Having 13-day rat embryo mid brain primary cells at our disposal we chose to culture them for two days in the presence and in the absence of retinoic acid. We had previously observed that the Hox2.3 probe recognizes rat transcripts identical in size to the mouse mRNAs (not shown) in embryos as well as in adult organs. We had also verified that the mouse Hox2.3 and En-1 (18) probes recognize similar amounts of transcript in the 13 day rat primary mid-brain cells cultured for two days, as in 13-day mouse embryonic midbrain (Fig.4 A, lane 2 vs. Fig. 4D, lane 2.

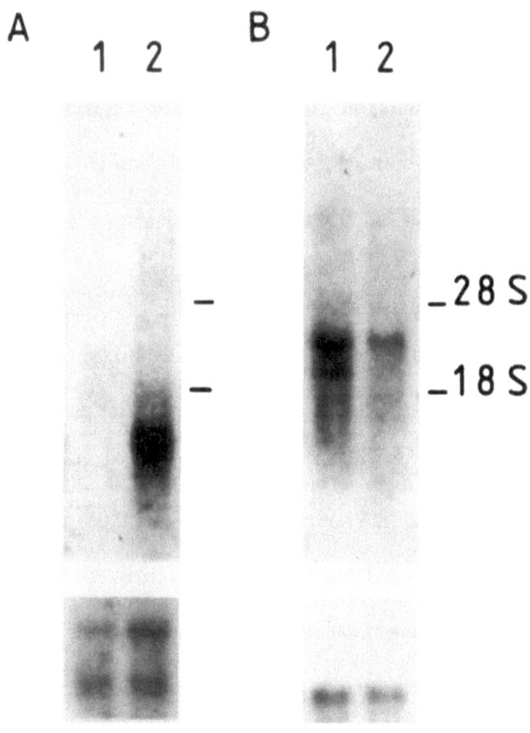

Fig. 5. Expression of five Hox genes in mouse embryonal tissues.
The diagram shows how the 13.5-day embryos were dissected. The number of the tissue sample corresponds to the number of the RNA samples loaded on the gels as indicated. Fractions 1-6 are mainly neural tissue: three brain and three spinal cord fractions. Other tissue fractions were the cranio-facial part of the head (fraction 7), the limb buds (8), the pooled organs from the viscera (gut, hart, lung, liver, kidney, and rudiments of genital organs; fraction 9), and the carcass (10, defined as what remained after the above-listed tissue fractions had been removed). Expression of Hox2.3 (panel A), Hox 1.1 (B), Hox 2.1 (C), and En-1 (D) and Hox1-4 (E) was analyzed by Northern hybridization of 10 microgram of total RNA per lane. Every RNA filter was successively hybridized with an homeobox probe (upper panel), the neurofilament NF 68 probe (middle panel), and the actin probe (lower panel).

No obvious morphological difference was noted between retinoic acid-treated and control cells. We compared the levels of expression of the En-1 gene, normally active in mouse mid-brain (see above and Ref. 21) and of the Hox 2.3 gene, normally expressed at a very low level in the embryonal mouse mid-brain (see above). This difference in level of expression is reflected by the levels found in normal (i.e. non-retinoic acid treated) primary cells. In the retinoic acid-treated cultures a strong induction of Hox2.3 transcripts is observed (Fig 5A, lane 2) whereas no increase in En-1 gene expression is detected (Fig. 5B, lane 2). As slightly less RNA was present on the filter in the case of the retinoic acid treated samples the apparent decrease in En-1 expression is probably not significant. For the same reason a slight increase in accumulation of neurofilament transcript in retinoic acid-treated cultures should be assumed.

The promoter region of the Hox2.3 gene

An obviously important aspect of the way regulator genes function, is the way they are regulated themselves. We have started experiments to investigate the transcriptional regulation of the Hox2.3 gene. This is being done mainly along two lines. Constructs consisting of a reporter gene linked to various portions of putative regulatory sequences will be transfected into cells and the activity of the reporter gene will be assayed. In addition interactions between these regulatory sites and transcription factors are being studied in vitro as will be shown below.

Location of a transcription start site of the Hox2.3 gene

To determine the transcription start site of the gene, we performed S1 nuclease protection analysis. A restriction fragment from a genomic clone bordered by a AvaI/SmaI site located in the protein coding portion of the gene and by a SacI site 257 bp upstream from the ATG codon was labeled using polynucleotide kinase. This probe was then hybridized with RNA from embryonal spinal cord known to contain a relatively high level of RNA transcribed from the Hox2.3 gene. As shown in Fig. 6, treatment of the hybrids with S1 nuclease resulted in a 207-bp band indicating that 87 bp upstream from the SmaI site there is a discontinuity between the mRNA sequence and the genomic sequence. Since the sequence found here does not resemble the consensus splice site found in eukaryotic spliced genes transcribed by RNA polymerase II we conclude that this is a transcription initiation site of the Hox2.3 gene. Our data do not exclude the possibility that there is an additional upstream transcription initiation site. Recently, Wright et al. (10) published the genomic organization of the Xenopus laevis gene which is homologous to Hox2.3 (XlHbox2; see above): it consists of three exons.

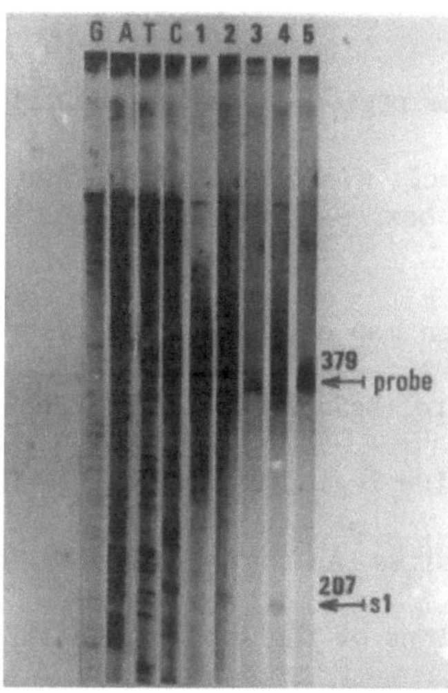

Fig. 6. S1 analysis in the upstream region of the Hox2-3 gene.
20 ug of embryonal spinal cord RNA was hybridized with the end-labeled ScSa382 (see Fig. 7) fragment. The hybrids were treated with S1-nuclease and the products analyzed by gel electrophoresis on denaturing poly acrylamide gels containing 8M urea. Lanes G-A-T-C: sequence ladders used as length markers. Lane 1, untreated probe; lanes 3 and 5, hybridization with E.coli tRNA; lanes 2 and 4, hybridization with 20 ug mouse spinal cord RNA. Lanes 2 and 3, digestion with 50 U of S1 nuclease; lanes 4 and 5, digestion with 90 U of S1 nuclease.

DNA sequences interacting with nuclear proteins

To see whether DNA fragments located near the transcription start site of Hox2.3 do specifically interact with transcription factors, or at least with proteins present in the nucleus, we performed gel retardation assays using these fragments and nuclear extracts. These extracts originated from the fibroblast-like cell line Fib9 which was found to constitutively express the Hox2.3 gene (J. D., unpublished).

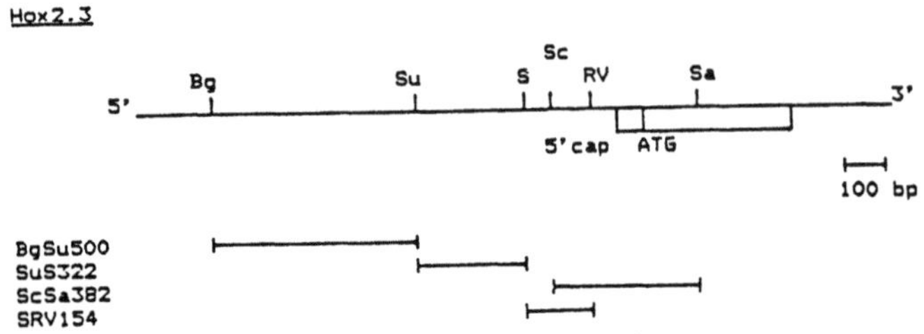

Fig. 7. Restriction map of the region around an identified transcription start site the Hox2.3 gene.
Bg= BglII; Su= Sau3AI; Sc= SacI; RV=EcoRV; S=SmaI. The first exon of the Hox2.3 gene is indicated by an open box.

Fig. 7 shows a restriction map of the Hox 2.3 upstream region. The four DNA fragments shown in this figure, BgSu500, SuS322, ScSa382, and SRV154, span a 1-kb region upstream from the coding region of the Hox2.3. They were assayed for the presence of protein binding sites. As shown in Fig. 8, significant complex formation was detected in incubations of the labeled ScSa382 fragment with Fib9 nuclear extracts. Addition of a 100-fold molar excess of pGem-Blue plasmid DNA did not change the result, whereas retardation was abolished by the presence of an excess of unlabeled ScSa 382. Fig. 8 shows that the SRV 154 fragment is specifically retarded in the same assay. Moreover, competition not only with the same fragment, but also with the overlapping ScSa 382 fragment prevents retardation of the SRV154 probe. The simplest explanation of this result is that the sequences shared by both fragments:

GTCGACAGCG GCCCTTAAGC CCCCAGCCCC AATCCGCAGA GCTCGGCCTT CCCATTCATT ATTGATCATA
TTTTATAAAT CCAACGCCAC ACAATTTTTT CCACATTACT GGGAGCCTCC GGGAGGCCGT CATACCATTG
GCCGAGGGGA TATC

are directly involved in the crucial interaction. It does not contain the sequence TCAATTAAATGA (24 and Desplan personal communication) which has been shown to be recognized by a number of homeodomain proteins, although there are a few sequences which resemble it.

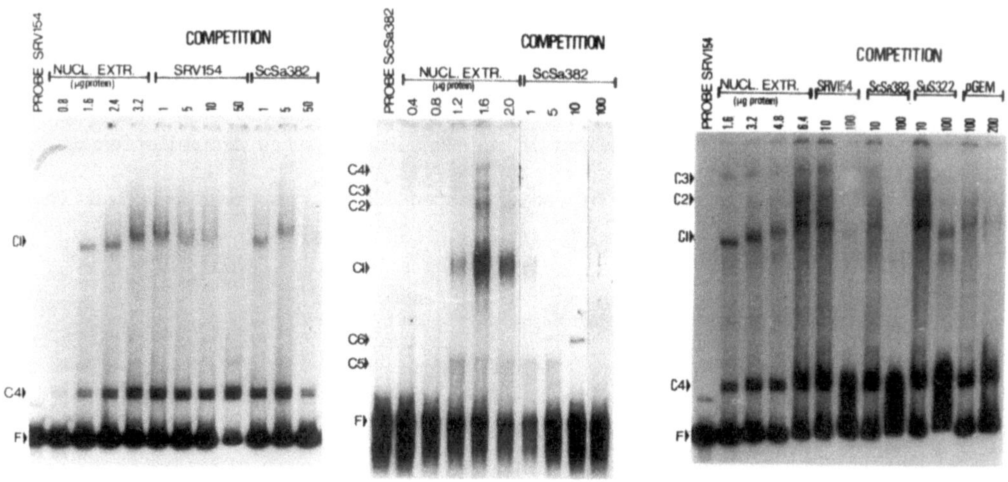

Fig. 8. Gel retardation analysis using fragments SRV154 and ScSa382 as probes. End-labeled fragments were incubated with different amounts of nuclear extract prepared from Fib9 cells (numbers above lanes indicate micrograms of protein) and electrophoresed in a 5% polyacrylamide gel. F indicates the position of the free probe. Several different protein-DNA complexes numbered C1-C6 were detected. We consider C1 as the most important specific complex. Competition experiments were performed by including non-labeled purified restriction fragment in the incubation. The molar excess (1-50) of cold over labeled fragment is indicated above the lanes.

Presence of the binding protein

We next asked whether the presence of the protein interacting with the SRV154 fragment correlates with the activity of the gene. We showed previously that C1003 embryocarcinoma cells, the line from which the FIB9 line is derived, do not contain detectable amounts of Hox2.3 mRNA. We therefore compared, by using retardation assays, the presence of proteins interacting with the Hox2.3 promoter in nuclear extracts from C1003 EC cells and C1003 cells treated for 24 hours with retinoic acid. As shown in Fig. 9, essentially the same retardation was observed with both extracts and with Fib9 extracts. An implication of this result is that it is not the Hox2.3 protein itself which is the responsible factor for the retardation observed. Binding of a homeodomain protein to the promoter of the gene encoding it, has been described several times, for instance for the Hox1-5 gene (25). The lack of correlation between this retardation and the presence of the Hox2.3 mRNA does not, however, exclude a connection between the retardation of the SRV154 fragment ad with the transcriptional regulation of the Hox2.3 gene since accumulation of homeobox gene transcripts in retinoic acid-treated EC cells may not be primarily a transcriptional event (26).

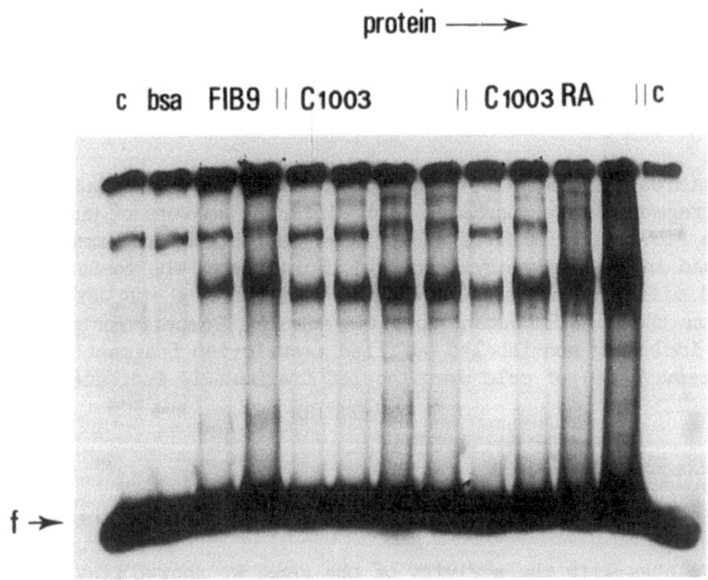

Fig. 9. Gel retardation analysis using nuclear extract from EC cells and differentiating EC cells.
The probe used was the SRV154 fragment. C, control, no nuclear extract added. Bsa, 10 ug of BSA added instead of nuclear extract. Retardation by 1 and 2 ug Fib9 extract is shown for comparison. Various amounts (1-8 ug) of nuclear extract from undifferentiated C1003 cells and C1003 cells treated for 24 hours with retinoic acid were used. Equal amounts of protein from both sources give a similar degree of retardation.

DISCUSSION

Considerablé evidence exists that homeobox-containing genes are in part regulated by their own products and by the products of other regulator genes. Studies of this "regulation of the regulator genes" is of obvious importance for an understanding of their role in pattern formation and the mechanistic features underlying it. We approach this problem by 1) studying the molecular basis of Hox2.3 differential expression in vivo, 2) studying the phenomenon of homeobox gene mRNA accumulation in EC cells as a result of induction of differentiation using retinoic acid 3) studying the interactions of nuclear proteins with the Hox2.3 promoter, assuming that these interactions are implicated in transcriptional control of the gene.

Retinoic acid induction

We wanted to make a link between the two following observations on the expression of homeobox containg genes:

1) The four Antennapedia-related and the En-1 homeobox genes considered in our studies similarly respond to the presence of retinoic acid during EC and ES cell differentiation in that high levels of transcript accumulate.

2) These genes are all expressed in the central nervous system of the developing mouse embryo, but each one in a specific region. We compared the distribution of transcripts from several homeobox genes in the 13.5-day mouse embryonal central nervous system, and found gene-specific, overlapping gradients of homeobox mRNAs along the longitudinal axis. An interesting possibility is, that the mechanisms leading to high levels of homeobox transcripts in the embryonal nervous system in vivo and to accumulation of homeobox mRNA in retinoic acid-induced EC cells in vitro would be in some way related. We therefore looked at the influence of retinoic acid on the expression of the homeobox genes in primary mid brain cells grown in vitro.

In these cells, the normally silent Hox2.3 gene is activated by retinoic acid, whereas expression of the normally active En-1 gene remains essentially unchanged. In another primary cell type (limb bud cells) no expression of these homeobox genes could be detected, either before or after retinoic acid treatment. It is tempting to speculate that the action of retinoic acid is to affect a negative control mechanism which, in the mid-brain, prevents Hox2.3 expression, but has been removed from the En-1 gene. Position -and perhaps time- dependent signals would be responsible for the level of expression of these genes in the mesencephalon. This positional information may relate to the gradients of neural proliferation, differentiation and migratory activity along the rostro-caudal axis at this stage of embryogenesis. Our speculative hypothesis can be tested by comparing expression levels of other homeobox genes in primary cells before and after retinoic acid treatment.

DNA-protein interactions

Understanding the mechanisms which control gene expression at the transcriptional level requires the characterization of the nucleotide sequences involved and the transcription complexes interacting with them. We are searching for sequences in the putative promoter region(s) of the Hox2.3 gene that are involved in interactions with transcription factors and report here specific binding of an as yet unidentified nuclear protein to a restriction fragment named SRV 154. This fragment contains a transcription initiation site. There are regions of extremely high sequence conservation with the *Xenopus laevis* Hox2.3 homologue XlHbox2 (Fig. 10). Surprisingly, these sequences in the Xenopus gene are present in cloned cDNAs, and therefore in transcribed mRNA. A possible explanation for this is that there is an additional upstream transcription start site for this gene. Our transcription mapping data do not exclude the existence of transcription starts upstream from the SacI site that borders the restriction fragment used in the S1 mapping experiment.

```
SEQUENCE CONSERVATION UPSTREAM FROM THE CODING REGIONS OF

Xlhox2    and    Hox2-3

CGTACATCTTACATCCCATGAAATATTCATCATATTTTATAAATTCAGCTCCACAACCAT
CAGAGCTCGGCCTTCCCATTCATTATTGATCATATTTTATAAATCCAACGCCAC ACAAT

TTTTT   TTTTCTTTCCACATTACAAGAGCTCG GTGGGGATTGGACGCCGGGAAATCA
TTTTTCCACATTACTGGGAGCCTCCGGGAGGCCGTCATACCATTGGCCGAGGGATATCA

CGTGTGGCGCAGTCACATGGTC GGGGAGGAAAAAGGGGTCCTTTTTGGTGTAAATCTG
CGTGGCCGGGGTCACGTGGTCAGAAGAGGAAAAAGGGGTCC TTTTGGTGTAAATCTG

GAGTCTAATTCTGTAATATATCACGGTACCTCGTAAAACCGACACTAAAAGGGCCCGGCC
GACTCTAATTCTGTAATATATCAAGGAATCTCGTAAAACCGACACTAAAACGTCCCCGAC

TACAAATCATTCGGCCGAATTATG
TACAAATCATCCGGCCAAATTATG
```

Fig. 10. Comparison of nucleotide sequences upstream from the protein coding regions of Hox2-3 and XlHbox2.
The XlHbox2 sequence (Ref. 10) is displayed in the upper line and the Hox2-3 sequence in the lower line. The ATG at the end of the compared stretches is the translation initiation codon. A line indicates the putative transcription initiation site in Hox2-3, established by S1 analysis.

Although it clear that the SRV 154 fragment is specifically bound by a protein present in nuclei of Fib9, C1003 and HeLa cells, the specific nature of this interaction has been difficult to assess, since we have not yet succeeded, for unknown reasons, in obtaining "footprints" demonstrating the exact nucleotide sequences involved in the interactions. Further studies addressing the origin of the interacting protein and its presence in cell nuclei isolated from various organs are now under way.

Acknowledgements.

We thank Drs. P. Gruss, B. Holland and G. Martin for providing us with plasmid probes, Dr. C. Stewart for the EK cell line CS-1, and Dr. K.A. Lawson for reading the manuscript.

REFERENCES

1. Gehring WJ (1987) Homeoboxes in the study of development. Science 236:1245–1252
2. Doe CQ, Hiromi Y, Gehring WJ, Goodman CS (1987) Expression and function of the segmentation gene fushi tarazu during Drosophila neurogenesis. Science 239:170–175
3. Hart CP, Awgulewitch A, Fainsod A, McGinnis W, Ruddle F (1985) Homeobox gene complex on chromosome 11: Molecular cloning, expression in embryogenesis and homology to a human homeo box cluster. Cell 43:9–18
4. Meijlink F, De Laaf R, Verrijzer P, Destrée O, Kroezen V, Hilkens J, Deschamps J (1987) A mouse homeobox on chromosome 11: Sequence and tissue-specific expression. Nucleic Acids Res. 15:6773–6786
5. Duboule D, Baron A, Maehl P, Galliot B (1986) A new homeo-box is present in overlapping cosmid clones which define the mouse HOX-1 locus. EMBO J. 5:1973–1980
6. Wolgemuth DJ, Engelmyer E, Duggal RN, Gizang-Ginsberg E, Mutter GL, Ponzetto C, Viviano C, Zakeri ZF (1986) Isolation of a mouse cDNA coding for a developmentally regulated, testis-specific transcript containing homeo box homology. EMBO J. 5:1229–1235
7. Verrijzer P, De Graaff W, Deschamps J, Meijlink F (1988) Nucleotide sequence of the Hox2.3 gene region. Nucleic Acids Res. 16:2729
8. Regulski M, McGinnis N, Chadwick R, McGinnis W (1987) Developmental and molecular analysis of Deformed; a homeotic gene controlling Drosophila head development. EMBO J. 6:767–777
9. Kessel M, Schulze F, Gruss P (1987) Primary structure and nuclear localization of a murine homeodomain protein. Proc. Natl. Acad. Sci. USA 84:5306–5310
10. Wright CVE, Cho KWY, Fritz A, Bürglin T, De Robertis EM (1987) A Xenopus laevis gene encodes both homeobox-containing and homeo-box-less transcripts EMBO J. 6:4083–4094
11. Simeone A, Mavilio F, Acampora D, Giampaolo A, Faiella A, Zappavigna V, D'Esposito M, Pannese M, Russo G, Boncinelli E, Peschle C (1987) Two additional homeobox gene, c1 and c8: Structure analysis and expression in human embryonic development. Proc. Natl. Acad. Sci. USA 84:4914–4918
12. Odenwald WF, Taylor CF, Palmer-Hill FJ, Friedrich V, Tani M, Lazzarini RA (1987) Expression of a homeodomain protein in noncontact-inhibited cultured cells and postmitotic neurons. Genes Devel. 1:482–496
13. Bürglin TR, Wright CVE, De Robertis EM (1987) Translational control in homebox mRNAs? Nature 330:701–702
14. Müller MM, Carrasco AE, De Robertis EM (1984) A homeo-box-containing gene expressed during oogenesis in Xenopus. Cell 39:157–162
15. Colberg-Poley AM, Voss SD, Chowdury K, Gruss P (1985) Structural analysis of murine genes containing homoeobox sequences and their expression in embryonal carcinoma cells. Nature 314:713–718
16. Hauser CA, Joyner AL, Klein RD, Learned TK, Martin GR, Tjian R (1985) Expression of homologous homeobox containing genes in differentiated human teratocarcinoma cells and mouse embryos. Cell 43:19–28
17. Jackson I, Schofield P, Hogan B (1985) A mouse homoeo box gene is expressed during embryogenesis and in adult kidney. Nature 317:745–748
18. Joyner AL, Kornberg T, Coleman KG, Cox DR, Martin GR (1985) Expression during embryogenesis of a mouse gene with sequence homology to the Drosophila engrailed gene. Cell 43:29–37
19. Deschamps J, De Laaf R, Joosen L, Meijlink F, Destrée O (1987) Abundant expression of homeobox genes in mouse embryonal carcinoma cells correlates with chemically induced differentiation. Proc. Natl. Acad. Sci. USA 84:1304–1308

20. Deschamps J, De Laaf R, Verrijzer P, De Gouw M, Destrée O, Meijlink F (1987) The mouse Hox2.3 homeobox-containing gene: regulation in differentiating pluripotent stem cells and expression pattern in embryos. Differentiation 35:21-30
21. Joyner AL, Martin GR (1987) En-1 and En-2, two mouse genes with sequence homology to the Drosophila engrailed gene: expression during embryogenesis. Genes Devel. 1:29-38
22. Krumlauf R, Holland PW, McVey JH, Hogan BLM (1987) Developmental and spatial patterns of expression of the mouse homeobox gene, Hox2.1. Development 99:603-617
23. Holland PWH, Hogan BLM (1988) Spatially restricted patterns of expression of the homeobox-containing gene Hox2.1. during mouse embryogenesis. Development 102:159-174
24. Desplan C, Theis J, O'Farrel PH (1985) The Drosophila developmental gene, engrailed, encodes a sequence-specific DNA binding activity. Nature 318:630-635
25. Fainsod A, Bogarad LD, Ruusala T, Lubin M, Crothers DM, Ruddle FH (1986) The homeo domain of a murine protein binds 5' to its own homeo box. Proc. Natl. Acad. Sci. USA 83:9532-9536
26. Colberg-Poley AM, Püschel AW, Dony C, Voss SD, Gruss P (1987) Post-transcriptional regulation of a murine homeobox gene transcript in F9 embryonal carcinoma cells. Differentiation 35:206-211

The Mouse <u>En-1</u> Gene Encodes a Nuclear Protein Expressed in the Nervous System of the Developing Embryo

Gail R. Martin and Michael A. Frohman

Department of Anatomy
University of California, San Francisco
San Francisco, California 94143

Introduction

The classical genetic approach to understanding how development is controlled is to first identify mutations that perturb the process, and then to study the mutant embryo phenotypes in order to gain some insight into the functions of the genes represented by the mutant alleles. Although this approach has been very successful in organisms well-suited to genetic analysis, such as the fruit fly (<u>Drosophila</u> <u>melanogaster</u>) and the nematode (<u>Caenorhabditis</u> <u>elegans</u>), attempts to identify genes that regulate development in mammals by this approach have been hindered by the difficulty of identifying mutations that affect early embryonic development. Moreover, once a particular mutation that affects development is identified, there is no straightforward means of isolating the mammalian gene represented by that mutant allele, except in those cases in which the mutation has been caused by insertion of foreign DNA (Jaenisch, et al., 1983; Wagner, et al., 1983; Woychik, et al., 1985).

An alternative means of identifying and isolating such genes is based on the assumption that there has been conservation during evolution of the basic mechanisms of embryonic development, and of the genes that control the process. If this premise is correct, it theoretically should be possible to isolate such genes from the mammalian genome by virtue of their sequence similarity with the genes that control development in other species. Although the validity of the premise that the mechanisms by which development proceeds are fundamentally the same in such different species as fruit flies and mice is still hotly debated (Raff and Raff; Brown and Greenfield, 1985), there is experimental evidence, discussed at length during this workshop, in support of the idea that the genes that control development in <u>Drosophila</u> may have been conserved during the evolution of vertebrates.

The specific approach we took to identifying genes that might control developmental processes in mammals was to ask whether the <u>engrailed</u> (<u>en</u>) gene of <u>Drosophila</u> has been conserved during mammalian evolution. This gene, first identified by traditional mutational analysis (Ecker, 1929), is involved in the processes that subdivide the insect embryo into

NATO ASI Series, Vol. H26
Cell to Cell Signals in Mammalian Development
Edited by S. W. de Laat et al.

separate developmental units called compartments. Each segment of the fly is composed of an anterior and a posterior compartment (Garcia-Bellido, et al. 1973), and in mutants that lack engrailed function, posterior compartments develop abnormal patterns and the compartment and segment borders are not maintained (Lawrence and Morata, 1976; Kornberg, 1981; Lawrence and Struhl, 1982). Molecular studies of the en gene have led to the identification of a gene adjacent to it, now designated invected (inv), which has extensive sequence homology with en (Poole, et al. 1985). Sequence comparisons of en and inv cDNA clones (Coleman, et al. 1987) have revealed two regions of homology between the genes. One region includes a stretch that encodes 17 contiguous amino acids that are identical in the two genes. This region is separated, by sequences encoding 5 (en) or 26 (inv) non-conserved amino acids, from a downstream 180 bp sequence known as the "homeo box," that is very similar in en and inv, but quite divergent from the homeo boxes found in more than 20 other Drosophila genes (Gehring and Hiromi, 1986). Additional homologous sequences encoding 31 amino acids are found immediately downstream of the homeo box. In situ RNA hybridization analysis of inv expression in developing Drosophila embryos has shown that, like en, early in development it is expressed selectively in the cells of the posterior compartments (Coleman, et al., 1987). These results suggest that the en and inv genes have related functions during embryogenesis. As yet, however, no mutations of inv are known that would suggest its role in development.

Using a Drosophila engrailed cDNA clone as a probe, we previously identified and isolated two mouse genes with significant sequence homology to the engrailed and invected genes of Drosophila. For both of these genes, named En-1 and En-2, we determined their stage and tissue specificities of expression during mouse embryogenesis by Northern blot hybridization analysis, and mapped them in the mouse genome (Joyner, et al., 1985 and Joyner and Martin, 1987). We describe here our efforts to localize expression of the En-1 gene in the developing mouse embryo using antisera raised against an En-1 polypeptide.

Results

Northern Blot Hybridization Analysis of En-l Expression During Mouse Embryogenesis

In our earliest study, we carried out a Northern blot hybridization analysis to determine the stages at which the mouse En-l gene is expressed during mouse embryogenesis (Joyner, et al., 1985). Assays were performed using poly $(A)^+$ RNA from PSA-1 teratocarcinoma cells, which serve as an in vitro model system for the mouse embryo at the peri-implantation stages of development (3.5 to 6.5 days of gestation; see Martin, 1980), and from whole mouse embryos at the mid- and late-gestation stages of development. We found that undifferentiated teratocarcinoma stem cells express a relatively low abundance 1.8 kb En-1 transcript. Following differentiation, an additional, larger (2.8 kb) En-1 transcript became detectable. Both transcripts were observed in RNA from embryos at 9.5 through 17.5 days of gestation. However, whereas the less abundant, smaller transcript appeared to be expressed at relatively constant levels in all samples tested, the levels of the more abundant, 2.8 kb transcript appeared to peak during the mid-gestation stages of development (Figure 1).

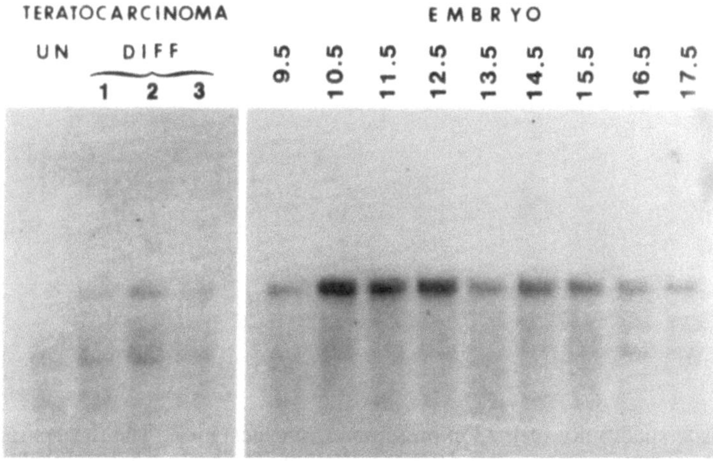

Figure 1. Stage-specificity of En-1 expression during mouse embryogenesis. A Northern blot of poly $(A)^+$ RNA (5 µg) isolated from PSA-1 teratocarcinoma stem cells maintained in the undifferentiated state (un) and at three stages of differentiation (diff 1,2,3), and from whole embryos at the days of gestation indicated, were analyzed by hybridization under conditions of high stringency to a single-stranded 140 bp En-1 homeo box-containing probe (see Joyner, et al. 1985). The readily detected mRNA is approximately 2.8 kb in length.

To examine the tissue-specificity of En-1 transcripts, conceptuses at 12.5 days of gestation were dissected into seven fetal and two extra-embryonic parts and total RNA was extracted from each sample. Northern blot hybridization analysis (Figure 2) showed that the 2.8 kb En-1 RNA was most abundant in the sample containing the posterior portion of the brain. This sample consisted primarily of midbrain, but also included cerebellum, pons and medulla. Little, if any En-1 RNA was detectable in the sample containing the anterior brain, but the 2.8 kb RNA was clearly detectable in the spinal cord, "face," limbs and carcass. Samples derived from the viscera appeared to be completely negative. Although there were several faint bands of hybridization in the lanes containing RNA from yolk sac and from placenta, neither of these extra-embryonic tissue samples contained detectable amounts of the 2.8 kb En-1 RNA. The 1.8 kb En-1 RNA observed in all poly$(A)^+$ RNA samples from teratocarcinoma cells and embryos was not observed in any of the total RNA samples from fetal tissues, possibly because it is expressed in most tissues at levels too low to detect under the conditions we employed. These data demonstrated that expression of the En-1 gene is localized to specific regions of the mid-gestation embryo.

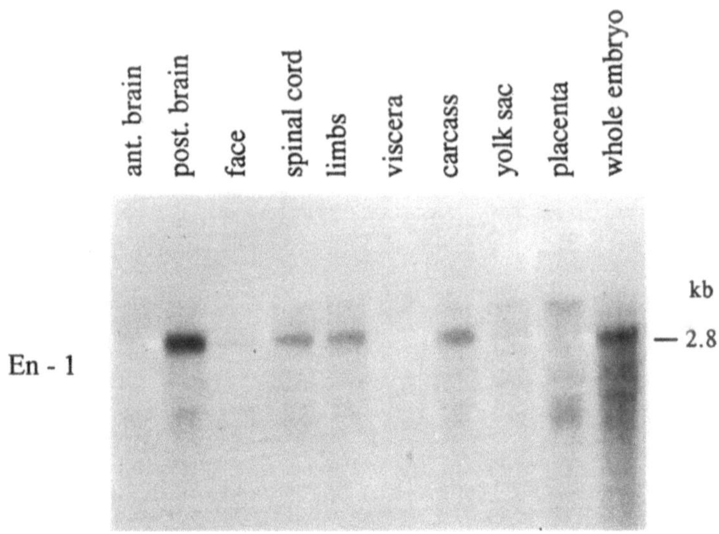

Figure 2. Tissue specificity of En-1 expression at mid-gestation. The tissues indicated were dissected from mouse fetuses at 12.5 days of gestation. The brain was removed from the head and divided into the anterior portion, containing the two frontal lobes, and the posterior portion consisting of midbrain, cerebellum, pons and medulla; the remainder of the head constituted the "face" sample. The spinal cord sample was free of the spinal column but included the spinal ganglia. The "viscera" sample contained a pool of internal organs, and the "carcass" sample contained what remained of the fetus after the other tissues were removed. In addition, yolk sac and placenta samples were dissected from the conceptuses. A Northern blot of total RNA (15 μg) from each tissue was hybridized with an En-1 cDNA insert probe labelled with ^{32}P by nick-translation.

Immunolocalization of the En-1 Protein in Mid-gestation Mouse Embryos

In order to obtain more precise information about the expression of En-1, we set out to raise an antibody that could be used to identify the specific cells in which the En-1 gene product is expressed in the embryo, and also to obtain information on the subcellular localization of this protein. Figure 3 shows a partial restriction map of the cloned mouse En-1 genomic region as well as a map of the longest En-1 cDNA clone available (1.2 kb). The region that shares significant sequence homology with the Drosophila en and inv genes, as well as with the mouse En-2 gene, and in which the homeo box is located, is indicated. To obtain an immunological reagent specific for the En-1 gene, we sought to raise an antibody against a portion of the protein product that is outside of this conserved region. We therefore isolated the 531 bp ClaI/AccI fragment of the En-1 gene immediately upstream of the conserved region (see Figure 3). This portion of the gene was found, by sequence analysis and other criteria, to contain a single long open reading frame that encodes part of the En-1 protein. This fragment was inserted in the appropriate reading frame into a bacterial expression vector system, which produced an ~ 22,000 dalton En-1 polypeptide. Using this partially purified polypeptide as an immunogen in rabbits, we raised a polyclonal antiserum. To make this serum more specific for the En-1 protein, it was purified by affinity chromatography using a column matrix to which the En-1 polypeptide had been coupled.

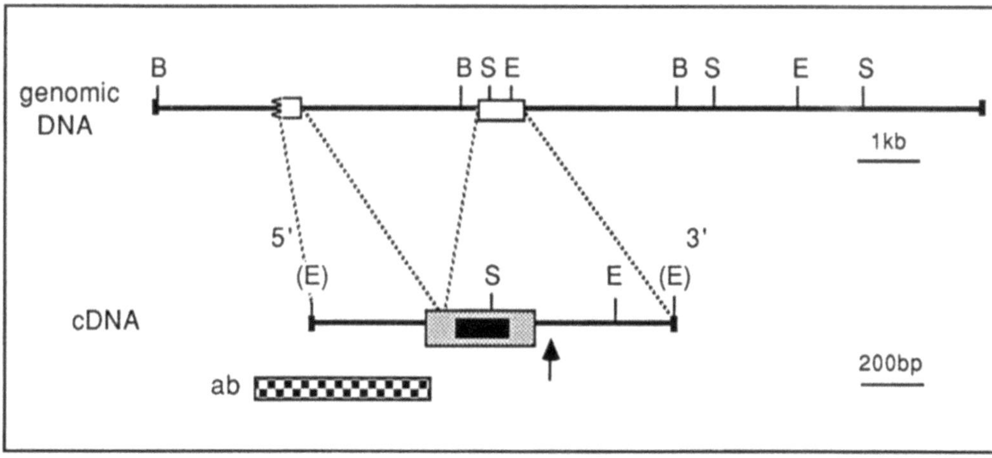

Figure 3. Restriction maps of En-l genomic and cDNA clones. A partial restriction map of the cloned En-l genomic region is shown. The open boxes represent the known positions of exons. Below is an illustration of the longest En-1 cDNA clone isolated thus far. The horizontal line indicates the DNA insert, the ends of which contain synthetic EcoRI sites (E). The stippled box represents the presently known limits of the "conserved" region, which is homologous to sequences in the Drosophila engrailed and invected genes and in the mouse En-2 gene; this region includes the homeo box, which is depicted as a filled rectangle. The vertical arrow indicates the end of the open reading frame that includes the homeo box. The direction of transcription of the gene was determined by hybridization of Northern blots with single-stranded DNA probes. The checkered box labelled **ab** represents the 531 bp ClaI/AccI genomic fragment that was inserted into a bacterial expression vector in order to obtain an En-1 polypeptide. It is drawn and positioned on the same scale as the cDNA clone to illustrate its relationship to the cDNA sequence. B, BamHI; H, HindIII;, Bg, BglII: S, SstI; E, EcoRI.

The anti-En-1 antiserum thus obtained was used to stain frozen sections of mouse embryos at various stages of development. In all cells that were clearly positive, the antibody staining was detected in the nucleus, thus demonstrating that En-1, like the Drosophila en gene to which it is related, encodes a nuclear protein. The pattern of expression of En-1, as detected by these immunolocalization studies, is summarized in the diagrams shown in Figure 4. Briefly, from about 10.5 days of gestation, the protein is readily detectable in the central nervous system, beginning in the anterior diencephalon, and extending posteriorly through the thalamus, pons, medulla (but excluding the cerebellum), and down the length of the spinal cord. The results are consistent with the findings described above, that En-1 RNA was most abundant in the posterior portion of the brain, and detectable in the spinal cord (Figure 2). Within the cord, expression appears to occur in a ventral to dorsal gradient, with expression in the ventral portion of the cord being strongest. Expression was not limited to the central nervous system,

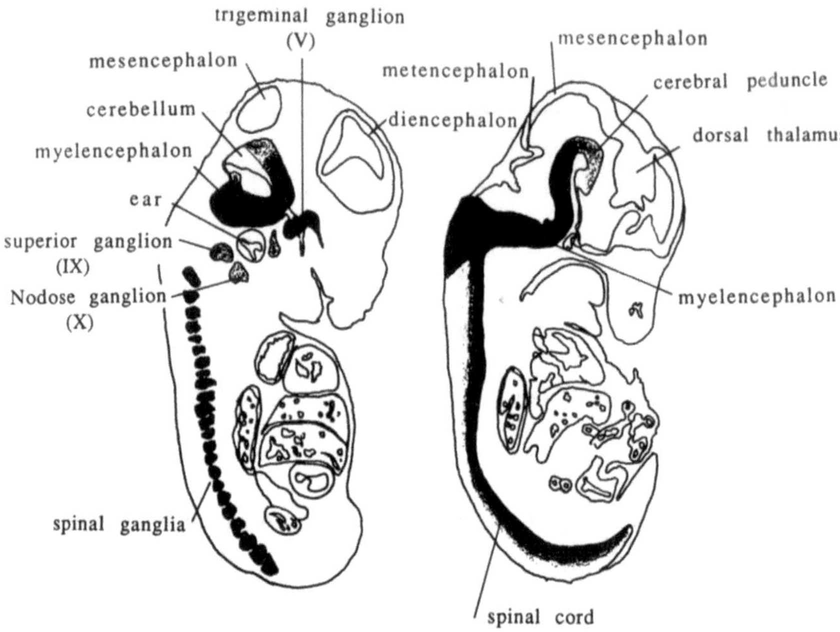

Figure 4. A schematic representation of the pattern of En-1 protein expression in the mid-gestation mouse embryo. The diagrams summarize the results of our immunolocalization studies of the En-1 protein in the mouse embryo at approximately 13 days of gestation. The shaded/black areas represent regions containing antibody-positive cells; the darker the shading, the more numerous or intensely staining the positive cells in that region. Left: parasagittal section; Right: mid-sagittal section.

but could also be detected in the peripheral nervous system. All spinal ganglia contained En-1-expressing cells, as did the trigeminal (V), facial/auditory (VII, VIII), geniculate (IX) and Nodose (X) cranial ganglia. Again, these observations are consistent with the finding that En-1 RNA was detected in the "face" and "carcass" samples, which contained cells from these ganglia. No expression of En-1 protein was detected in any cells outside of the nervous system. This is consistent with our finding that En-1 RNA was not present in the samples of viscera. In fact, the only discordance between our immunolocalization and Northern blot hybridization studies was our inability to detect cells in the developing limbs that contained En-1 protein. One possible explanation is that the En-1 protein may not be expressed in any cells whose nuclei are in the limbs, but rather that the En-1 RNA we detected was present in the axons of neurons whose nuclei are located in the spinal cord or spinal ganglia. Alternatively, it is possible that En-1 expression is under post-transcriptional control and the protein product is not produced in cells in the limb that transcribe the RNA.

In all regions of the nervous system that were positive, En-1 could be detected in a variable proportion of the cells. This proportion appeared to decrease with increasing gestational age, as did the intensity of En-1 staining in individual cells in a given region. No other temporal changes in the pattern of En-1 protein expression were apparent. Cells that are positive are morphologically indistinguishable from their negative neighbors.

Discussion

We have previously identified a mouse gene, En-1, using a cDNA clone of the Drosophila engrailed gene as a probe (Joyner, et al., 1985). Further analysis has revealed that the region of sequence similarity in En-1 and en has also been conserved in another mouse gene, En-2 (Joyner and Martin, 1987), as well as the Drosophila invected gene (Poole, et al., 1985; Coleman, et al. 1987). This region, which is located at the 3' ends of the genes, can code for 107 amino acids, including a centrally located 60 aa residue homeo box. Of these amino acid residues, 78 (73%) are identical in all four genes. Upstream of this region the genes diverge significantly, although there does appear to be one additional short stretch of sequence similarity in all four genes (unpublished data).

The conserved sequences represent a carboxy-terminal protein domain with a molecular mass of approximately 12,000 daltons. The existence of such extensive similarity suggests that the conserved domain in the products of all four genes performs a similar biochemical function. Although this function may be related to that of other homeo box-containing genes, it is likely to have some unique features specified by the conserved sequences outside of the homeo box. Although the two mouse genes might be the homologs of the Drosophila engrailed and invected genes with respect to their specific biochemical functions (e.g. DNA binding), it remains to be determined whether En-1 and En-2 play some role in the control of embryogenesis in mammals that is in any way analogous to the developmental functions of their counterparts in Drosophila.

With respect to En-1, the Northern blot hybridization data shown here demonstrate that the gene is expressed in differentiating PSA-1 teratocarcinoma cells, implying that it is also expressed early in normal embryonic development, during periods when fundamental cell lineage decisions are being made and when the basic embryo body plan is being established. Thus the En-1 gene could conceivably play a role in these processes, as does the engrailed gene is Drosophila. What that role might be remains obscure, since we were unable to identify cells in early embryos that express the En-1 protein product. This could be due either to insensitivity of our antibody localization technique, or possibly because the En-1 protein is not produced in the early embryo.

The results of our immunolocalization study clearly demonstrate that the En-1 protein is expressed in specific cells of the central and peripheral nervous system of the mid-gestation embryo, suggesting that the gene may play some role in the development of the nervous system. The observation that only a fraction of the cells in En-1-positive regions of the mouse embryonic nervous system appear to express the En-1 protein, raises the intriguing possibility that the presence of this protein provides those cells with information necessary for their differentiation along a particular neuronal pathway. Expression of both engrailed and invected has also been detected in specific cells of the nervous system of developing Drosophila embryos (Ingham, et al., 1986; Weir and Kornberg, 1985; Coleman, et al. 1987), further suggesting that there might be some common function in the different species. Although these are attractive ideas, at present they remain highly speculative.

Although the pattern of expression we observed for En-1 is unique, it has certain features in common with the patterns of expression of other murine homeo box-containing genes. Each of the genes in this category that has been studied in any detail, has been found to be expressed in some portion of the central nervous system (Awgulewitsch, et al., 1986; Toth, et al., 1987, Dony and Gruss, 1987; Gaunt, 1987; Utset, et al. 1987; Holland and Hogan, 1988; Brier, et al., 1988; LeMouellic, et al., 1988; Davis, et al., 1988). However, expression of En-1 appears to occur more extensively along the length of the nervous system than that of any other homeo box-containing gene. An even more striking difference is that all of the other homeo box-containing genes, with the exception of En-2 (Davis, et al., 1988), are also expressed in cells outside the nervous system, whereas En-1 and En-2 are not. As yet, the significance of these findings is unknown.

Our studies have identified En-1 as a candidate for a gene that plays a role in the development of the mammalian nervous system, and possibly has functions earlier in development as well. To further explore its role in embryogenesis, we plan to carry out studies aimed at determining the consequences of perturbing the expression of En-1 during embryogenesis. In particular, we are working to create animals homozygous for a mutant En-1 gene that is unable to express the normal protein product. This should bring us full circle, back to the classical genetic approach, whereby the study the mutant embryo phenotypes may provide some insight into the function of the wild-type En-1 gene in the developing mammalian embryo.

Acknowledgments

This work was supported by funds from NIH grant HD-20959. M.A.F. was supported by a post-doctoral fellowship from the American Cancer Society.

References

Awgulewitsch, A., Utset, M.F., Hart, C.P., McGinnis, W. and Ruddle, F.H. (1986). Spatial restriction in expression of a mouse homeo box locus within the central nervous system. Nature 320, 328-335.

Breier, G., Dressler, G.R. and Gruss, P. (1988). Primary structure and developmental expression pattern of Hox 3.1, a member of the murine Hox 3 homeobox gene cluster. EMBO J., in press.

Coleman, K.G., Poole, S.J., Weir, M., Soeller, W.C. and Kornberg, T. (1987). The invected gene of Drosophila: sequence analysis and expression studies reveal a close kinship to the engrailed gene. Genes and Devel. 1, 19-28.

Davis, C.A., Rossant, J. and Joyner, A.L. (1988). Expression of the homeo box-containng gene En-2 delineates a specific region of the developing mouse brain. Genes and Development, in press.

Dony, C. and Gruss, P. (1987). Specific expression of the Hox 1.3 homeo box gene in murine embryonic structures originating from or induced by the mesoderm. EMBO J. 6, 2965-2975.

Ecker, R. (1929). The recessive mutant engrailed in Drosophila melanogaster. Hereditas 12, 217-222.

Garcia-Bellido, A., Ripoll, P. and Morata, G. (1973). Developmental compartmental-ization of the wing disc of Drosophila. Nature New Biol. 245, 251-253.

Gaunt, S.J. (1987). Homeobox gene *Hox-1.5* expression in mouse embryos: earliest detection by in situ hybridization is during gastrulation. Development 101, 51-60.

Gehring, W.J. and Hiromi, Y. (1986). Homeotic genes and the homeobox. Ann. Rev. Genet. 20, 147-173.

Holland, P.W. and Hogan, B.L.M. (1988), Spatially restricted patterns of expression of the homeobox-containng gene Hox-2.1 during mouse embryogenesis. Development 102, 159-174.

Ingham, P., Martinez-Arias, A., Lawrence, P., and Howard, K. (1985). Expression of engrailed in the parasegment of Drosophila. Nature 317, 634-636.

Jaenisch, R., Harbers, K., Schnieke, A., Lohler, J., Chumakov, I., Jahner, D., Grotkopp, D. and Hoffman, E. (1983). Germline integration of Moloney murine leukemia virus at the Mov-13 locus leads to recessive lethal mutation and early embryonic death. Cell 32, 209-216.

Joyner, A.L. and Martin, G.R. (1987). En-l and En-2, two mouse genes with sequence homology to the Drosophila engrailed gene: expression during embryogenesis. Genes and Devel. 1: 29-38.

Joyner, A.L., Kornberg, T., Coleman, K., Cox, D., and Martin, G.R. (1985). Expression during embryogenesis of a mouse gene with sequence homology to the Drosophila engrailed gene. Cell 43, 29-37.

Kornberg, T. (1981). Engrailed: a gene controlling compartment and segment formation in Drosophila. Proc. Natl. Acad. Sci. USA 78, 1095-1099.
Lawrence, P.A., and Morata, G. (1976). Compartments in the wing of Drosophila: a study in the engrailed gene. Dev. Biol. 50, 321- 337.
Lawrence, P.A., and Struhl, G. (1982). Further studies of the engrailed phenotype in Drosophila. EMBO J. 1, 827-833.
Le Mouellic, H., Condamine, H. and Brulet, P. (1988). Pattern of transcription of the homeo-gene Hox-3.1 in the mouse embryo. Genes and Devel., in press.
Martin, G.R. (1980). Teratocarcinomas and mammalian embryogenesis. Science 209, 768-776.
Poole, S.J., Kauvar, L.M., Drees, B., and Kornberg, T. (1985). The engrailed locus of Drosophila: structural analysis of an embryonic transcript. Cell 40, 37-43.
Raff, E.C. and Raff, R.A.; Brown, S.D.M. and Greenfield, A.J. (1985). Possible functions of the homeo box. Nature 313, 185-186.
Toth, L.E., Slawin, K.L., Pintar, J.E. and Nguyen-Huu, M.C. (1987). Region-specific expression of mouse homeobox genes in the embryonic mesoderm and central nervous system. Proc. Nat. Acad. Sci. USA 84, 6790-6794.
Utset, M.F., Awgulewitsch, A., Ruddle, F.H. and McGinnis, W. (1987). Region-specific expression of two mouse homeo box genes. Science 235, 1379-1382.
Wagner, E.F., Covarrubias, L., Stewart, T.A. and Mintz, B. (1983). Prenatal lethalities in mice homozygous for human growth hormone gene sequences integrated in the germ line. Cell 35, 647-655.
Weir, M. and Kornberg, T. (1985). Patterns of engrailed and fushi tarazu transcripts reveal novel intermediate stages of Drosophila segmentation. Nature 318, 433-439.
Woychik, R.P., Stewart, T.A., Davis, L.G., D'Eustachio, P. and Leder, P. (1985). An inherited limb deformity created by insertional mutagenesis in a transgenic mouse. Nature 318, 36-40.

The Hox 1.3 Homeo Box Gene Encodes a Sequence Specific DNA Binding Phosphoprotein

Ward F. Odenwald, James Garbern, Heinz Arnheiter, Elizabeth Tournier-Lasserve, Hilary A. Overton* and Robert A. Lazzarini
Laboratory of Molecular Genetics, NINCDS, NIH,
Bethesda, Maryland 20892
*NERC Institute of Virology
Mansfield Road, Oxford, OX1 3SR, UK

Introduction

The murine Hox 1.3 homeo box gene is the third member of the Hox 1 cluster, located on chromosome 6 (Colberg-Poley et al., Búcan et al.). Analysis of the Hox 1.3 cDNA and genomic sequences indicate that its major transcript of 1.85 Kb (detected in both embryonic and adult tissues) encodes a 270 amino acid homeo domain protein (Odenwald et al.). Our previous studies suggested that the Hox 1.3 protein present in cultured fibroblasts may be post-translationally modified (i.e., aberrant migration and multiple immunoreactive forms in SDS gels). In order to obtain sufficient amounts of the Hox 1.3 protein to study its biology, we have employed the Baculovirus expression system. In addition to high levels of protein expression, this eukaryotic system has been shown to perform correct post-translational modification of nuclear proteins (Miyamoto et al.).

Results and Discussion

The Hox 1.3 protein encoding cDNA sequences (present in the major transcript) were inserted behind the polyhedrin promoter of the recombinant Baculovirus and used to infect host Spodoptera frugiperda cells. Twenty hours post-infection, phase-dense intranuclear masses appear and by 40 hours, they occupy more than 50% of the nuclear volume. Immunofluorescent studies with either Hox 1.3 peptide specific, rabbit polyclonal or mouse monoclonal antibodies identify these structures as Hox 1.3 accumulations (Figure 1). Western analysis of protein extracts obtained from nuclei 20 hours post-infection and fractionated on one dimensional SDS gels reveals five distinct forms of Hox 1.3 immunoreactive protein. No cross-reacting proteins are detected in extracts obtained from uninfected, wild-type virus or unrelated recombinant virus infected cells (Figure 2). The five forms migrate through 12.5% polyacrylamide SDS gels with apparent molecular weights ranging from approximately 35 to 43 Kd.

NATO ASI Series, Vol. H26
Cell to Cell Signals in Mammalian Development
Edited by S. W. de Laat et al.

Pulse-chase experiments followed by immunoprecipitation reveal that the ^{35}S labeled methionine appears first in the fastest migrating form and is chased into the slower migrating proteins (Figure 3a). Comparison of these immunoprecipitates to those obtained from cells incubated with ^{32}P orthophosphate demonstrates that the slower migrating forms are phosphoproteins (Figure 3b). Alkaline phosphatase treatment of ^{35}S methionine labeled proteins demonstrates that the slower mobilities are the result of different phosphorylation events (Figure 3c). Alkaline treatment alone (100 mM NaOH for 30 minutes) does not alter the SDS gel pattern suggesting that Hox 1.3 is not poly ADP ribosylated. Endogenous Hox 1.3 proteins, present in the nuclei of primary embryonic fibroblasts, co-migrate with the

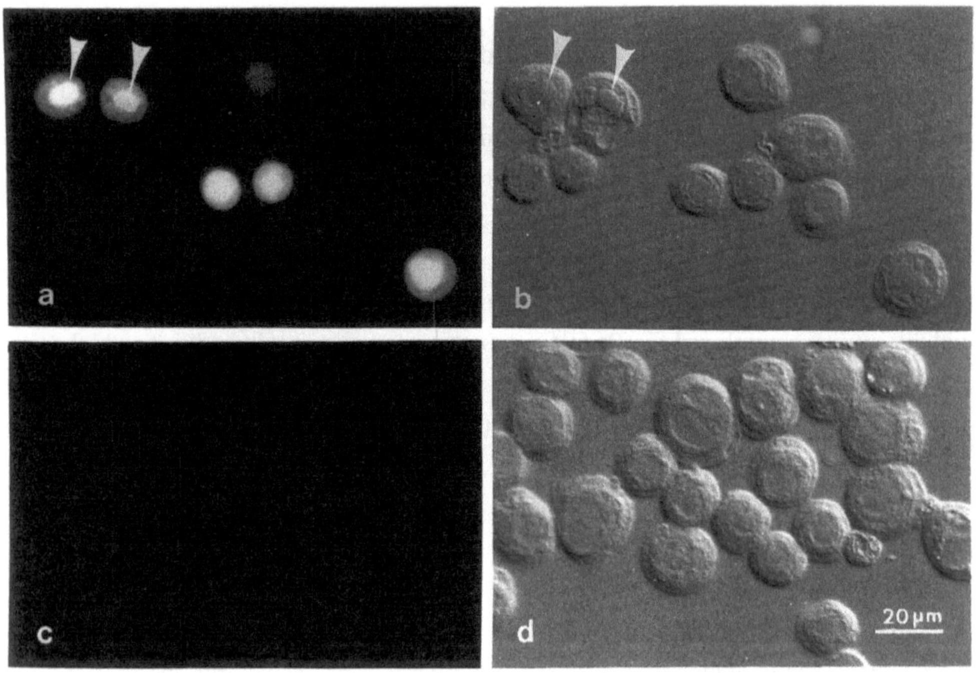

Figure 1: The Hox 1.3 protein accumulates in the nuclei of S. Frugiperda cultured cells infected with the recombinant Baculovirus. Immunofluorescence and differential interference contrast microscopy (D.I.C.) views (a and b) of host cells 20 hours post-infection with a low multiplicity of infection. Note that Hox 1.3 protein co-localizes with intra nuclear inclusions (arrows). Control immunostaining and D.I.C. views (c and d) demonstrating that uninfected cells lack the Hox 1.3 epitope. Cells infected with wild-type or unrelated recombinant viruses also show no immunostaining (data not shown).

recombinant proteins (Figure 4) suggesting that similar post-translational processing occurs in mammalian cells. Differential salt elution of Hox 1.3 proteins from fibroblast nuclei suggests that phosphorylation increases the protein's nuclear binding affinity (Figures 4 and 5).

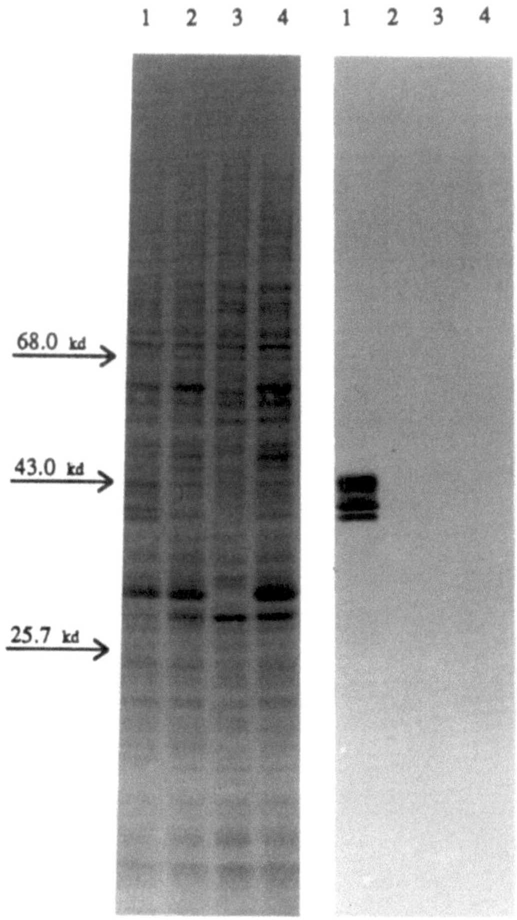

Figure 2: Immunoblot analysis of crude nuclear extracts demonstrates that Hox 1.3 immunoreactive proteins are unique to extracts from cells infected with the Hox 1.3 recombinant Baculovirus. Crude nuclear extracts were prepared according to the Dignam et al. protocol from Hox 1.3 recombinant virus infected cells (36 hrs., p.I) lane 1; wild-type Baculovirus infected cells (36 hrs., p.I.) lane 2; uninfected cells lane 3; and an unrelated recombinant virus infected cells (also 36 hrs., p.I.) lane 4. The proteins were resolved on a 12.5% SDS polyacrylamide gel and electroblotted onto nitrocellulose. Left half, Amido black staining. Right half, Immunoblot with Hox 1.3 peptide specific monoclonal antibodies. The same immunostaining profile was observed with our peptide specific polyclonal antibodies (data not shown).

It is possible that homeo domain proteins act as regulators of transcription by binding DNA in a sequence specific manner. To ascertain if Hox 1.3 protein(s) bind to specific DNA sequences, we have developed a procedure which enables one to screen large blocks of genomic sequences for protein-DNA binding sites. Unlabeled DNA restriction fragments

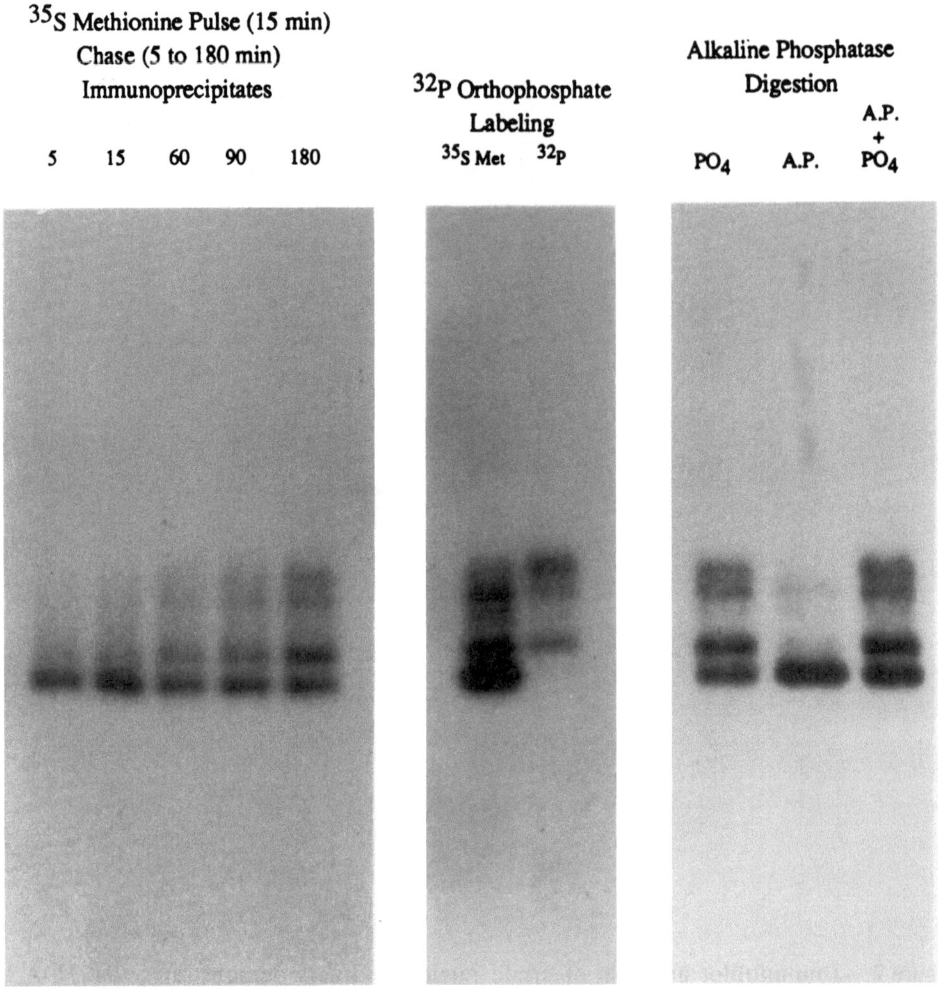

Figure 3: Multiple phosphorylation events alter the mobility of <u>Hox 1.3</u> in SDS polyacrylamide gels. (a) Pulse-chase experiments reveal that the multiple immunoreactive <u>Hox 1.3</u> proteins arise from the fastest migrating form. (b) Comparison of ^{35}S methionine and ^{32}P orthophosphate labeled <u>Hox 1.3</u> immunoprecipitated proteins demonstrates that the slower migrating forms of <u>Hox 1.3</u> are phosphoproteins. (c) Disappearance of the <u>Hox 1.3</u> phosphoproteins after treatment with alkaline phosphatase suggests that their slower mobilities are due to phosphorylation.

are separated by agarose electrophoresis, transferred to membranes (Immobilon, Millipore Corp.) while in their native, double-stranded form and probed with DNA binding proteins. After removal of unbound proteins, the position of the DNA binding protein is determined by immunochemical staining. Using this Southern cis/trans blot protocol with nuclear extracts containing <u>Hox 1.3</u> proteins, we have recently identified one such site which binds

Recombinant and Endogenous

Hox 1.3 Proteins

1 2 3

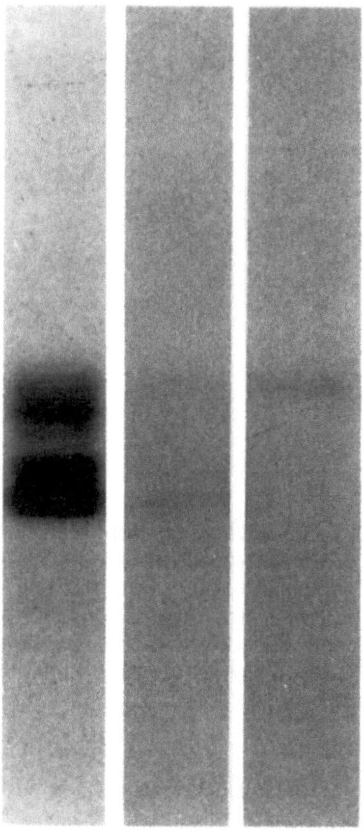

Figure 4: Autoradiographs of ^{35}S methionine labeled immunoprecipitated proteins resolved on a 12.5% SDS polyacrylamide gel. Lane 1 contains the Hox 1.3 recombinant proteins. Lane 2 contains Hox 1.3 immunoreactive proteins present in a 600 mM NaCl wash of nuclei from cultured mouse primary embryonic fibroblasts. Lane 3 contains Hox 1.3 immunoreactive proteins present in these fibroblasts nuclei after the 600 mM NaCl wash (solubilized by RIPA buffer). Note the comigration of the fibroblast proteins with those of the recombinant. Also note, that a significant amount of the slower migrating (presumably phosphorylated) fibroblast protein is not eluted from the nuclei during the 600 mM NaCl wash.

Hox 1.3 protein(s) under stringent conditions (250 mM KCl) (data not shown). This binding site is located upstream from the Hox 1.3 protein encoding sequences in a region which contains blocks of non-colinear sequence homology with the Hox 2.1 gene(Krumlauf et al., Garbern et al.). DNase I protection analysis reveals a binding site "a-b" located 131 bp upstream from the 5' end of the 1.85 Kb transcript (see Figures 6 and 7). Binding to these sequences is detected with the least amount of the Hox 1.3 protein containing extract and is not observed in our controls which contained 6 to 8 times more protein. Part of these protected sequences (region b) share homology with sequences protected by other Antennapedia (Antp) class homeo domain proteins. The Drosophila Ultrabithorax (P. Beachy et al., unpublished), Antp (W. Gehring et al., unpublished, presented at the Advances in Gene Technology: The Molecular Biology of Development meeting, Miami, Florida, February 1987), even-skipped and Deformed (M. Levine et al.,

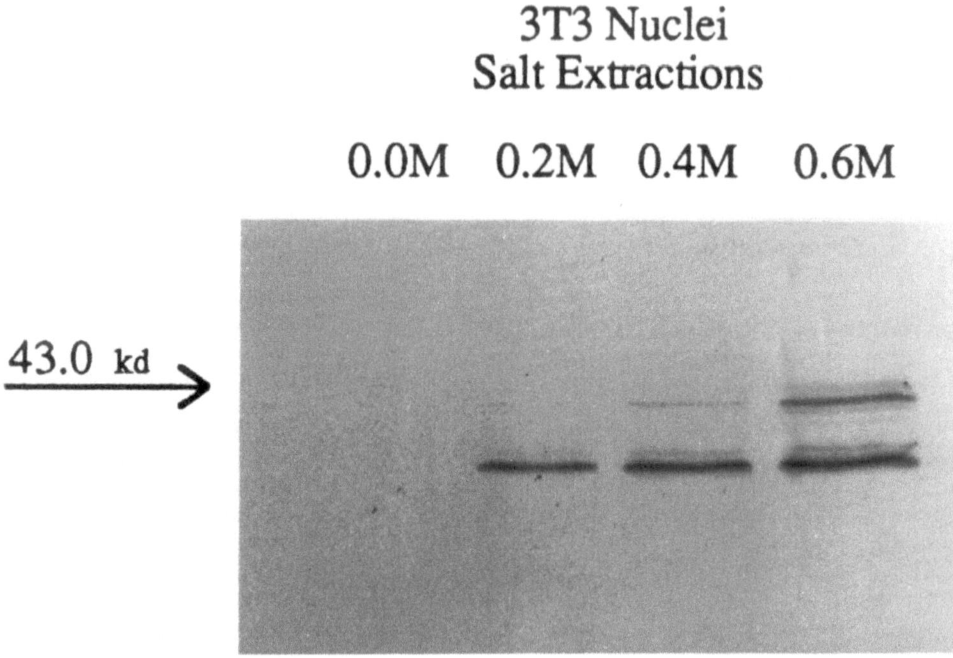

Figure 5: Differential salt elution of Hox 1.3 proteins from 3T3 nuclei. Nuclei prepared from nonconfluent 3T3 fibroblasts were aliquoted and exposed to one of four different NaCl concentrations. The proteins in these salt washes were resolved in a 12.5% SDS polyacrylamide gel, electroblotted onto nitrocellulose and immunostained with Hox 1.3 specific antibodies. The observed differential elution profile of the Hox 1.3 proteins suggests that the slower migrating forms may have a higher affinity for nuclear binding sites.

unpublished, presented at the NIH bi-centennial meeting, Bethesda, Maryland, October 1987), and the Xenopus Xhox 1a (G. Oliver et al., unpublished, presented at the Advanced NATO Workshop, Les Arcs, France, February 1988) all share a core protected sequence of ATTA. Protected sequences c, d and e (detected only when higher amounts of the Hox 1.3 extracts were used) do not share sequence homology with each other or with regions a and b and may represent binding sites protected by other factors present in the Hox 1.3 recombinant Baculovirus infected S. Frugiperda extract.

Conclusions and Speculations

Expression of Hox 1.3 protein encoding sequences in the Baculovirus expression system results in the synthesis of Hox 1.3 nucleophilic phosphoproteins. These multiple

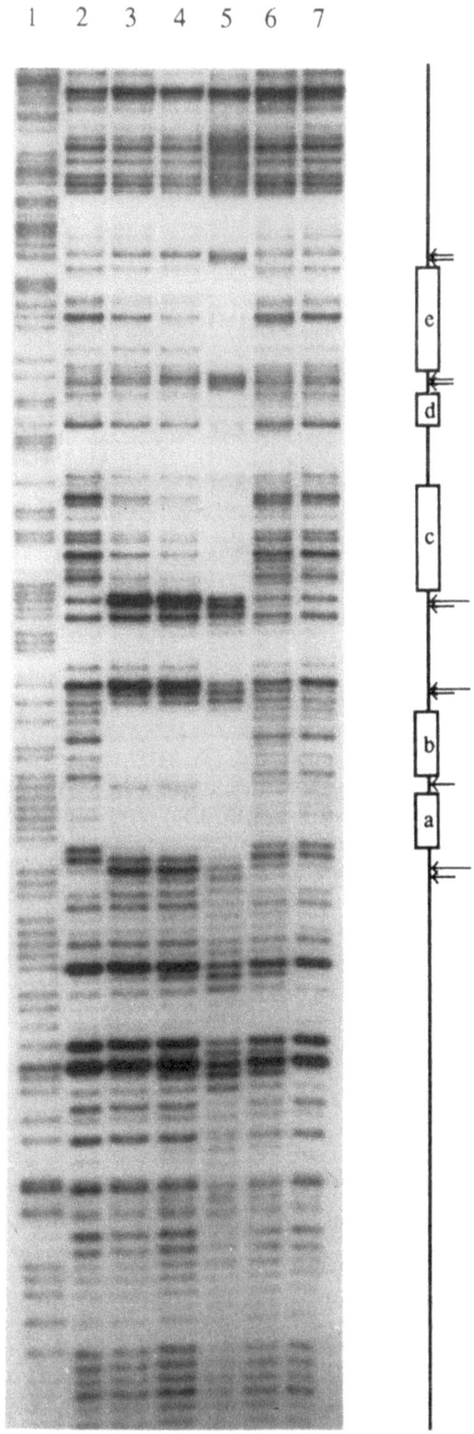

Figure 6: DNase I protection analysis of Hox 1.3 upstream sequences by nuclear extracts from S. Frugiperda cells. A 310 amino bp genomic fragment which contains 250 bp of Hox 1.3 upstream sequences from the lung major cap site was employed (see accompanying figure). Lane 1, Maxam-Gilbert G+A purine ladder. Lane 2, DNase I digest of one ng of end labeled fragment with 1.5 μg poly dI·dC. Lane 3, as in lane 2, but pre-incubated with 1.2 μg of crude extract from Hox 1.3 recombinant virus infected cells. Lane 4, as in lane 3, but pre-incubated with 2.4 μg of extract from the Hox 1.3 recombinant infected cells. Lane 5, as in lane 3, but pre-incubated with 9.6 μg of the recombinant containing extract. Lane 6, as in lane 2, but pre-incubated with 8 μg of crude extract from wild-type infected cells. Lane 7, as in Lane 2, but pre-incubated with 10 μg of crude extract from uninfected cells. In the flanking cartoon, boxes represent protected regions and arrows indicate DNase I hypersensitive sites induced by the recombinant extracts. Note that protected regions a and b are detected in the digests which contain 1.2 μg of the recombinant extracts while regions c, d and e show significant protection only with 9.6 μg of the extract.

Alignment of DNase I Footprints With
Upstream Homologies Between Hox 1.3 and Hox 2.1

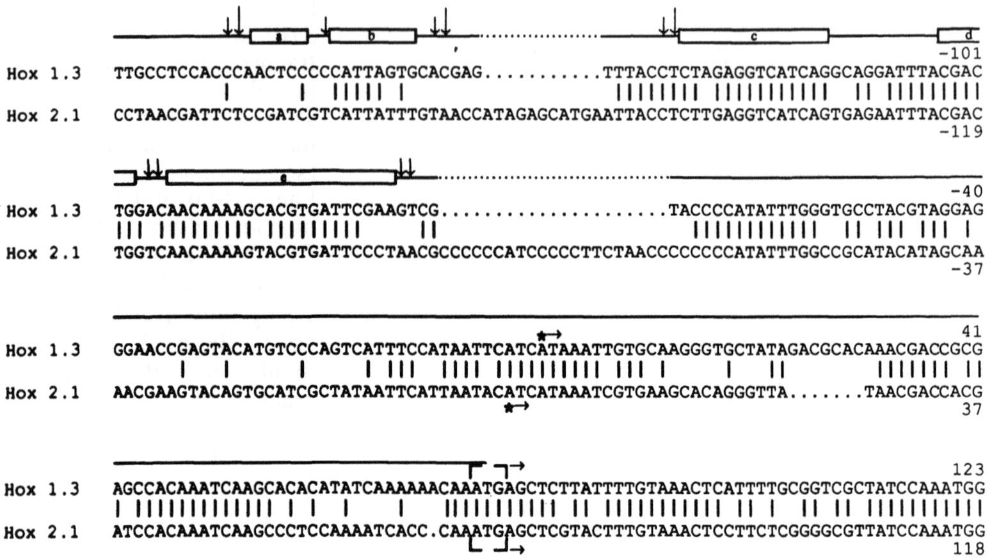

Figure 7: Optimal alignment of Hox 1.3 and Hox 2.1 upstream sequences reveals blocks of non-collinear homologies. One of these regions (Hox 1.3, -109 to -77; Hox 2.1, -127 to -95) shares 67% sequence homology with a human Ig germ line heavy chain enhancer element (Rabbitts et al.). Note that this region is coincident with DNase I protected regions c, d and e. Sequences homologous to protected region b are also present in multiple copies in the heavy chain enhancer element. Adult lung mRNA major cap sites are denoted by stars and the translation start codons are bracketted.

forms, resolved on one dimensional SDS polyacrylamide gels, co-migrate with Hox 1.3 proteins present in the nuclei of cultured fibroblasts. Differential salt elution of Hox 1.3 proteins from fibroblasts nuclei suggests that phosphorylation modulates the proteins function by increasing its binding affinity.

Nuclear extracts from Hox 1.3 recombinant virus infected cells contain proteins which bind to different regions upstream of the major Hox 1.3 transcript (as determined by DNase I protection assays). All but one of these protected regions are conserved between Hox 1.3 and Hox 2.1. The lack of sequence homology among the five binding sites suggests that different factors are responsible for these footprints. One of these footprints shares homology with the core sequence present in the binding sites of other Antp class homeo domain proteins and most likely represents part of the Hox 1.3 binding site. Protection in this region is observed with the lowest concentration of extract while eight times more extract is required to reveal the other footprints. DNase I protection is not observed when equivalent amounts of extracts from wild-type virus infected cells or uninfected cells are used. Perhaps the binding of the Hox 1.3 protein is required for the binding of these other factors which recognize cis-elements conserved not only between Hox 1.3 and Hox 2.1 but also between insects and mammals. The observed homology between these upstream sequences and a transcriptional enhancer element, the Ig heavy chain enhancer, further suggests that Hox 1.3 protein may function as a trans-acting transcriptional regulatory factor which in addition to its own gene, may modulate the expression of others.

References

Búcan M, Yang-Feng T, Colberg-Poley AM, Wolgemuth DJ, Guenet J-L, Francke U, Lehrach H (1986) Genetic and cytogenetic localization of the homeo box containing genes on mouse chromosome 6 and human chromosome 7. EMBO J 5:2899-2905.

Colberg-Poley AM, Voss SD, Chowdhury K, Gruss P (1985) Structural analysis of murine genes containing homeo box sequences and their expression in embryonal carcinoma cells. Nature 314:713-718.

Dignam JD, Lebowitz RM, Rioeder RG (1983) Accurate transcription initiation by RNA polymerase II in a soluble extract from isolated mammalian nuclei. Nucl Acids Res 11: 1475-1489.

Garbern J, Krumlauf R, Odenwald W, Lazzarini R Sequence and expression homology of two homeo box genes on different chromosomes. In preparation.

Krumlauf R, Holland PWH, McVey JH, Hogan BLM (1987) Developmental and spatial patterns of expression of the mouse homeo box gene, Hox 2.1. Development 99:603-617.

Odenwald WF, Taylor CT, Palmer-Hill FJ, Friedrich V, Tani M, Lazzarini, RA (1987) Expression of a homeo domain protein in non-contact inhibited cultured cells and post-mitotic neurons. Genes and Development 1:482-496.

Miyamoto C, Smith GE, Farrell-Towt J, Chizzonite R, Summers MD, Ju G (1985) Production of human c-myc protein in insect cells infected with a baculovirus expression vector. Mol Cell Biol 5:2860-2865.

Rabbitts TH, Forster A, Baer R, Hamlyn PH (1983) Transcription enhancer identified near the human C-mu immunoglobulin heavy chain gene is unavailable to the translocated c-myc gene in a Burkitt lymphoma. Nature 306:806-809.

Analysis of transcription of the murine homeobox gene Hox 1.3

James Garbern, Ward Odenwald, Elizabeth Tournier-Lasserve, and Robert Lazzarini

Laboratory of Molecular Genetics

National Institute of Neurological and Communicative Disorders and Stroke

National Institutes of Health

Bethesda, Maryland 20892, USA

Introduction

Development in Drosophila is critically dependent on the expression of classes of regulatory genes which establish the polarity of the organism (rev. Nüsslein-Volhard et al.), define the segmentation (rev. Scott and Carroll), determine individual segmental characteristics (rev. Gehring), and establish neuronal identities (Doe et al.). These genes were discovered by the effects of mutations which disturbed embryogenesis. Many of the genes which take part in these processes in Drosophila have been characterized at the molecular level and, surprisingly, were found to share a 180 base pair region which is called the homeobox (Gehring). The homeobox encodes a sequence specific DNA binding region, the homeodomain, of these regulatory proteins which presumably modulate the expression of other genes in a coordinated fashion (Gehring; Scott and Carroll). Tremendous excitement was generated by the discovery that the homeobox is highly conserved not only between many developmentally important Drosophila genes, but also among many other organisms including sea urchins (Dolecki et al.), fish (Eiken et al.), frogs (Müller et al.), mice (McGinnis et al., Colberg-Poley et at. 1985a) and humans (Acampora et al.).

While development in these higher organisms follows much different plans than does Drosophila there is suggestive evidence for regulatory roles for homeobox genes in these species as well. These genes are expressed very early during development in all organisms studied and generally have specific spatial patterns of expression, especially along the neuraxis (review Fienberg et al.). Their protein products are localized

NATO ASI Series, Vol. H26
Cell to Cell Signals in Mammalian Development
Edited by S. W. de Laat et al.

to the nucleus (Odenwald et al. and Kessel et al.), and more recently have been shown to have sequence specific DNA binding properties (Fainsod et al. and Odenwald this meeting). As yet, there are no reports of developmental anomalies caused by mutations of homeobox genes in higher organisms.

Mice have at least 20 homeobox containing genes (as does Drosophila) which are clustered on at least 4 chromosomes (Martin et al.). Our laboratory has previously described Hox 1.3, which lies in a cluster of at least 6 other homeobox genes on chromosome 6 (Odenwald et al.). The gene organization for the major Hox 1.3 transcript is shown in fig. 1. This tran-

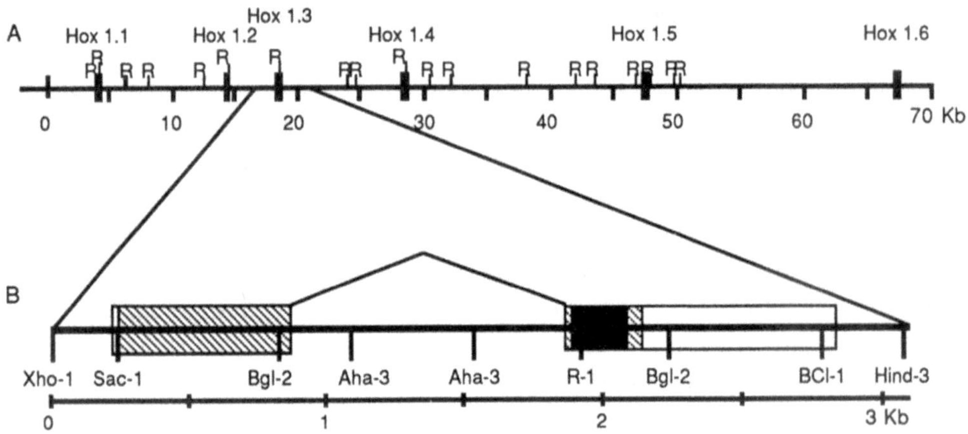

Fig. 1 (A). The murine Hox-1 cluster with EcoRI restriction sites as compiled by Colberg-Poley et al. 1985b and Duboule et al. (B) Transcription unit map of the Hox 1.3 1.85 Kb transcript. The solid boxes denote the homeobox. The hatched boxes contain the open reading frame and the open boxes are untranslated parts of the exons.

script is generated by splicing two exons which are separated by a 960 bp intron. The deduced open reading frame encodes a protein of 270 amino acids which is rich in serine, proline, and glycine and also has several domains (PEST regions - regions more than 10 residues in length flanked by basic amino acids and enriched in proline, glutamate, aspartate, serine and threonine) which are characteristic of rapidly degraded intracellular proteins (Rogers et al.). There is a 792 bp 3' untranslated region which contains sequences similar to those found by Shaw and Kamen and which confer susceptibility to RNAse degradation (Shaw and Kamen). Our prior studies of Hox 1.3 gene expression determined the levels of transcription of the gene in various embryonic and adult tissues (fig. 2). In most tissues examined, transcripts of approximately 1.85, 4, 8 and 9 Kb are found, irrespective of whether hybridization probes from the first exon or the 3' untranslated region are used. In addition to the

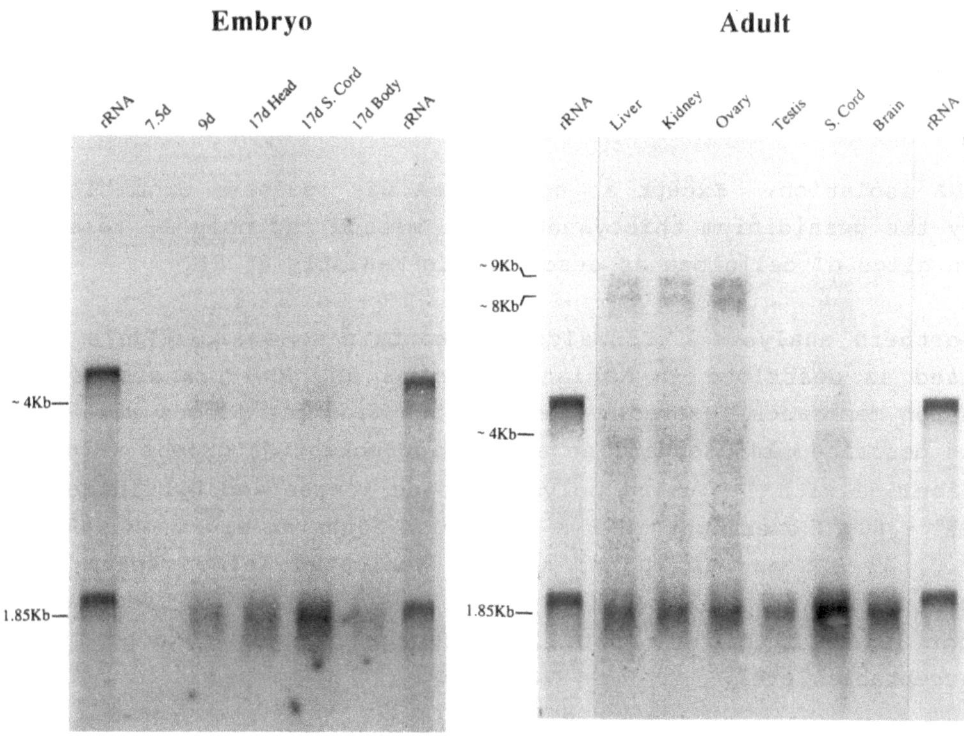

Fig. 2. Northern analysis of poly A+ RNA from embryonic and adult mouse tissues. 1 µg poly A+ RNA from the indicated embryonic tissues was analyzed and 10 µg poly A+ RNA from the adult tissues were used. Exposure times were 3 and 7 days for the embryo RNA and adult RNA blots, respectively. ^{14}C-labelled rRNA was used as molecular weight markers.

embryonic tissues expressing higher levels of the gene than the corresponding adult tissues, the relative proportions of the differently sized transcripts are also different. There is also apparent tissue specific variation in the levels of the larger transcripts with respect to the predominant 1.85 Kb form.

Because of the presumption that homeobox genes regulate development we have undertaken studies to understand their own transcriptional regulation with the hope of finding genes which reside at a higher level in the hierarchy of gene control during development. In order to define potential upstream regulatory regions of the Hox 1.3 gene we have determined the transcriptional initiation sites of the 1.85 Kb transcript and have begun studies to determine the structure of the 4 Kb transcript.

Materials and Methods

RNA isolation. Except as noted, RNA was isolated from tissues by the guanidinium thiocyanate-CsCl method and poly A+ selected on oligo dT cellulose as described in Maniatis et al.

Northern analysis. Formaldehyde containing agarose gels were used as described in Maniatis' manual and RNA transferred to nylon membranes (Biorad). Restriction fragment probes were used as described in Odenwald et al. Oligonucleotide probes were end labelled with ^{32}P using polynucleotide kinase and hybridized at 55° - 60° C overnight in 2 X SSC, 10 X Denhardt's, 20 mM NaPi pH 7.0, 1% SDS and 100 µg/ml sheared sonicated salmon sperm DNA. Blots were washed in 1 X SSC at the hybridization temperature for 10 to 20 minutes until the background was reduced to an acceptable level.

Primer extension. RNA was annealed in 0.4M NaCl, 10 mM PIPES pH 6.7 at 55° C for 12-16 h to an end labeled oligonucleotide primer which was then extended with AMV reverse transcriptase. The products were analyzed on a 6% polyacrylamide sequencing gel with labeled molecular weight standards or a parallel sequencing reaction of a Hox 1.3 template primed with the same but unlabeled oligonucleotide.

cDNA library preparation. Poly A+ RNA from spinal cords pooled from Balb/c mouse embryos of 13 - 15 days gestation was isolated and an Okayama-Berg cDNA library prepared, all as described in Okayama, H. et al. The vector primer and oligo dG-tailed linker DNA fragments were generous gifts of Dr. Hiroto Okayama.

DNA sequencing. Single and double stranded DNA templates were sequenced by the dideoxynucleotide method using AMV reverse transcriptase or Sequenase (modified T7 DNA polymerase (Tabor and Richardson)) using the Promega and United States Biochemicals protocols, respectively.

Results and discussion

The major cap site for the 1.85 Kb Hox 1.3 transcript in mouse lung was found to reside at position -74 from the initiation codon (figure 3). Other transcription initiation sites at positions -44 to -47 are used less efficiently. There are no canonical TATA or CAAT boxes in the sequences immediately 5' from this site. Initiation tends to occur at CAPy sequences. Most embryonic spinal cord Hox 1.3 transcripts of approximately 1.85 Kb in length also initiate within 74 nucleotides of the start codon, but a small fraction utilizes slightly more upstream cap sites (data not shown) and are represented by the faint hybridization signal in the 1.85 - 1.9 Kb region seen on hybridization of Northerns with the 5' oligonucleotide probe. This same probe, however, hybridizes with the same intensity to the 4 Kb transcript, of both embryonic spinal cord and adult lung, as does the 3' probe. Of interest, the mouse homeobox gene hox 2.1, which resides on chromosome 11, shares several blocks of homology upstream of its initiation codon with Hox 1.3, including the regions of its cap site and blocks both upstream and downstream from this region (Garbern, J., Krumlauf, R., Odenwald, W. et al. in preparation). These may represent sequences important for transcriptional and/or translational regulation of these genes' expression (Soeller et al. and Bürglin et al.).

From the embryonic mouse spinal cord cDNA library, 24

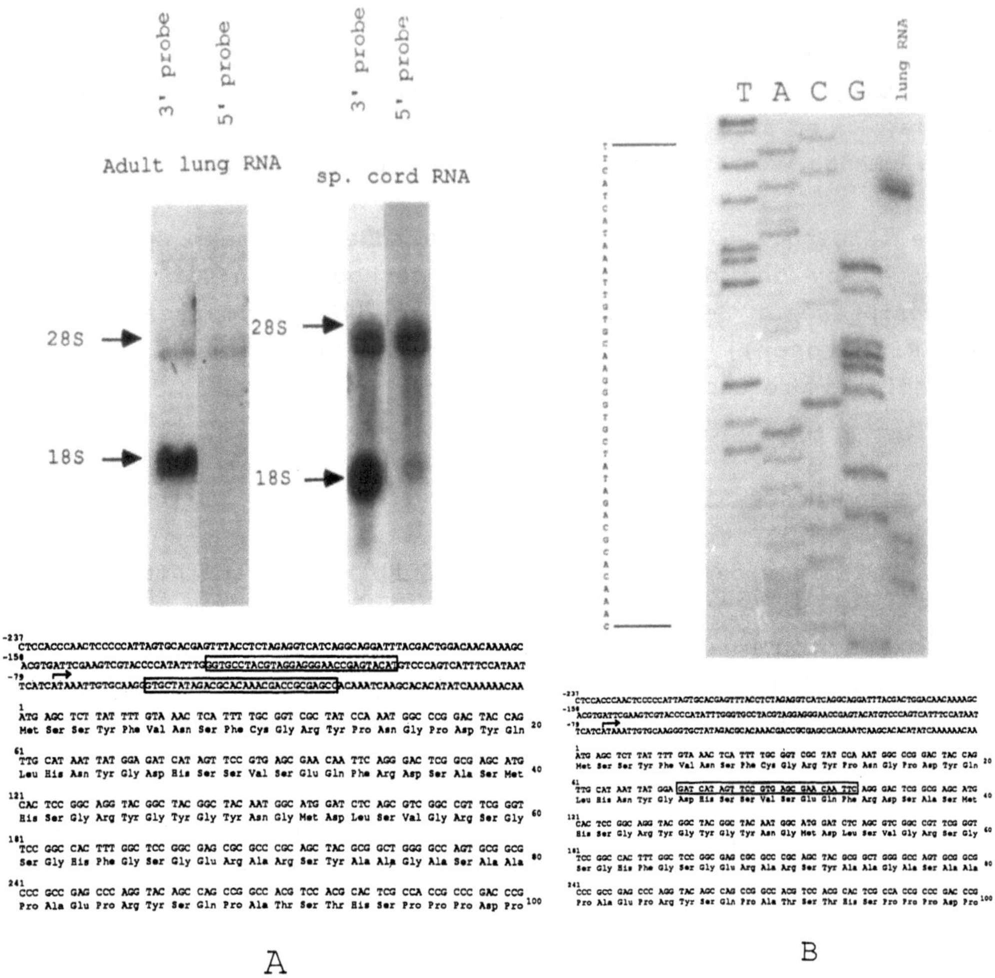

Figure 3 (A) Cap site determination for the 1.85 Kb transcript. A) Differential oligonucleotide hybridization poly A+ RNA to two oligonucleotides (shown below), lane 1: hybridized with the 3' probe; lane 2: hybridized with the 5' probe; lane 3: hybridized with the 3' probe; lane 4: hybridized with the 5' probe. Lanes 1 and 2 contain adult mouse lung RNA, 5µg; lanes 3 and 4 contain 6 µg fetal (E13-15) mouse spinal cord RNA. Lower: The locations of the oligonucleotides are enclosed in boxes.
Fig 3 (B) Primer extension analysis of adult lung poly A+ RNA. The primer is designated below. A sequence ladder of a Hox 1.3 genomic template primed with the same primer and extended in the presence of dideoxynucleotides was used for molecular weight determination of the extension products. The sequencing lanes are labelled as the complement of the dideoxynucleotide used in each reaction. Primer extension analysis (not shown) of embryonic spinal cord RNA indicated that cap sites up to about 100 bases upstream of that indicated can be utilized, though at much lower efficiencies, and these further upstream initiating transcripts are detected with the 5' oligonucleotide in the Northern of spinal cord RNA in 3A. Labeled PhiX174 DNA digested with HaeIII was used for molecular weight determination.

independent positive clones were selected using an oligonucleotide probe for the 5' end of the 1.85 Kb transcript. The longest insert of the clones is about 3.5 Kb and sequence analysis shows it has spliced out the 960 bp intron. From sequence analysis the 3' end of this clone coincides with that of the 1.85 Kb transcript. We have sequenced the entire 5' end of this cDNA and found that it is an approximately 1.6 Kb extension colinear with the genomic sequence from the 5' end ofthe 1.85 Kb transcript. From primer extension analysis (data not shown) we have deduced that the 4 Kb transcript is initiated approximately 220 bp upstream of the end of this cDNA clone, which would place the cap site about 200 bases from the end of the Hox 1.2 homeobox (Colberg-Poley et al 1985b). See fig. 4 for the comparison of transcription units for the 1.85 Kb and ~4 Kb mRNAs.

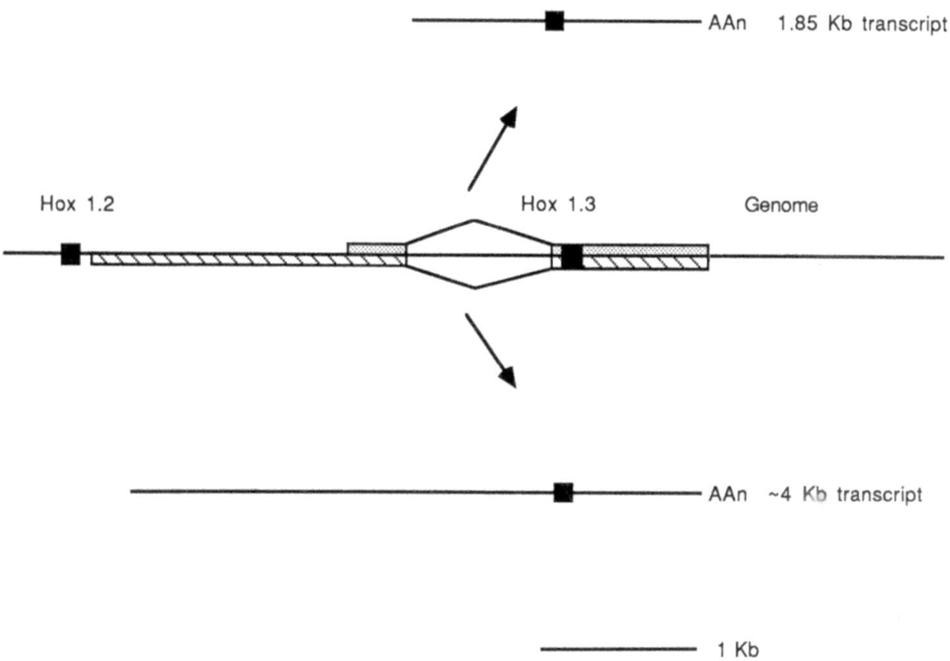

Fig. 4. Transcription map of the 1.85 Kb and 4 Kb transcripts. The predicted cap site for the 4 Kb transcript lies 198 bases downstream from the end of the Hox 1.2 homeo box (Colberg-Poley, 1985b). Exons are denoted by shaded boxes. The homeobox is denoted by a solid box.

Sequence analysis of the longer transcript reveals a potential upstream extension of 333 bases of the open reading frame utilized by the shorter transcript, which could lead to a protein of 381 amino acid residues being expressed (figure 5). While we lack direct proof that this potential reading frame is used, supporting evidence can be derived from our sequence analysis of the human cognate of Hox 1.3 (see Tournier-Lasserve, E. et al., this symposium). There is extensive and high similarity between these genes, including regions encoding the predicted intron and 3' untranslated trailer. The sequences in these two regions are not in register with the open reading frame of their homolog. In contrast, the nucleotide sequence of the human gene is 98% identical with that of the mouse gene for at least 200 bases upstream of the initiation codon and is in frame with that of the mouse. Two of four nucleotide sequence differences in the 235 bases upstream of the initiation codon occur at the wobble position and would not change the amino acid sequence. Homeodomain containing proteins are thought to bind to specific DNA sequences through this domain and to interact with other proteins or other DNA sequences through other domains. An amino terminus extended protein would have the ability to bind to the same DNA sequences as the smaller form, but through its extension be able to modify or take on other interactions. Therefore, choice of alternative promotors would dictate which protein form is made. Also, if the protein binding sites in the 300 bases upstream of the 1.85 Kb transcript AUG codon represent cis elements, the possibility is raised that a eukaryotic DNA sequence may be utilized both as an open reading frame and as a transcriptional regulatory region. The extreme length of the 5' region is unusual and the 4 Kb transript may represent a partially processed transcript, however, other genes, notably cellular oncogenes, have long 5' untranslated sequences. These may contain regulatory sequences important for transcription (Soeller WC et al.), RNA stability or translational control (Kozak).

```
   1 CGTGCGATCGGCAGCTGACGGCCTCACAATTGGTACGTCCTAATGGAACTGCGAGGGAAATGCAATAATTTTGCCATAATGGGCTGTAACCTCAATTCGA

 101 CCCCGGCCCTTGAAGCCCGAGGTCGGAAGCGAAGCG ATG CGC CCA GTC TCC AGC GGG TGG CGC TCG AGT CCG ACT GAA CGG CGG
                                          Met Arg Pro Val Ser Ser Gly Trp Arg Ser Ser Pro Thr Glu Arg Arg

 185 CAA CGG GTG GCG GGG ACG CGC CCA GGG CGC GCG CGC CAC CCC TCT TGC CTC CAC CCA ACT CCC CCA TTA GTG CAC
     Gln Arg Val Ala Gly Thr Arg Pro Gly Arg Ala Arg His Pro Ser Cys Leu His Pro Thr Pro Pro Leu Val His

 260 GAG TTT ACC TCT AGA GGT CAT CAG GCA GGA TTT ACG ACT GGA CAA CAA AAG CAC GTG ATT CGA AGT CGT ACC CCA
     Glu Phe Thr Ser Arg Gly His Gln Ala Gly Phe Thr Thr Gly Gln Gln Lys His Val Ile Arg Ser Arg Thr Pro

 335 TAT TTG GGT GCC TAC GTA GGA GGG AAC CGA GTA CAT GTC CCA GTC ATT TCC ATA ATT CAT CAT AAA TTG TGC AAG
     Tyr Leu Gly Ala Tyr Val Gly Gly Asn Arg Val His Val Pro Val Ile Ser Ile Ile His His Lys Leu Cys Lys

 410 GGT GCT ATA GAC GCA CAA ACG ACC GCG AGC CAC AAA TCA AGC ACA CAT ATC AAA AAA CAA ATG AGC TCT TAT TTT
     Gly Ala Ile Asp Ala Gln Thr Thr Ala Ser His Lys Ser Ser Thr His Ile Lys Lys Gln Met Ser Ser Tyr Phe

 485 GTA AAC TCA TTT TGC GGT CGC TAT CCA AAT GGC CCG GAC TAC CAG TTG CAT AAT TAT GGA GAT CAT AGT TCC GTG
     Val Asn Ser Phe Cys Gly Arg Tyr Pro Asn Gly Pro Asp Tyr Gln Leu His Asn Tyr Gly Asp His Ser Ser Val

 560 AGC GAA CAA TTC AGG GAC TCG GCG AGC ATG CAC TCC GGC AGG TAC GGC TAC GGC TAC AAT GGC ATG GAT CTC AGC
     Ser Glu Gln Phe Arg Asp Ser Ala Ser Met His Ser Gly Arg Tyr Gly Tyr Gly Tyr Asn Gly Met Asp Leu Ser

 635 GTC GGC CGT TCG GGT TCC GGC CAC TTT GGC TCC GGC GAG CGC GCC CGC AGC TAC GCG GCT GGG GCC AGT GCG GCG
     Val Gly Arg Ser Gly Ser Gly His Phe Gly Ser Gly Glu Arg Ala Arg Ser Tyr Ala Ala Gly Ala Ser Ala Ala

 710 CCC GCC GAG CCC AGG TAC AGC CAG CCG GCC ACG TCC ACG CAC TCG CCA CCG CCC GAC CCG CTG CCC TGC TCA GCG
     Pro Ala Glu Pro Arg Tyr Ser Gln Pro Ala Thr Ser Thr His Ser Pro Pro Pro Asp Pro Leu Pro Cys Ser Ala

 785 GTG GCC CCC TCG CCC GGC AGC GAC AGC CAC CAC GGC GGG AAA AAC TCC CTG GGC AAC TCC AGC GGC GCC TCG GCC
     Val Ala Pro Ser Pro Gly Ser Asp Ser His His Gly Gly Lys Asn Ser Leu Gly Asn Ser Ser Gly Ala Ser Ala

 860 AAC GCC GGC AGC ACC CAC ATC AGC AGC AGA GAG GGG GTT GGC ACG GCG TCC GCA GCC GAG GAG GAC GCC CCT GCC
     Asn Ala Gly Ser Thr His Ile Ser Ser Arg Glu Gly Val Gly Thr Ala Ser Ala Ala Glu Glu Asp Ala Pro Ala

 935 AGC AGC GAG CAG GCG GGC GCC CAG AGC GAG CCG AGC CCG GCG CCG CCC GCT CAG CCC CAG ATC TAC CCC TGG ATG
     Ser Ser Glu Gln Ala Gly Ala Gln Ser Glu Pro Ser Pro Ala Pro Pro Ala Gln Pro Gln Ile Tyr Pro Trp Met

1010 CGC AAG CTG CAC ATT AGT CAC GAC AAT ATA GGT GGC CCA GAA GGC AAA AGG GCC CGG ACG GCC TAC ACT CGC TAC
     Arg Lys Leu His Ile Ser His Asp Asn Ile Gly Gly Pro Glu Gly Lys Arg Ala Arg Thr Ala Tyr Thr Arg Tyr

1085 CAG ACC CTG GAG CTG GAG AAA GAA TTC CAC TTC AAC CGC TAC CTG ACC CGC CGA AGA AGG ATC GAA ATA GCT CAT
     Gln Thr Leu Glu Leu Glu Lys Glu Phe His Phe Asn Arg Tyr Leu Thr Arg Arg Arg Arg Ile Glu Ile Ala His

1160 GCC CTT TGC CTC TCC GAG AGA CAA ATT AAA ATC TGG TTC CAA AAC AGG AGG ATG AAG TGG AAA AAA GAT AAT AAG
     Ala Leu Cys Leu Ser Glu Arg Gln Ile Lys Ile Trp Phe Gln Asn Arg Arg Met Lys Trp Lys Lys Asp Asn Lys

1235 CTG AAA AGC ATG AGT ATG GCC GCG GCA GGG GGG GCT TTC CGC CCC TGA GCATCTGAGCGGCCAAAGTACTGAGCAGTAGTAGCC
     Leu Lys Ser Met Ser Met Ala Ala Ala Gly Gly Ala Phe Arg Pro ***

1319 GGGCAGCTCTCTGTAGTGTCAGTACTAAGGTGACTTTCTGAAACTCCCCTTGTGTTCCTTCTGTGAAGAAGCCCTGTTCTCGTTGCCCTAATTCATCTTT

1419 TAATCATGAGCCTGTTTATTGCCATTATAGCGCCTGTATAAGTAGATCTGCTTCTGTTCATCTCTTTGTCCTGAATGGCTTTGTCTTGAAAAAAAAATAG

1519 ATGTTTTAACTTATTTATATGAAGCAAGCTGTGTTACTTGAAGTAACTAAAACAAAAAAAAAAAAAAAAGAAAAGAGAAAAAAAAACTACTCACACAAAA

1619 AGCCCCCCCACCTCTTTAGTGCCAATGTTGTGTGTTGCACTTGAGTTCTTTAATGTGCATGTACGTGGAAGTGTTCCTGTCTCAATAGCTCCAAGCTGTT

1719 AAAGATATTTTTATTCAAACTACCTATATTCCTTGTGTAATTAATGCTGTTGTAGAGGTGACTTGATAAGACACAAATTAACTTGTTCAACGTGTAGTGG

1819 CTAGTGGCTCTGTGACGAAAACTGTGACTCCAAGCGGTGTGTCCCTGCGTGCCTTTGTAGGACCCTTTGCACGAACTCTGGAAGTGGCTCTTATAAGCGC

1919 AGTTCAGTGATGTATGTTTTTGTGAAAAAGTTACAAATATTGTCCAAGTCTGGCTGTTTAAGCAAACTGTGATCAGCTTTTTTTTTTTTTGTATTTGTT

2019 TTTAAGGAAAAAAAACACTGACTGGAAACAAAACAAAATAAACTTTCTATTGTAA
```

Fig. 5. Partial sequence of the 4 Kb Hox 1.3 cDNA. This shows only the region including the potential extension of the open reading frame. Note there is a stop codon at position 77 (TAA). The cap sites and initiation codon of the 1.85 Kb transcript are indicated, by an arrows and box, respectively; the intron is marked by the inverted triangle, the homeobox is bracketed and the polyadenylation signal is in bold type.

References

Acampora D, Pannese M, d'Esposito M, Simeone A, Boncinelli E (1987) Human homoeobox-containing genes in development. Human Reproduction 2:407-414

Bürglin TR, Wright CVR, De Robertis EM (1987) Translational control in homeobox mRNAs? Nature 330:701-702

Colberg-Poley, AM, Voss SD, Gruss P (1985a) Structural analysis of murine genes containing homoeo box sequences and their expression in embryonal carcinoma cells. Nature 314:713-718

Colberg-Poley AM, Voss SD, Chowdhury K, Stewart CL, Wagner EF, Gruss P (1985b) Elustered homeo boxes are differentially expressed during murine development. Cell 43:39-45

Doe CQ, Hiromi Y, Gehring WJ, Goodman CS (1988) Expression and function of the segmentation gene fushi-tarazu during Drosophila neurogenesis. Science 239:170-175

Dolecki GJ, Wannakrairoj S, Lum R, Wang G, Riley HD, Carlos R, Wang A Humphreys T (1986) Stage-specific wxpression of a homeo box-containing gene in the non-segmented sea urchin embryo. EMBO J 5:925-930

Duboule D, Baron A, Mähl P, Galliot B (1986) A new homeo-box is present in overlapping cosmid clones which define the mouse Hox-1 locus. EMBO J 5:1973-1980

Eiken HG, Njølstad PR, Molven A, Fjose A (1987) A zebrafish homeobox-containing gene with embryonic transcription. Biochem Biophys Res Comm 149:1165-1171

Fienberg AA, Utset MF, Bogarad LD, Hart CP, Awgulewitsch A, Ferguson-Smith A, Fainsod A, Rabin M, Ruddle FH (1987) Homeo box genes in murine development. Curr Topics Dev Biol 23:233-256

Gehring WJ (1987) Homeo boxes in the study of development. Science 236:1245-1252

Kessel M, Schulze F, Fibi M, Gruss P (1987) Primary structure and nuclear localization of a murine homeo domain protein. Proc Natl Acad Sci USA 84:5306-5310

Kozak M (1987) An analysis of 5'-noncoding sequences from 699 vertebrate RNAs. Nucl Acids Res 15:8125-8148

McGinnis W, Hart CP, Gehring WJ, Ruddle FH (1984) Molecular cloning and chromosome mapping of a mouse DNA sequence homologous to homeotic genes of Drosophila. Cell 38:675-680

Maniatis T, Fritsch E, Sambrook J (1982) Molecular cloning - A laboratory manual. Cold Spring Harbor Lab., Cold Spring Harbor, New York

Martin GR, Boncinelli E, Duboule D, Gruss P, Jackson I, Krumlauf R, Lonai P, McGinnis W, Ruddle F, Wolgemuth D (1987) Nomenclature for homeo-box-containing genes. Nature 325:21-22

Müller MM, Carrasco AE, De Robertis EM (1984) A homeo-box-containing gene expressed during oogenesis in Xenopus. Cell 39:157-162

Nüsslein-Volhard C, Frohnhöfer HG, Lehmann R (1987) Determination of anteroposterior polarity in Drosophila. Science 238:1675-1681

Odenwald WF, Taylor CF, Palmer-Hill FJ, Friedrich V, Tani M, Lazzarini RA (1987) Expression of a homeo domain protein in non contact-inhibited cultured cells and post mitotic neurons. Genes and Development 1:482-496

Okayama H, Kawaichi M, Brownstein M, Lee F, Yokota T, Arai K (1987) High efficiency cloning of full length cDNA; Construction and screening of cDNA expression libraries from mammalian cells. Methods Enzymology 154:3-28

Rogers S, Wells R, Rechsteiner M (1986) Amino acid sequences common to rapidly degraded proteins: the PEST hypothesis. Science 234:364-368

Scott MP, Caroll SB (1987) The segmentation and homeotic gene network in early Drosophila development. Cell 51:689-698
Shaw G, Kamen R (1986) A conserved AU sequence from the 3' untranslated region of GM-CSF mRNA mediates selective mRNA degradation. Cell 46:659-667
Soeller WC, Poole SJ, Kornberg T (1988) In vitro transcription of the Drosophila engrailed gene. Genes and Development 2:68-81
Tabor S, Richardson CC (1987) DNA sequence analysis with a modified bacteriophage T7 DNA polymerase. Proc Natl Acad Sci USA 84:4767-4771

The Human Cognate of the Murine Hox 1.3 Homeo Box Gene is Almost Identical to its Murine Counterpart

Elizabeth Tournier-Lasserve, Ward F. Odenwald, James Garbern, and Robert A. Lazzarini
Laboratory of Molecular Genetics, NINCDS, NIH
Bethesda, Maryland 20892

Introduction

Development is based on spatially and temporally regulated expression of the genetic information contained in embryonic cells. In Drosophila melanogaster, at least three classes of genes are involved in the regulation of expression of genes for the establishment of the body pattern and segment identity. Among them, the homeotic genes responsible for the segment identity have been shown as being clustered in two major loci: the Antennapedia complex (Ant-C) and the Bithorax complex (BX-C) (Gehring, W., Review, 1985). A highly conserved sequence of 180 bp, the homeo box, has been observed repeatedly in this class of genes. This homeo box encodes a 60 amino acid protein domain, the homeo domain. These homeo box sequences have recently been shown in the genomes of a number of species including mice and humans.

The role of homeo box containing genes in mammals is still completely unknown. In the mouse, at least 15 homeo box genes have been described. They are organized in at least four major clusters. Two of these clusters, the Hox 1 cluster on chromosome 6 and the Hox 2 cluster on chromosome 11, contain at least six homeo boxes each (Baron et al., 1987; Hart et al., 1985; Odenwald et al., 1987; Coldberg-Poley et al., 1985).

To explore which regions could be functionally important in these genes, we have compared the primary structures of the murine gene and its human cognate looking at the conservation of coding sequences, but also untranslated regions and the introns.

We report here the striking conservation of the 270 amino acid predicted protein encoded by the human cognate of Hox 1.3 (Hum Hox 1.3), which differs from the murine one by only seven amino acids. Not only is the coding sequence of this gene conserved, but also the untranslated regions and the intron.

Materials and Methods

Screening of the Human Genomic Library, Subcloning and Sequence Analysis

A human recombinant genomic λEMBL3 library (prepared in our laboratory) was

NATO ASI Series, Vol. H26
Cell to Cell Signals in Mammalian Development
Edited by S. W. de Laat et al.

screened under stringent conditions with a cDNA probe derived from the 5' murine sequence flanking the murine Hox 1.3 box ("5' probe"). Two positive recombinant λ clones (λ26 and λ32) were then hybridized with two cDNA probes derived respectively from the murine Hox 1.3 and the 3' untranslated region ("Hox 1.3 box probe" and "3' probe"). A 3.5 kb (Sal I-Eco RI) DNA fragment from λ32 showing a positive hybridization signal with the 5', 3' and Hox 1.3 box probes and a 6 kb (Sal I) DNA fragment from λ26 showing a positive signal with the 5' and the Hox 1.3 box probes have been subcloned into the Blue scribe plasmid vector (Stratagene). The two strands of the Hox 1.3 human cognate gene were sequenced by the dideoxy nucleotide chain termination technique (Sanger et al., 1977) using primers synthesized on an Applied Biosystems 380A DNA synthesizer. Sequences were analyzed using the sequence analysis software package of the genetics computer group (Devereux, Haeberli and Smithies, Nucleic Acid Research 12(1):387-395, 1984) on a VAX computer.

Southern Blot Analysis of λ DNA Clones

DNA from λ26 and λ32 was prepared by standard procedures (Maniatis et al., 1982). DNA λ clones were restricted with different enzyme combinations. DNA fragments were separated on a 1% agarose gel and transferred to a zeta probe membrane (Biorad). Different probes have been used:

- Four oligonucleotide probes were derived from the murine sequences flanking immediately at the 3' end the respective boxes of the Hox 1.3 ("oligo 1.3"), Hox 1.1 ("oligo 1.1"), Hox 1.2 ("oligo 1.2") and Hox 1.4 ("oligo 1.4") genes.

- Another oligonucleotide probe ("oligo generic") was prepared from the most conserved sequence in the different boxes of the murine Hox 1 cluster.
- A genomic Pvu II-Sac I DNA digested fragment ("Hox 1.1 pM6 probe") from the murine cloned DNA (kindly provided by Dr. Colberg-Poley) has also been used.

Different stringency conditions have been utilized:

- When using oligo probes 1, 2, 3, and 4, hybridization was performed overnight at 55°C in 2 x SSC without formamide, 10 x Denhardt's solution and 1% SDS and the blots washed at 55°C in 1 x SSC.

- When using the generic oligo probe, hybridization was performed at 42°C and the blot washed in 1 x SSC at 42°C.

- When using the Hox 1.1 PM6 probe, the hybridization was performed at 37°C, in 43% formamide and the blot was washed for a half hour at 55°C in 1 x SSC.

Results and Discussion

Comparison of the Human and Murine Protein Coding Regions

The human sequence is shown in Figure 1 and the comparison between the human and murine sequences is shown on the dot matrix of Figure 2. To facilitate the discussion in comparing these two sequences, we will use the terms intron, untranslated regions, deletions... in reference to the murine sequence.

Assuming conservation of the intron position, the 810 bp reading frame between the ATG triplet at position 1 and the TGA stop codon at position 1773 would encode a 270 amino acid protein strikingly similar to the murine protein. The homeo box nucleotide sequence is 93% identical to the murine Hox 1.3 homeo box. The amino acid sequences of the two homeo domains are identical in mouse and man.

The homeo domain of Hum Hox 1.3 is also identical to the human c10 homeo domain. c10 is the human homolog of murine Hox 2.1 (Hauser et al., 1985; Acampora, 1987; Hart et al., 1985); consequently,the homology described between genes of the Hox 1 and 2 cluster in the mouse genome seems to exist also in the human genome. When comparing the nucleotide sequence of the human Hox 1.3 homeo box to the other homeo boxes previously described in human genes, we observed that the similarity was 88% with c1, 82% with c8, 90% with c13 and 83% with c2.2 (Simeone et al., 1987; Hauser et al., 1985; Acampora et al., 1988). The pentapeptide "Ile Tyr Pro Trp Met" a motif which is conserved in several Mouse, Xenopus and Drosophila homeo domain proteins is also present in Hum Hox 1.3. Downstream from the homeo domain, the 16 amino acids preceding the stop codon are also identical in the two species. When comparing the 5' coding sequences between the first ATG triplet and the exon intron boundary, 94% identity at the nucleic acid level is observed. The predicted protein sequence encoded by this open reading frame would have exactly the same length as the murine protein, i.e., 270 amino acids and differ by only 7 amino acids. These differences reside in the central region of the protein (Figure 1). Interestingly, this central region is not conserved between the murine genes Hox 2.1 and Hox 1.3. Although generally the similarity found in the homeo domain of the other described homeo box genes drops off outside the homeo box, such a conservation has been described between the murine Hox 2.3 gene and its human counterpart c1 (Meijlink et al., 1987).

```
-248 ACCCCCATCGCCTCCACCCAACTCCCCTATTAGTGCACGAGTTTACCTCTAGAGGTCATCAGGCAGGATTTACGACTGGACAACAAAAGCACGTGATTCG

-148 AAGTCGTACCCCATATTTGGGTGCCTACGTAGGAGGGAACCAAGTACATGTCCCAGTCATTTCCATAATTCATCATAAATTGTGCAAGGGTGCTATAGAC

 -48 GCACAAACGACCGCGAGCCACAAATCAAGCACACATATCAAAAAACAA ATG AGC TCT TAT TTT GTA AAC TCA TTT TGC GGT CGC TAT
                                                      Met Ser Ser Tyr Phe Val Asn Ser Phe Cys Gly Arg Tyr     13

  40 CCA AAT GGC CCG GAC TAC CAG TTG CAT AAT TAT GGA GAT CAC AGT TCC GTG AGC GAG CAA TTC AGG GAC TCG GCG
     Pro Asn Gly Pro Asp Tyr Gln Leu His Asn Tyr Gly Asp His Ser Ser Val Ser Glu Gln Phe Arg Asp Ser Ala     38

 115 AGC ATG CAC TCC GGC AGG TAC GGC TAC GGC TAC AAT GGC ATG GAT CTC AGC GTC GGC CGC TCG GGC TCC GGC CAC
     Ser Met His Ser Gly Arg Tyr Gly Tyr Gly Tyr Asn Gly Met Asp Leu Ser Val Gly Arg Ser Gly Ser Gly His     63

 190 TTT GGC TCC GGA GAG CGC GCC CGC AGC TAC GCT GCC AGC GCC AGC GCG GCG CCC GCC GAG CCC AGG TAC AGC CAG
     Phe Gly Ser Gly Glu Arg Ala Arg Ser Tyr Ala Ala [Ser] Ala Ser Ala Ala Pro Ala Glu Pro Arg Tyr Ser Gln     88

 265 CCG GCC ACG TCC ACG CAC TCT CCT CAG CCC GAT CCG CTG CCC TGC TCC GCC GTG GCC CCC TCG CCC GGC AGC GAC
     Pro Ala Thr Ser Thr His Ser Pro [Gln] Pro Asp Pro Leu Pro Cys Ser Ala Val Ala Pro Ser Pro Gly Ser Asp     113

 340 ACG CAC CAC GGC GGG AAA AAC TCC CTA AGC AAC TCC AGC GGC GCC TCG GCC GAC GCC GGC AGC ACC CAC ATC AGC
     [Thr] His His Gly Gly Lys Asn Ser Leu [Ser] Asn Ser Ser Gly Ala Ser Ala [Asp] Ala Gly Ser Thr His Ile Ser     138

 415 AGC AGA GAG GGG GTT GGC ACG GCG TCC GGA GCC GAG GAG GAC GCC CCT GCC AGC AGC GAG CAG GCG AGT GCG CAG
     Ser Arg Glu Gly Val Gly Thr Ala Ser [Gly] Ala Glu Glu Asp Ala Pro Ala Ser Ser Glu Gln Ala [Ser] Ala Gln     163

 490 AGC GAG CCG AGC CCG GCG CCG CCC GCC CAA CCC CAG ATC TAC CCC TGG ATG CGC AAG CTG CAC ATA AGT CAT G▽GT
     Ser Glu Pro Ser Pro Ala Pro Pro Ala Gln Pro Gln Ile Tyr Pro Trp Met Arg Lys Leu His Ile Ser His     187

 565 AAAGCCAGCCTTTTTCTAAATCCACGCGACCGCGGGGACGCGGCTCTCGGTCCCCCCTCTCTTCTCTGCCGCCCTCTCCCAGTCTCTCTTGGTCTCATTT

 665 CTCCCAGCCCTGCGGAGCTCTCCTTGCCGTTCTCCCTCCCCTGCCTTCCTCCTTCCTCCTCTTCTTTGCGAGCACCCTGCGCTCACAAATAGTGGGGAA

 765 ATGGGCGTTCTCTGGGACAGTTTAGACGTTGGAAGGGGGAGGAAGCAAAAAACCCCTCTGGAACCCCACGCCTTGGGACGCGCTCCCGGGTCAGGCCAGC

 865 CGAGCAAGGCGCAGAGAGGTAGAGGATGGCTGTAGCAGCCGTGAATCGGGCTTGTCACGGCGGATAATTTATGAGGAGGGCTACGCTGGGAAACAGCGT

 965 TACTAATTACAGCCCCCAAAAAGGGGCTTGGGGGAAAGAATCGAGGCGAGACCTGCAGGATTCTGAATTTTGGGGGCAGGAGGGAGAGAGAAGGAAAGGG

1065 AAGAAAAAGAAAACAGGCTCCCCAACCCTGCAGGCTGGAAACGGGAGGCGGCTCTCGGGGCTGGAACTTTGAGGGAGGGTGACCCGAAGGCCACTTGGGC

1165 GCTCAGGAAAGGGCCTTGCTTCCTGGGTTTCTGTGCGGTGGGCAGCCTGGGAGGGCTGTGCCTCCCGATCGGGGNNCGCCCGGGGCAGGGCGGAGGGGGC

1265 AGGAGAGGGGCCAGGGAAAGCCGGAGTCCGCCGGGACACGGCCCCAGCCTCAGATGGGCAGATTGTTCCCAGGGTCCAAATCGTATTGTTTTCTTTCTAG

1365 AAAGGAAGAGAGAAGGAAATTCGGGAGGGGTGTGCGGGCTGGTAGGCAGAACTTGTTGAGCTTTTCGCCTGGGTTCCCTGCTCATGACCCAAGCTTGTCC

1465 CCCTGGCGGACTTTGGAAGACAGGAGTTGGTGGCTAAACCGCTGACTTTTCTATTGCAG▽AC AAC ATA GGC GGC CCG GAA GGC AAA AGG
                                                                Asp Asn Ile Gly Gly Pro Glu Gly Lys Arg     197

1553 GCC CGG ACG GCC TAC ACG CGC TAC CAG ACC CTG GAG CTG GAG AAG GAG TTC CAC TTC AAC CGT TAC CTG ACC CGC
     Ala Arg Thr Ala Tyr Thr Arg Tyr Gln Thr Leu Glu Leu Glu Lys Glu Phe His Phe Asn Arg Tyr Leu Thr Arg     222

1628 AGA AGG AGG ATT GAA ATA GCA CAT GCT CTT TGC CTC TCC GAG AGA CAA ATT AAA ATC TGG TTC CAA AAC CGG AGA
     Arg Arg Arg Ile Glu Ile Ala His Ala Leu Cys Leu Ser Glu Arg Gln Ile Lys Ile Trp Phe Gln Asn Arg Arg     247

1703 ATG AAG TGG AAA AAA GAT AAT AAG CTG AAA AGC ATG AGC ATG GCC GCG GCA GGA GGG GCC TTC CGT CCC TGA GTAT
     Met Lys Trp Lys Lys Asp Asn Lys Leu Lys Ser Met Ser Met Ala Ala Ala Gly Gly Ala Phe Arg Pro ***     270

1779 CTGAGCGTTTAAAGTACTGAGCAGTATTAGCGGATCCCGCGTAGTGTCAGTACTAAGGTGACTTTCTGAAACTCCCTTGTGTTCCTTCTGTGAAGAAGCC

1879 CTGTTCTCGTTGCCCTAATTCATCTTTTAATCATGAGCCTGTTTATTGCCATTATAGCGCCTGTATAAGTAGATCTGCTTTCTGTTCATCTCTTTGTCCT

1979 GAATGGCTTTGTCTTGAAAAAAAATAGATGTTTTAACTTATTTATATGAAGCAAGCTGTGTTACTTGAAGTAACTATAACAAAAAAAGAAAAGAGAAAAA

2079 AAAAACACACAAAAAGTCCCCCTTCAATCTCGTTTAGTGCCAATGTTGTGTGTTGCACTCAAGTTGTTTAACTGTGCATGTGCGTGGAAGTGTTCCTGTC

2179 TCAATAGCTCCAAGCTGTTAAAGATATTTTTATTCAAACTACCTATATTCCTTGTGTAATTAATGCTGTTGTAGAGGTGACTTGATGAGACACAACTTGT

2279 TCGACGTGTAGTGACTAGTGACTCTGTGATGAAAACTGTGACTCCAAGCGGTGTGTCCCTGCGTGCCTTTATAGGACCCTTTGCACGAACTCTGGAAGTG

2379 GCTCTTATAAGCGCAGCTTCAGTGATGTATGTTTTTGTGAACAAAGTTACAAATATTGTCCAAGTCTGGCTGTTTTAAGCAAACTGTGATCAGCTTTTTT

2479 TTTTTTTTTTTTTTTGTATTTGTTTTAAGGAAAAAATACTGACTGGAACAAA[AATAAA]CTTTCTATTGTAAGTTCTCTTGGTCTGGTTTGTGCCAAA

2579 TAGTGAGCGGCTCTGTCTGCTTTTCTGTCTGTCTGTGCAGTCTTGGAAGCTGTTGGGTCTGAGGCTACCTGAGCAGATGACCTGTGCAGGGAGACCTCAT

2679 ACCAACACTGTCCCATCGCTTCCCTACCTCTGACCCATTGCAAAGTTCAGGGCATAAGGTGGAAAAAGCTGTAGGCTGTTCCAAAGCCCCAGAAGACCTG

2779 TCCATCTCTGAGGAAACCAAGTTAACTTGCTGGGTACAAAAAXGAGAGAAGAGC
```

Figure 1: Genomic DNA sequence of the Hum <u>Hox 1.3</u> gene. Numbering begins at the predicted translation start codon. The boundaries of the predicted intron (by analogy with the murine gene) are denoted by arrowheads. The homeo box is underlined with a solid line. A consensus polyadenylation signal is boxed with a thick line. The seven differences in the predicted amino acid sequence between the human and the murine genes are boxed.

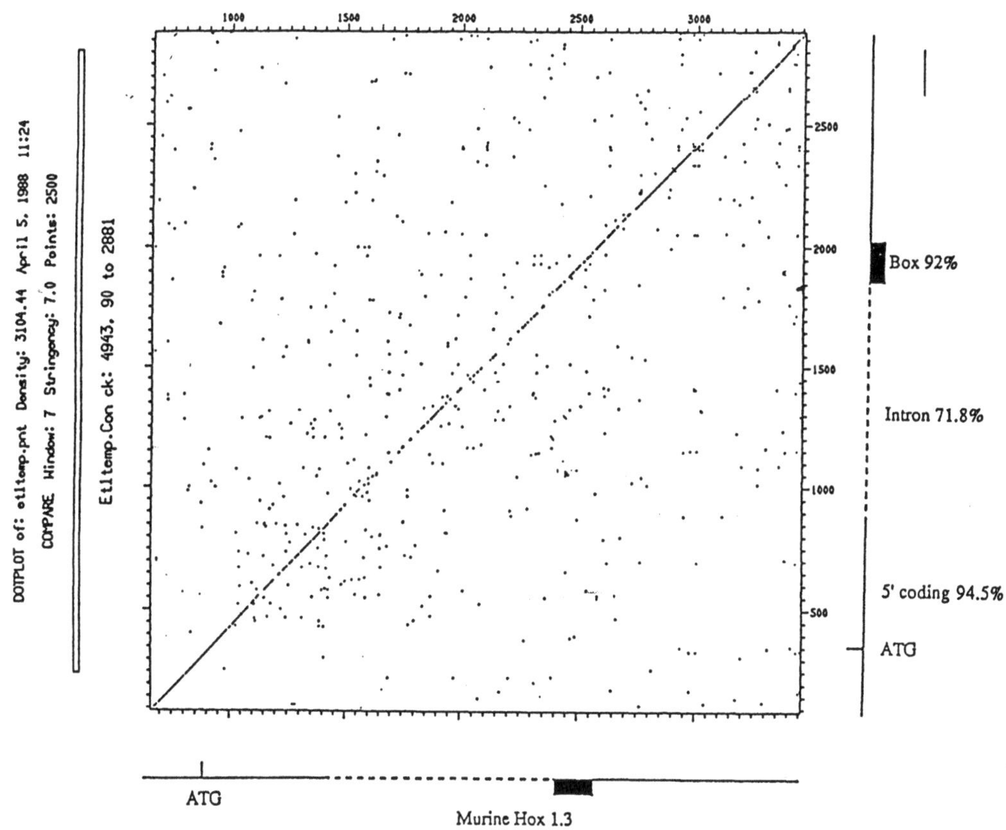

Figure 2: Dot matrix comparison of the human cognate of Hox 1.3 and its murine counterpart. The human nucleotide sequence (pos. -248 to +2833) is compared to the murine sequence (Odenwald et al.) under very stringent parameters. Segments of seven nucleotides are compared and a dot indicates identity of seven. The human and murine sequences are displayed vertically and horizontally, respectively. The black boxes show the homeo boxes. The hatch lines code for the murine intron and the predicted human intron. The predicted translation initiation codons are denoted as ATG. Conservation in the various regions are indicated by the percentages of homology.

Comparison of the 5' and 3' Untranslated Regions

Upstream the first ATG codon, there is a 250 bp long stretch of DNA that differs from the mouse sequence at only four positions. In the human sequence as in the murine one, the open reading frame described above extends beyond the first ATG triplet in the 5' direction. We cannot exclude the possibility that these sequences are not part of a longer exon. To explore this possibility, we have raised antibodies directed against a predicted peptide encoded by this sequence (J. Garbern, unpublished). On the other hand, the extreme

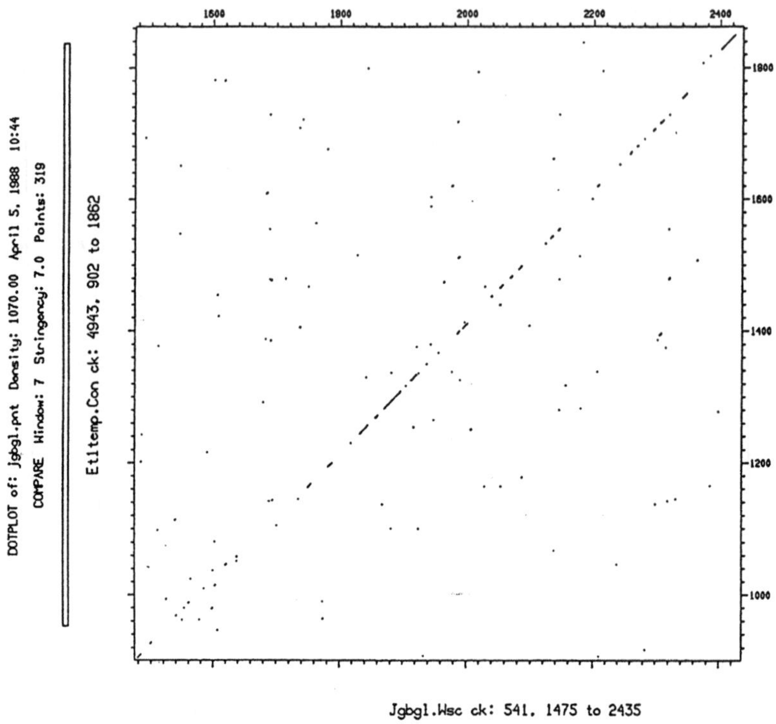

Figure 3: Dot matrix comparison of the murine and human introns. The murine intron and the predicted human intron are compared using the same stringency parameters as above.

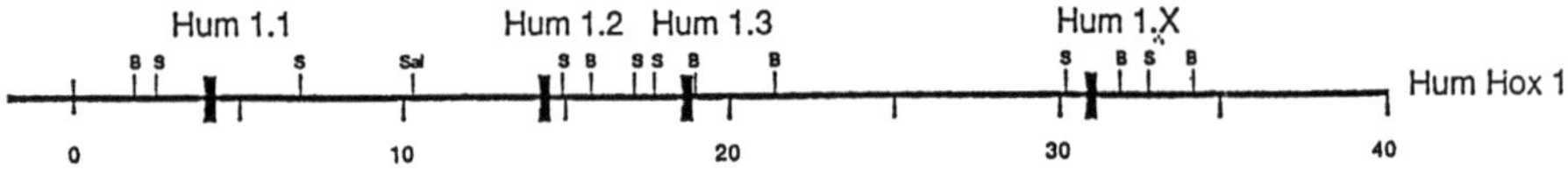

Figure 4: Genomic restriction map of the human homeo box region studied and homology to the murine Hox 1 cluster. The top shows a simplified restriction map of the genomic DNA region surrounding the human cognate of Hox 1.3. The two overlapping recombinant phage inserts (λ26 and λ32) are shown below this map. The square boxes identify the location of the respective Hum 1.1, Hum 1.2, and Hum 1.3 boxes. A fourth homeo box designated Hum 1.x has been located approximately 11 kb downstream from the Hum 1.3 box. The bottom shows the Eco RI restriction map of murine Hox 1 cluster compiled from different authors (Coldberg-Poley et al., Duboulé et al., Martin et al., Odenwald et al.).

conservation of this 5' sequence may be related to a conserved regulatory element in the two species.

In the 3' region downstream of the coding sequence, the nucleic acid similarity is over 90%. The reading frame is not maintained due to insertions or deletions in the human sequence relative to the murine sequence. Between these differences, multiple stretches of perfect identity exist. The conservation of these regions may suggest that these sequences also play an important role in the functioning of these genes.

Comparison of the Intron Sequences

The striking similarity between the two introns is shown in the dot matrix of Figure 3. The human gene contains consensus eukaryotic splice donor and acceptor sites at the same position as in the mouse. The length of the two introns are identical, 960 bp. Although the intron sequences are more divergent than the coding sequences, the sequence conservation reaches 72%. We find many stretches of perfect conservation. One stretch of 111 bp shows a 90% identity with the murine sequence. The large number of conserved bases in these two introns is unusual. However, although less striking, such a similarity has been observed in the introns of the engrailed genes between two Drosophila species which evolutionary distance is estimated to be 60 million years (Kassis et al.).

Conservation of the Clustering of Hum Hox 1.3

The restriction map of the human Hox 1 cluster is shown in Figure 4. Southern blot hybridization with the Hox 1.3 box probe and a "generic" oligonucleotide probe derived from the most conserved region of the homeo boxes in the murine Hox 1 cluster (cf Material and Methods) identified four regions of homeo box homology on λ26 and λ32 phage clones. These two clones were overlapping by 400 bp in the 5' coding regions, and the human Hox 1.3 box was included in the λ32 clone. To further characterize the other boxes present in this cluster, we used, in Southern blotting experiments, oligonucleotide probes derived from the murine sequences flanking immediately at their 3' end the respective boxes of Hox 1.3, Hox 1.1, Hox 1.2 and Hox 1.4 boxes and also a genomic Pvu II-Sac I DNA digested fragment from the murine Hox 1.1 cloned DNA. None of these probes contained any box sequences.

Under stringent conditions, we detected a strong hybridization signal with the Hox 1.1 probe and the oligonucleotide Hox 1.2 with the two respective boxes of λ26. No hybridization signal was detected with the Hox 1.4 oligonucleotide probe on the fourth box included in λ32, but because of the length of these oligos (approximately 20 bp) and the stringency conditions that we used, minor sequence differences between the murine and human genes could lead to absence of signal. Therefore, we cannot exclude that this fourth box would be part of the human cognate of Hox 1.4. We did not map the human Hox 1 locus on human chromosome ourselves, but Rabin et al. (1986) using a mouse probe specific for a flanking region of the Hox 1 cluster mapped it to chromosome 7 (7p14-p21) and demonstrated conserved linkage between the two species.

The comparative analysis of the sequences of the homeo box gene members of the Hox 1 and Hox 2 clusters in the mouse might indicate a close evolutionary relationship between these genes. For example, the Hox 1.3 and 2.1 boxes are identical at the amino acid level and a strong similarity is also observed in their 5' noncoding sequences. These similarities have also been described between 2.2 and 1.2, 2.3 and 1.1. The human Hox 1.3 domain is also identical to the c10 homeo domain which is the human cognate of Hox 2.1 and the human Hox 1.2 box and the murine Hox 2.2 box are 92% identical.

Using the same oligonucleotide Hox 1.2 as a primer we sequenced 162 out of 180 bp of the Hum Hox 1.2 box. This box showed a 93% identity at the nucleic acid level with the murine Hox 1.2 gene and only 77% to the Hox 1.3 box.

A clustered chromosomal organization has been observed in most mammalian homeo box genes. In the mouse, at least 6 homeo boxes have been described in the Hox 1 cluster on chromosome 6. The organization of the human Hox 1 cluster is similar to the murine one. Strikingly, the spacing between the different boxes of the cluster is conserved between these two species and the genes show the same relative order. The reason for this conserved organization is unknown. However, the multiple mRNA transcripts of the Hox 1.3 genes indicate that it has a complex transcriptional pattern. The Hox 1.3 gene uses alternative transcriptional initiation sites to generate two of its transcripts (Garbern, NATO meeting). Interestingly, one of these sites lies near the upstream homeo box of Hox 1.2. Perhaps this proximity has biological signficance that maintains the clustering of the genes over long periods of evolutionary time.

Conclusion

We have identified, isolated and characterized the human cognate of the murine homeo box gene, Hox 1.3 (Odenwald et al.). Two overlapping recombinant phage clones have been isolated from a human genomic library using a mouse Hox 1.3 probe. The sequence of the human gene reveals a surprising degree of homology with the murine gene. At the nucleic acid level there is a 93% identity in the homeo boxes and the amino acid sequences of these two homeo domains are identical. The human sequence is also highly conserved both in the 5' and 3' domains. There are only seven amino acid differences between the predicted sequences of both proteins. Even more surprising is the fact that part of the sequence corresponding to the murine intron is also conserved.

Using different oligonucleotide probes, we have been able to show that the human cognate is part of a cluster containing at least four homeo boxes just as the murine Hox 1.3 gene is a part of the Hox 1 locus. The homeo box adjoining the human Hox 1.3 cognate was sequenced and shows 94% identity at the nucleotide level with the murine Hox 1.2 gene. Under stringent conditions, a Hox 1.1 probe gives a very strong hybridization signal with the 5' portion of one lambda clone, strongly indicating that one of the other homeo boxes that we have identified is the human cognate of Hox 1.1.

These results provide direct evidence for a highly conserved structure and organization of the human cognate of the murine Hox 1.3 gene as well as a conservation of the clustering of this gene with other homeo box genes. Our preliminary evidence suggests that even the order of the genes along the chromosome has been preserved. The importance of this clustered organization is indicated by its conservation over 45 million years of evolution (the divergence of rodent and man).

We are currently studying the expression pattern at the mRNA level of this human cognate in different normal and pathological tissues.

Bibliography

Acampora D, Pannese M, D'Esposito M, Simeone, A, Boncinelli E (1987) Human homeo box-containing genes in development. Human Reproduction 2:407-414.

Baron A, Featherstone MS, Hill RE, Hall A, Galliot B, Duboule D (1987) Hox 1.6: A mouse homeo box containing gene member of the Hox 1 locus complex. EMBO J 6:2977-2986.

Coldberg-Poley AM, Voss SD, Chowdhury K, Grüss P (1985) Structural analyis of murine genes containing homeo box sequences and their expression in embryonal carcinoma cells. Nature 314:713-718.

Gehring WS (1985) The homeo box: A key to the understanding of development? Cell 40:3-5.

Hart CP, Awgulewitsch A, Fainsod A, McGinnis W, Ruddle F (1985) Homeo box gene complex on mouse chromosome 11: Molecular cloning, expression in embryogenesis and homology to a human homeo box locus. Cell 43:9-18.

Hart CP, Fainsod A, Ruddle F (1987) Sequence analysis of the murine Hox 2.2, 2.3, and Hox 2.4 homeo boxes: Evolutionary and structural comparisons. Genomics 1:182-195.

Hauser CA, Soyner Al, Klein RD, Learned TK, Martin GR, Tjian R (1985) Expression of homologous homeo box containing genes in differentiated human teratocarcinoma cells and mouse embryos. Cell 43:19-28.

Kassis JA, Poole SJ, Wright DK, O'Farell PH (1986) Sequence conservation in the protein coding and intron regions of the engrailed transcription unit. EMBO J 5:3583-3589.

Levine M, Rubin GM, Tsian R (1984) Human DNA sequences homologous to a protein coding region conserved between homeotic genes of Drosophila. Cell 38:667-673.

Mavilio F, Simeone A, Giampaolo A, Faiella A, Zappavigna V, Acampora D, Poiana G, Russo G, Peschle C, Boncinelli E (1986) Differential and stage related expression in embryonic tissues of a new human homeo box gene. Nature 324:664-667.

Meijlink F, Delaaf R, Verrijzer P, Destree O, Kroezen V, Hilkens J, Deschamps J (1987) A mouse homeo box containing gene on chromosome 11: Sequence and tissue specific expression. Nucl Acids Res 15:6773-6786.

Odenwald, WF, Taylor CF, Palmer-Hill FS, Friedrich V, Tani M, Lazzarini RA (1987) Expression of a homeo domain protein in noncontact-inhibited cultured cells and post-mitotic neuron. Genes and Development 1:482-496.

Rabin M, Ferguson-Smith A, Hart C, Ruddle F (1986) Cognate homeo box loci mapped on homologous human and mouse chromosomes. Proc Natl Acad Sci 83:9104-9108.

Simeone A, Mavilio F, Acampora D, Giampaolo A, Faiella A, Zappavigna V, D'Esposito M, Pannese M, Russo G, Boneinelli E, Peschle C (1987) Two human homeo boxes c1 and c8: Structure analysis and expression in embryonic development. Proc Natl Acad Sci 84:4914-4918.

Simeone A, Mavilio F, Bottero L, Giampaolo A, Russo G, Faiella A, Boncinelli E, Peschle C. (1986) A human homeo box gene specifically expressed in spinal cord during embryonic development. Nature 320:763-765.

Temporal and Cellular Specificity of Expression of the Murine Homeobox-Containing Gene Hox-1.4

Debra J. Wolgemuth, Margaret Mostoller, Carolyn M. Viviano, Zahra F. Zakeri, Yasuhisa Iwaoki, and George L. Mutter
Department of Genetics and Development
The Center for Reproductive Sciences
Columbia University College of Physicians and Surgeons
630 West 168th Street
New York, NY 10032 USA

ABSTRACT

The homeobox is a 183 bp DNA sequence which was originally identified in *Drosophila* genes that are important in development. Its conservation in the genomes of diverse groups of animals, including mammals, provides a means for isolating potential development-regulating genes. Our studies have focused on the isolation and characterization of the structure and expression of the mouse homeobox-containing gene Hox-1.4. Hox-1.4 is a member of the Hox-1 cluster of six homeobox-containing genes covering a region of ~80 kb on mouse chromosome 6. Hox-1.4 is unique in that its expression is restricted to the testis in the adult animal in a cell type-specific manner. The transcripts are found at comparable levels in meiotic prophase spermatocytes, early spermatids, and in cytoplasmic fragments of elongated spermatids and residual bodies. Northern blot hybridization analysis revealed that two different sized Hox-1.4 transcripts are detected in germ cells. The larger of the transcripts is found in post-meiotic cells. A third size class of Hox-1.4 transcripts are seen in embryos as early as 9.5 days of gestation. *In situ* hybridization analysis of Hox-1.4 expression in day 13.5 embryos reveals localizaton in the dorsal myelencephalon and the high cervical spinal column. We are currently: (a) investigating the molecular basis of the differences in Hox-1.4 transcript size; (b) examining the expression of other members of the Hox-1 locus during male and female germ cell development; and (c) using transgenic mice carrying Hox-1.4 constructs to identify the regulatory elements necessary for the testicular and embryonic expression.

NATO ASI Series, Vol. H26
Cell to Cell Signals in Mammalian Development
Edited by S. W. de Laat et al.

INTRODUCTION

A 183 bp DNA sequence known as the homeobox has been identified in genes which are known to be important in Drosophila development (McGinnis et al., 1984; Scott and Weiner, 1984). The homeobox domain is conserved in the genomes of animals as evolutionarily divergent as insects, sea urchins, frogs, and man (Gehring and Hiromi, 1986). Its conservation in the genomes of mammals provides a means for isolating potential development-regulating genes from these organisms. This is particularly useful since identifying mammalian development regulating genes is complicated by their large genome size, poorly defined genetics, and long developmental life span.

We used a cDNA clone containing the homeobox region of the Antennapedia gene of Drosophila to screen a mouse testis cDNA library (Wolgemuth et al., 1986). A homeobox-containing clone, pHBT-1, was isolated and identified as representing the mouse homeobox-containing gene Hox-1.4 (Martin et al., 1987). The gene has been subsequently identified in a genomic cosmid clone (Duboule et al., 1986) and again as a cDNA clone from a mouse testis library (Rubin et al., 1986). Hox-1.4 is a member of a group of at least six homeobox-containing genes mapping to mouse chromosome 6 (Bucan et al., 1986; Duboule et al., 1986). These genes are clustered in a region of <80 kb (Duboule et al., 1986; Bucan et al., 1986), known as the Hox-1 locus (Martin et al., 1987).

Although all the Hox-1 genes have been shown to be expressed in the developing mouse embryo (Colberg-Poley et al., 1985; Rubin et al., 1986; Duboule et al., 1986; Wolgemuth et al., 1987; Duboule et al., 1987; Gaunt et al., 1986; Odenwald et al., 1987), Hox-1.4 is unique with respect to its apparent tissue-restricted expression in the adult animal. Hox-1.4 is expressed at high levels in the adult testis, specifically in germ cells which have entered meiosis (Wolgemuth et al., 1986; 1987). The studies presented herein report observations on the cellular and temporal specificity of Hox-1.4 expresssion in the embryo and in the adult testis, and refer to current studies in which we are attempting to identify important regulatory elements responsible for the embryonic and adult expression.

MATERIALS AND METHODS

Swiss Webster mice were used as the source of all adult and embryonic tissues unless otherwise noted. Adult tissues were stored at -70°C until processed for RNA extraction. Enriched populations of testicular germ cells were prepared according to Wolgemuth et al. (1985). Embryos were obtained from naturally mated females; observation of a vaginal plug denoted day 0.5 of gestation (Wolgemuth et al., 1987). Conceptuses at various stages of development, from 6.5 to 14.5 days of embryogenesis, were recovered, placed in phosphate buffered saline, and dissected free of uterine tissues. Embryonic and extraembryonic compartments were separated in conceptuses from day 10.5 of development to later stages. For earlier stages, the entire conceptus was used for isolation of RNA.

RNA was isolated according to Cathala et al., 1983, poly(A)-selected (Aviv and Leder, 1981), and electrophoresed according to our previously published procedures (Ponzetto and Wolgemuth, 1985). All hybridization and washing conditions were at high stringency, with the final wash at 0.1x SSC (1x SSC = 0.15 M NaCl, 0.015 M Na citrate) at 65°C.

Embryos of day 13.5 of gestation were processed for histological analysis by making an incision in the abdomen and in the region of the ventricles and fixing in 4% paraformaeldehyde. Specimens were subjected to dehydration and paraffin embedding exactly as described in Mutter and Wolgemuth (1987), except that the rehydrated paraffin sections were post-fixed in 4% paraformaldehyde for 10 minutes after proteinase K digestation.

The genomic probe p2181 B2 4a (Duggal et al., 1987) was used to detect Hox-1.4 transcripts. The purified insert was nick translated (Weinstock et al., 1978) or RNA probes were synthesized with ^{35}S-UTP from linearized p2181 B2 4a inserted into pGem 3 (Promega Biotech) in a sense or antisense orientation. Control hybridization for the *in situ* hybridizations included labeled RNA generated from linearized templates for the Ava I-Hind III c-*mos* insert (Mutter and Wolgemuth, 1987). Control hybridizations contained an equal number of counts as the experimental.

RESULTS AND DISCUSSION

Expression of Hox-1.4 in adult tissues

Hox-1.4 was originally identified as the cDNA clone pHBT-1 which had been isolated from a mouse testis cDNA library (Wolgemuth et al., 1986). A survey of several adult mouse tissues revealed that Hox-1.4 was expressed at high levels, if not uniquely, in the testis. Transcripts of ~1.4 kb in length are readily detected in RNA from adult testis (Figure 1). To identify the particular cell types within the

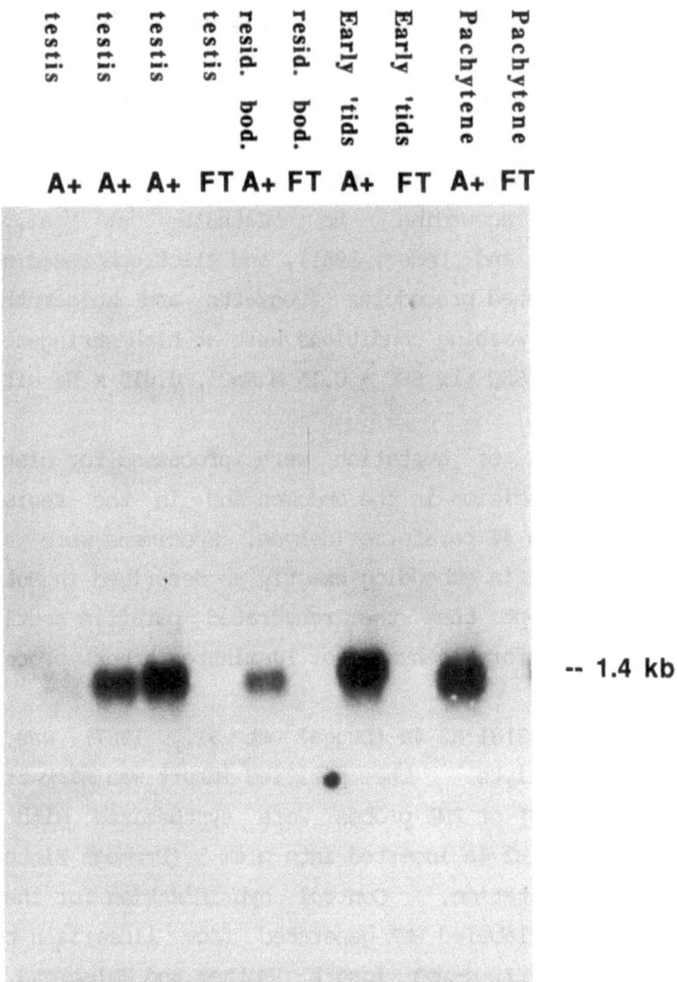

Figure 1. Hox-1.4 transcripts detected in male mouse germ cells. RNA from enriched populations of spermatogenic cells and from intact testis was analyzed by Northern blot hybridization using ^{32}P-labeled Mbo II-Eco Rl fragment of pHBT-1 (Wolgemuth et al., 1986). Total testis RNA samples contain 1, 4, and 8 ug of poly(A)$^{+}$ RNA respectively. All poly(A)$^{-}$ samples contain 30 ug of RNA. The pachytene and early spermatid fractions each contain 5 ug and the residual body sample contains 1.3 ug of (A)$^{+}$ RNA, respectively. Exposure, 5 days.

testis which express Hox-1.4, a combination of genetic, developmental, and cell enrichment approaches has been used. Hox-1.4 transcripts are not found in testes from mice homozygous for the atrichosis mutation, which lack germ cells (Wolgemuth et al., 1987). This suggests that the gene is expressed in the germ cell lineage. Next, RNA was isolated from testes at various stages of embryonic and postnatal development, which lack specific spermatogenic cell types depending upon the age of the animal (Bellve et al., 1977). Hox-1.4 transcripts were only detected in testes containing meiotic germ cells (Wolgemuth et al., 1986; Rubin et al., 1986). Finally, enriched populations of meiotic prophase spermatocytes, early spermatids, and a mixture of cytoplasmic fragments from elongating spermatids and residual bodies were obtained and examined for the presence of Hox-1.4 transcripts (Figure 1). The ~1.4 kb transcripts were apparent in RNA from all three populations of cells, at approximately equivalent levels. The less intense hybridization detected in the RNA from cytoplasmic fragments and residual bodies relative to the other samples is most likely due to the fact that only about one-third the amount of RNA was loaded in this lane.

We had observed repeatedly that the Hox-1.4 testicular transcript migrated as a rather broad band. We therefore determined that the testicular transcripts are actually composed of two discretely migrating species of 1.35 and 1.45 kb in length by electrophoresis on high percentage formaldehyde gels (Wolgemuth et al., 1987). Experiments are currently in progress to determine the molecular basis of this heterogeneity.

Embryonic Expression

Virtually all the mammalian homeobox genes identified to date are known to be expressed in the developing mid-gestation embryo (e.g. Snow, 1987). We therefore isolated RNA from embryonic day 11.5 and 12.5 embryos and determined if Hox-1.4 transcripts were present. As noted in Figure 2, Hox-1.4 is expressed in the day 11.5 embryos and yields a transcript of 1.7 kb, distinct in size from the testicular transcripts. To assess the temporal pattern of expression of Hox-1.4 expression in development, we isolated RNA from embryos (or conceptuses) of gestational ages days 6.5 to 16.5. The results of this analysis are shown in Figure 3 and in Wolgemuth et al. (1987). Hox 1.4 transcripts

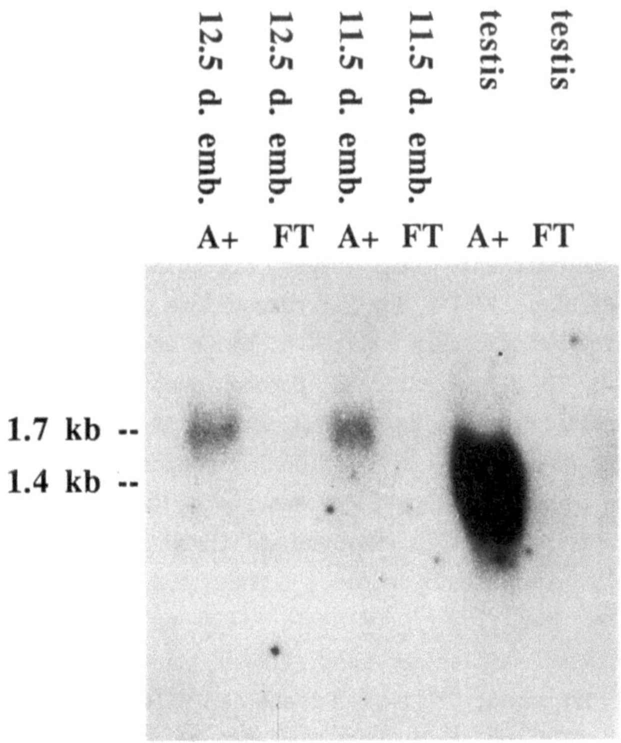

Figure 2. Northern blot hybridization analysis of testicular versus embryonic Hox-1.4 transcripts.
RNA was isolated from day 11.5 and 12.5 embryos and mature testes and poly(A)-selected. Poly(A)$^+$ samples each contained 5 ug; poly(A)$^-$ samples contained 30 ug. Hybridization was with Hox-1.4 clone p2181 B2 4a (Duggal et al., 1987), 10 day exposure.

are readily detected in day 9.5 embryos. The transcripts continue to be present at day 14.5 of development (Figure 3), and even as late as day 16.5 of development, the latest stage examined (Wolgemuth et al., 1987). Careful examination of the autoradiograms revealed the presence of low levels of Hox-1.4 transcripts as early as day 8.5.

To examine the distribution of Hox-1.4 transcripts in the embryos, we have utilized both Northern blot hybridization analysis of RNA from dissected embryonic compartments and *in situ* hybridization analysis of histological sections of midgestation embryos. Seven embryonic compartments were dissected from day 12.5 embryos as follows: anterior brain, posterior brain, face, spinal cord, viscera, limbs, and carcass. RNA was isolated from these samples as well as from placenta and yolk sac of the corresponding embryos and examined by Northern blot

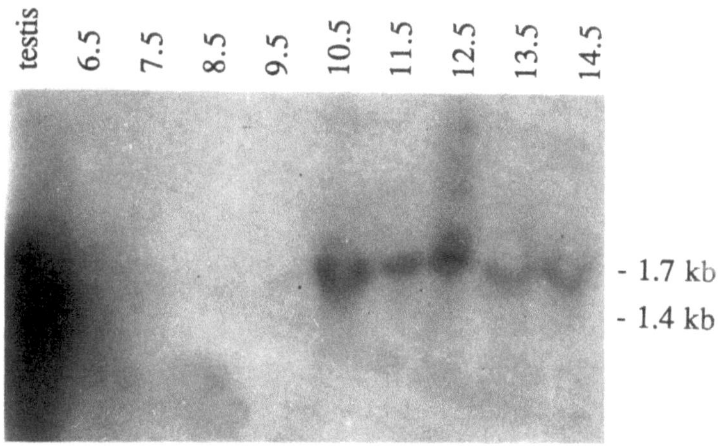

Figure 3. Northern blot hybridization analysis of Hox-1.4 expression in developing embryos.
Total RNA was extracted from Swiss Webster embryos at days 6.5, 7.5, 8.5, 9.5, 10.5, 11.5, 12.5, 13.5, 14.5 of gestation and adult testis. RNAs were electrophoresed in denaturing 0.8% agarose:2.2 M formaldehyde gels and processed for Northern blot hybridization analysis with ^{32}P-labeled Hox 1.4 clone p2181 B2 4a (Duggal et al., 1987). Hybridization conditions were at high stringency. 30 ug of total RNA was loaded for each sample. Exposure, 7 days.

hybridization for the presence of Hox-1.4 transcripts. In day 12.5 embryos, the most abundant expression of Hox-1.4 was seen in the spinal cord (Wolgemuth et al., 1987).

To determine more precisely the localization of Hox-1.4 mRNAs within the central nervous system, histological sections were prepared from embryos at day 13.5 of gestation, according to our previously described protocols (Mutter and Wolgemuth, 1987), and hybridized with antisense orientation ^{35}S-labeled Hox 1.4 RNA probes. Control hybridizations included antisense c-mos RNA probes to sections containing embryos and testis (data not shown). The results of this analysis are depicted in Figures 4 and 5. Localization is readily observed in the myelencephalon (Figure 4B, 4D) and to a lesser extent, in the high cervical region of the spinal cord (Figure 5B). Within the myenlencephalon, labeling was more intense over dorsal versus ventral neuronal cells. In the high cervical spinal cord however, the labeling appeared to shift to a more ventral position.

The significance of this shift in localization is not clear;

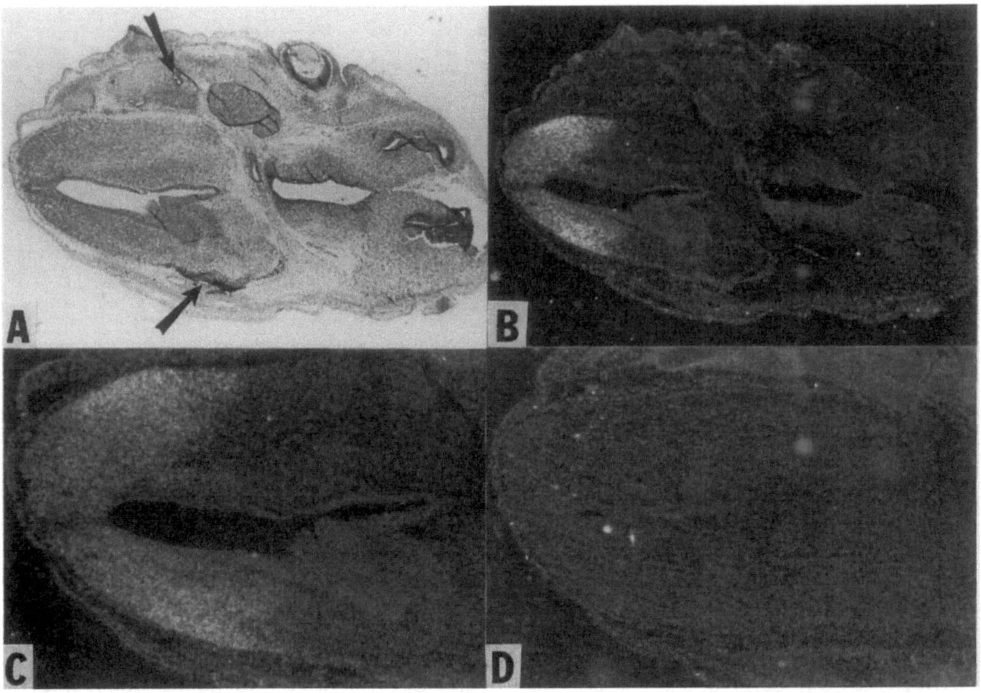

Figure 4. Localization of Hox-1.4 transcripts in day 13.5 embryos by in situ hybridization.
Transverse embryo sections were hybridized in situ with ^{35}S-labeled Hox-1.4 antisense probe (4A, 4B, 4C), or antisense c-mos probe (4D, negative control). Autoradiographic signal is visible as bright grains under a darkfield illumination (4B, 4C) in the region of the dorsal myelencephalon. The dorsal aspect of the embryos is to the left, and arrows indicate the otic vesicles adjacent to the myelencephalon. Exposure 177 days.

however Le Mouellic et al. (1988) have shown by in situ hybridization analysis both anterioposterior and dorsoventral changes in Hox 3.1 transcript localization. Hox-3.1 transcript first appears in all posterior tissues of the 8.5 day embryo. As development progresses, the signal becomes slightly more anterior and specifically localized to an area of the neural tube posterior to the third cervical vertebra. During this time changes in dorsoventral transcript localization also occur. However, the pattern of expression which will be maintained throughout the rest of prenatal development is established by day 11.5. These results emphasize the importance of examining the pattern of

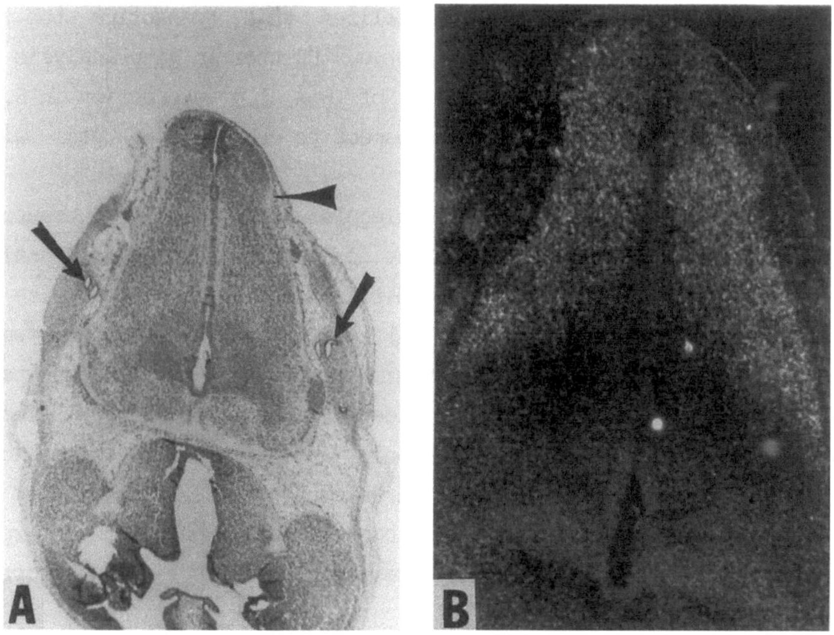

Figure 5. Distribution of Hox-1.4 transcripts in the high cervical spinal cord region of day 13.5 embryos.
Slightly oblique transverse sections of a 13.5 day embryo hybridized with ^{35}S-labeled antisense Hox-1.4 RNA probe, visualized under transmitted light (5A) or darkfield (5B). The dorsal aspect is at the top of the picture, and the section shows the transition from myelencephalon to spinal cord, delineated by an arrowhead. Note the otic vesicles (arrows) adjacent to the myelencephalon. Autoradiographic signal is distributed in the dorsal portion of the myelencephalon and extends to the ventral portion of high cervical spinal cord. Exposure, 177 days.

expression of homeo box-containing genes in the mouse during progressive stages of embryonic development. .

SUMMARY AND FUTURE DIRECTIONS

We have shown that three differently sized transcripts are produced from the murine homeobox-containing gene Hox-1.4. The two smaller sized transcripts (1.35 and 1.45 kb) are detected in male germ cells that have entered into and progressed beyond meiosis. The 1.7 kb message is most abundant in the spinal cord of the mid-gestation embryo. *In situ* hybridization experiments have localized this transcript to the myelencéphalon of the 13.5 day embryo. Further in situ analysis will charäcterize the temporal specificity of Hox 1.4 expression in 8.5 to 16.5 day embryo, particularly with respect to changing patterns within the central nervous system.

Experiments to determine the molecular basis of Hox 1.4 transcript heterogeneity are in progress. RNase protection and RNase H analyses are being used to determine if the differences in transcript size are due to alternate splicing or poly (A) tail length. In addition, we are developing transgenic mouse lines that carry the Hox 1.4 gene tagged with SV40 sequences in an effort to identify the 5' regulatory elements involved in controlling the embryonic and testicular expression of Hox-1.4. Transgenic constructs in which Hox 1.4 is placed under the control of stage- and cell-specific promoters will be used to explore possible functions of Hox 1.4 by examining the effect of inappropriate expression of the gene.

ACKNOWLEDGEMENTS

This work was supported by NIH grant HD18122. We wish to thank George Grills for help in preparation of the illustrations.

REFERENCES

Aviv, H, Leder P (1972) Purification of biologically active globin messenger RNA by chromatography on oligo thymidylic acid-cellulose. Proc Natl Acad Sci USA 69:1408-1412

Bellve AR, Cavicchia JC, Millette CF, O'Brien DA, Bhatnagar YM, Dym M (1977) Spermatogenic cells of the pre-puberal mouse. J Cell Biol 74:68-85

Bucan M, Yang-Feng T, Colberg-Poley AM, Wolgemuth DJ, Guenet JL, Francke U, Lehrach H (1986) Genetic and cytogenetic localisation of the homeobox genes on mouse chromosome 6 and human chromosome 7. EMBO J 5:2899-2905

Cathala G, Savouret JF, Mendez B, West BL, Karin M, Martial JA, Baxter JD (1983) A method for isolation of intact, translationally active ribonucleic acid. DNA 2:329-335

Colberg-Poley AM, Voss SD, Chowdhury K, Steward CL, Wagner EF, Gruss P (1985) Clustered homeo boxes are differentially expressed during murine development. Cell 43:39-45

Duboule D, Baron A, Mahl P, Galliot B (1986) A new homeo-box is present in overlapping cosmid clones which define the mouse HOX-1 locus. EMBO J 5:1973-1980

Duggal RN, Zakeri ZF, Ponzetto C, Wolgemuth DJ (1987). Differential expression of the c-abl proto-oncogene and the homeo box-containing gene Hox-1.4 during mouse spermatogenesis. In: Cell Biology of the Testis and Epididymis. Ed: M.C. Orgebin-Crist. Annals of The New York Academy of Sciences, in press.

Gaunt SJ, Miller JR, Powell DJ, Duboule D (1986) Homeobox gene expression in mouse embryos varies with position by the primitive streak stage. Nature 324:662-664.

Gehring WJ, Hiromi Y (1986) Homeotic genes and the homeobox. Ann Rev Genet 20:147-173

Le Mouellic H, Condamine H, Brulet P (1988) Pattern of transcription of the homeo gene Hox-3.1 in the mouse embryo. Genes and Devel 2:125-135.

Martin GR, Boncinelli E, Duboule D, Gruss P, Jackson I, Krumlauf R, Lonai P, McGinnis W, Ruddle F, Wolgemuth DJ (1987) Nomenclature for homeobox-containing genes. Nature 325:21-22

McGinnis W, Garber RL, Wirz J, Kuroiwa A, Gehring WJ (1984) A homologous protein coding sequence in Drosophila homoeotic genes and its conservation in other metazoans. Cell 38:403-409

Mutter GL, Wolgemuth DJ (1987) Distinct developmental patterns of c-mos protooncogene expression in female and male mouse germ cells. Proc Natl Acad Sci 84:5301-5305

Odenwald WF, Taylor CF, Palmer-Hill FJ, Friedrich V, Tani M, Lazzarini RA (1987) Expression of a homeo domain protein in noncontact-inhibited cultured cells and postmitotic neurons. Genes and Dev 1:482-496

Ponzetto C, Wolgemuth DJ (1985) Haploid expression of a unique c-abl transcript in the mouse male germ line. Mol Cell Biol 5:1791-1794

Rubin MR, Toth LE, Patel MD, D'Eustachio P, Nguyen-Huu MC (1986) A mouse homeo box gene is expressed in spermatocytes and embryos. Science 233:663-668

Scott MP, Weiner AF (1984) Structural relationships among genes that control development: sequence homology between the Antennapedia, Ultrabithorax and fushi tarazu. Proc Natl Acad Sci USA 81:4115-4119

Snow, MH (1986) New data from mammalian homeobox-containing genes. Nature 324:618-619.

Weinstock R, Sweet R, Weiss M, Cedar H, Axel R (1978) Intragenic DNA spacers interrupt the ovalbumin gene. Proc Natl Acad Sci USA 75:1299-1303

Wolgemuth DJ, Gizang-Ginsberg E, Engelmyer E, Gavin BJ, Ponzetto C (1985) Separation of mouse testis cells on a Celsep apparatus, and their usefulness as a source of high molecular weight DNA or RNA. Gamete Res 12:1-10

Wolgemuth DJ, Engelmyer E, Duggal RN, Gizang-Ginsberg E, Mutter GL, Ponzetto C, Viviano C, Zakeri ZF (1986) Isolation of a mouse cDNA coding for a developmentally regulated, testis-specific transcript containing homeo box homology. EMBO J 5:1229-1235

Wolgemuth DJ, Viviano CM, Gizang-Ginsberg E, Frohman MA, Joyner AL, Martin GR (1987) Differential expression of the mouse homeobox-containing gene Hox-1.4 during male germ cell differentiation and embryonic development. Proc Natl Acad Sci USA 84:5813-5817

Murine Homeo-Genes : Some Aspects of their Organisation and Structure

Denis DUBOULE, Brigitte GALLIOT, Agnès BARON and Mark S. FEATHERSTONE
Laboratoire de Génétique Moléculaire des Eucaryotes du CNRS,
Unité 184 de Biologie Moléculaire et de Génie Génétique de l'INSERM,
Faculté de Médecine
11 rue Humann
67085 STRASBOURG Cédex
France

Introduction

The mammalian genome contains at least 30 sequences related to the *Drosophila Antennapedia* (Antp) homeo-box (see references below). Most, if not all, of these sequences are present in the coding regions of actively transcribed genes (1-23) and thus define a homeo-gene family of as yet unknown function. It is supposed, by analogy with similar genes in insects (24, 25 and refs. therein), and by structural comparisons with prokaryotic DNA-binding domains (26), that such homeo-genes are involved in a regulatory network during ontogenesis through a DNA-binding activity. This gene family might therefore be of considerable importance in vertebrate morphogenesis, pattern formation and the specification of cell lineages. In this paper, we present some aspects of the structure of the homeo-gene family and homeo-genes themselves. Organisational features apparent at both levels may provide insight into their evolution, phylogenetic relationships, and possible modes of action. These aspects will be considered separately and exemplified by the situation as it is known in the house mouse.

1) Mammalian homeo-genes are clustered.

All murine genes reported so far which contain an Antp-related homeo-box are found within clusters (2, 4, 14, 19, 21, 22, 27; see Fig. 1). The Hox-1 and Hox-2 complexes contain at least 7 homeo-genes each, whereas the Hox-3, 5 and 6 complexes contain a minimum of 2 to 3 genes. These clusters, located on different chromosomes (4, 7, 18, 21, 22, 28, 29), seem to be similarly arranged in humans (29, 30) and other

NATO ASI Series, Vol. H26
Cell to Cell Signals in Mammalian Development
Edited by S.W. de Laat et al.

species (31). This grouping of homeo-box-containing genes is clearly reminiscent of their organization in insects (32), though such clusters are likely more numerous in vertebrates (27, 30). This particular arrangement may simply reflect the mechanism by which the gene family was generated during evolution (see below). Alternatively, this physical linkage might be required for the correct expression of each member with respect to its functional position within the gene network. This could be achieved by the presence of cis-acting regulatory elements, though there is as yet no evidence for this in the vertebrates.

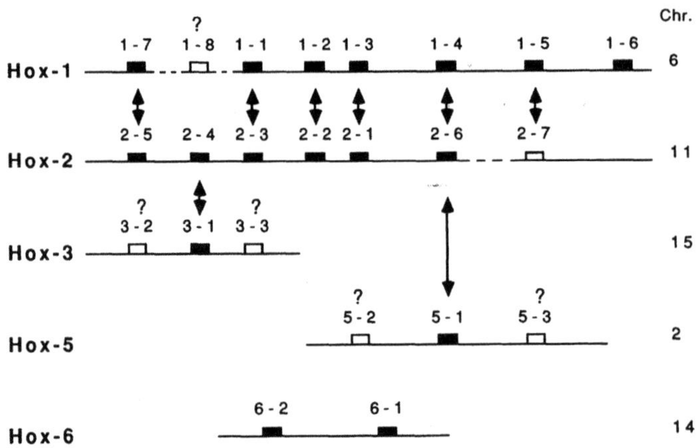

Fig. 1 : Schematic representation of the structure and chromosomal locations of murine homeo-gene complexes. The dark rectangles are reported homeo-boxes whereas white ones represent presumed (Hox-1.8) or partially characterized homeo-boxes. The distance between the different genes are approximate and interrupted lines indicate a presumptive mapping only. The arrows denote highly similar homeo-boxes which define homeo-gene subfamilies (see the text). The Hox-1 complex : data from Duboule et al. (1986), Colberg-Poley et al. (1985b), McGinnis et al. (1985), Rubin et al. (1987), Odenwald et al. (1987); Hox-2 : Hart et al. (1985, 1987), Lonai et al. (1987), Krumlauf et al. (1987) and personal communication, Meijlink et al. (1987); Hox-3 : Awgulewitsch et al. (1986), Breier et al. (1986), Gruss, personal communication; Hox-5 : Featherstone et al. (1988) and Hox-6 : Sharpe et al. (1988). The nomenclature is that proposed in Martin et al. (1987).

2) Homeo-box sub-families.

A comparison of homeo-box sequences to the Antp homeo-box (33) reveals positions of amino acid (aa) variability (25, 34). Homeo-genes can thus be divided into sub-groups based on identical aa changes within these variable regions. For example, 9 out of 11 aa substitutions found in found in Hox-1.4 with respect to Antp are also found in the Hox-5.1 homeo-box, whereas Hox-1.3 and Hox-2.1 show identical changes relative to Antp (see Fig. 2).

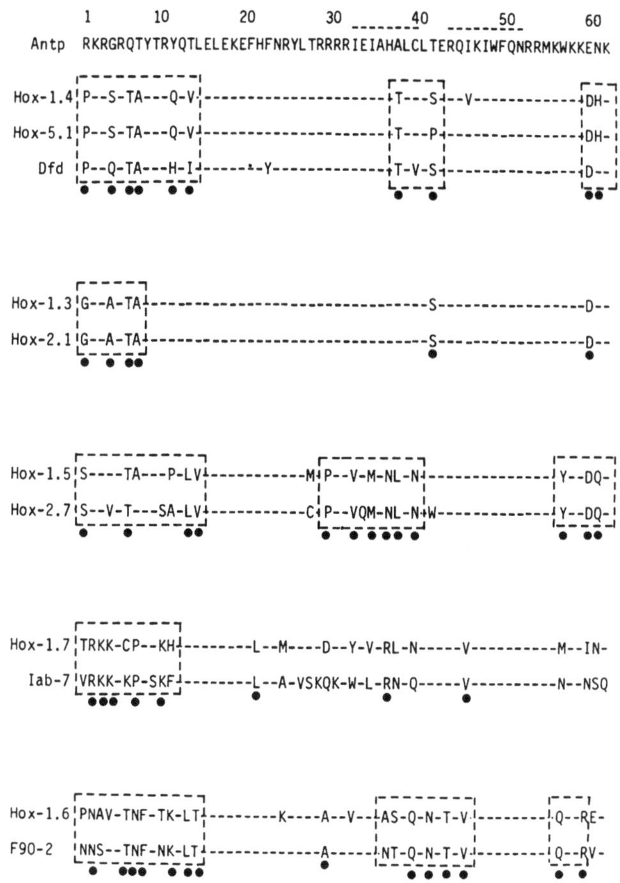

Fig. 2 : Homeo-box sub-families. One letter-code protein sequence of different murine and *Drosophila* homeo-boxes. The Antp sequence is used as a reference. Only those aa different from the Antp sequence are indicated. The number, quality and positions of these differences define sub-families as exemplified by the Hox-1.4-like, Hox-1.3-like and Hox-1.5-like sub-families. The dots indicate identical aa substitutions and the dashed rectangles show that the changes, which define a given sub-family, occur in similar regions of the homeo-domain (variable regions). The two short dashed lines on the top of the figure indicate the position of the putative helix-turn-helix DNA-binding structure (see below). Data are from the following sources : Antp, McGinnis et al. (1984); Hox-1.4, Duboule et al. (1986); Hox-5.1, Featherstone et al. (1988); Hox-1.3, Odenwald et al. (1987); Hox-2.1, Jackson et al. (1985); Hox-1.5, McGinnis et al. (1985) and our unpublished work; Hox-2.7, Lonai et al. (1986); Hox-1.7, Rubin et al. (1987); iab-7, Regulski et al. (1985); Hox-1.6, Baron et al. (1987); F90-2, Hoey et al. (1986); Dfd, Regulski et al. (1987).

A similar situation is observed for Hox-1.5 and Hox-2.7, and Hox-1.1 and Hox-2.3 as well (see Hart et al., 1987, arrows in Fig. 1). Furthermore, homeo-boxes within the same sub-family are located at the same relative positions along the Hox-1 and Hox-2 complexes. Thus, these two clusters can be aligned with respect to the ordering of homologous homeo-box sequences on the chromosome (Fig. 2). This situation likely reflects the large scale duplication of a unique cluster which may itself have been the result of duplications and divergence from an original homeo-gene (34). In mammals, such large scale duplications seem to have occured more than once, since sets of homologous genes are found in other complexes (21). For example, the Hox-5.1 homeo-gene shows high conservation with the Hox-1.4 gene and its Hox-2 counterpart, Hox-2.6 (R. Krumlauf, personal communication; 21). Likewise, the Hox-3.1 gene is highly similar to Hox-2.4 (whose putative Hox-1 counterpart, Hox-1.8, has not yet been detected). Both the Hox-3 and Hox-5 complexes contain additional homeo-boxes (19, 21) which may also have counterparts in the Hox-1 and Hox-2 complexes. It will also be of interest to learn whether unmapped members of homeo-gene sub-families, such as two genes related to Hox-1.6 (B. Robert and R. Hill, personal communication), are all located at the same relative positions with respect to adjacent homeo-genes.

The fact that divergent variable regions are conserved within sub-families of homeo-genes suggests that they are functionally important domains. In at least three cases, a marked divergence from Antp within these variable regions results in a striking similarity to other *Drosophila* homeo-box genes (Fig. 2). Thus Deformed (Dfd; 35) can be considered with the Hox-1.4-like sub-family, iab-7 (32) with Hox-1.7, and F90-2 (36) with Hox-1.6. In this last case, all the F90-2 aa positions which diverge from the Antp sequence also diverge in the Hox-1.6 sequence which has three more variable positions (4, 24, 32). Of 20 changes between F90-2 and Antp (∿ 35%), 14 involve identical aa substitutions in Hox-1.6, the two homeo-domains being more than 85% identical (see Fig. 2). These non-random substitutions revealed by inter- and intra-species comparisons reinforce the idea that these variable regions may be involved in the specifications of various homeo-domain sub-functions conserved between insects and vertebrates, even though the global functions of homeo-proteins in these various species might be different. For example, a region of variability is observed at the position of the putative α-helix 2 (26; Fig. 2) which might therefore affect the binding characteristics of the protein to DNA. Thus, it is possible that both the Hox-1.6 and F90-2 proteins display similar binding properties, as may Hox-1.4 and Dfd, though their respective functions are likely to be different.

3) Other regions of similarity between homeo-genes.

A second region of similarity among homeo-gene sub-families is a short sequence encoding a penta- or hexa-peptide. This conserved penta-peptide (CP) has been found in virtually all homeo-proteins described so far (37, refs. above and in Fig. 4).

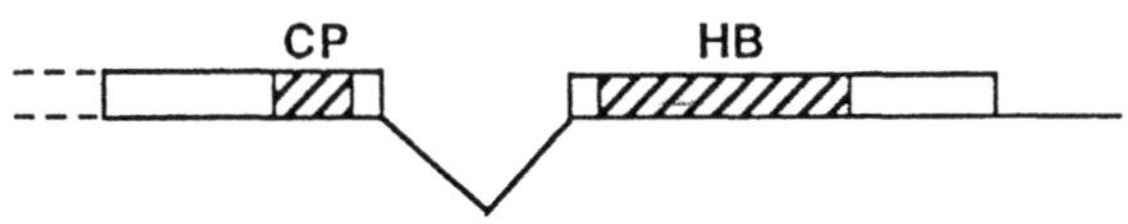

Fig. 3 : General structure of vertebrate homeo-genes. The homeo-box (HB) containing exon codes for the C-terminal part of the protein. Translation generally terminates 15 to 50 codons 3' from the homeo-box. A splice site (originally described for Hox-1.4) found 10 to 40 nucleotides upstream from the homeo-box is typically used to bring the CP adjacent to the homeo-box (see the text). A single 5' exon (containing the CP) is often found, though several additional 5' exons have been reported.

The CP is encoded in the 3' end of the exon preceding the homeo-box exon, except in a few cases. This CP is consistantly linked to the homeo-box via a splicing event. [Interestingly, in the case of the Ultrabithorax (Ubx) gene, two mini-exons preceding the homeo-box-exon, and which can be spliced in or out, do not encode this short aa sequence. Instead, it is found in the exon located 5' to these mini-exons and is therefore present in most of the different types of Ubx proteins (38, 39)]. The alignment of protein sequences allows the inclusion of the CP in the region of homology surrounding the homeo-box. Different authors (13, 40) have proposed a "consensus" sequence for the CP such that Val, Ile or Leu would occupy the first position, followed by Tyr^2 or Phe^2, Pro^3, Trp^4, Met^5, Arg/Lys^6. Though this sequence contains a "core", (Pro^3) Trp^4 Met^5 (see Fig. 4), it seems that the different residues surrounding this "core" might again be specific for given sub-families rather than deriving from a consensus. For example, the Hox-1.1-like and Hox-1.3-like sub-families show almost identical sequences. The very conserved Ile^{1} and Arg^{6} are, however, not present in Hox-1.4 or Hox-5.1 which have a Val^{1} and Lys^{6}. The Hox-1.6 CP sequence is more divergent than others since it does not even show the Pro^{3} but rather a Asp^{3}, Phe^{2} and Thr^{1}. As for the variable sub-domains of the homeo-box, this very divergent Hox-1.6 CP (10) can be found in a *Drosophila* gene, since caudal (cad; 41) also shows an "atypical" Tyr^{1} and the same Phe^{2}, Asp^{3} and Lys^{6}.

Hox-1.4	...	Val	Tyr	Pro	Trp	Met	Lys	...
Hox-5.1	...	Val	Tyr	Pro	Trp	Met	Lys	...
Hox-1.1	...	Ile	Tyr	Pro	Trp	Met	Arg	...
Hox-2.3	...	Ile	Tyr	Pro	Trp	Met	Arg	...
Hox-2.1	...	Ile	Phe	Pro	Trp	Met	Arg	...
Hox-1.3	...	Ile	Tyr	Pro	Trp	Met	Arg	...
Hox-1.6	...	Thr	Phe	Asp	Trp	Met	Lys	...
caudal	...	Tyr	Phe	Asp	Trp	Met	Lys	...

Fig. 4 : Examples of conserved pentapeptide (CP) sequences. The "core" Trp Met sequence is boxed. Sequences are from : Hox-1.4 (our unpublished work); Hox-5.1, Featherstone et al. (1988); Hox-1.1, Kessel et al. (1987); Hox-2.3, Meijlink et al. (1987); Hox-2.1, Krumlauf et al. (1987); Hox-1.3, Odenwald et al. (1987); Hox-1.6, Baron et al. (1987); caudal, Mlodzik et al. (1985).

Like the homeo-domain, aa substitutions in the CP which are conserved within sub-families might determine functional sub-specificities. Alternatively, the presence of a Met codon within the "core" could permit the initiation of a short protein, in the cases where the splicing pattern would not allow the translation of the full-length native protein (10). This protein would almost exclusively consist of the homeo-domain and therefore have a DNA binding activity in the absence of additional N-terminal domains. Though such proteins would be of great interest as far as the regulation of homeo-gene function is concerned, their existence remains to be demonstrated.

	1	2	3	4	5	6	7	8	9	10	11	12	13	14	15	16	17
1-3	Met	Ser	Ser	Tyr	Phe	Val	Asn	Ser	Phe	Cys	Gly	Arg	Tyr	Pro	Asn	Gly	Pro
2-1	Met	Ser	Ser	Tyr	Phe	Val	Asn	Ser	Phe	Ser	Gly	Arg	Tyr	Pro	Asn	Gly	Pro
							Val									Gly	
1-1	Met	Ser	Ser	Ser	Tyr	Tyr	▽Asn	Ala	Leu	Phe	Ser	Lys	Tyr	Thr	Ala	▽Ala	Ser
2-3	Met	Ser	Ser	Leu	Tyr	Tyr	Asn	Ala	Leu	Phe	Ser	Lys	Tyr	Pro	Ala	Ala	Ser
3-1	Met	Ser	Ser	Tyr	Phe	Val	Asn	Pro	Leu	Phe	Ser	Lys	Tyr	Lys	Gly	Gly	Glu
5-1	Met	Ser	Ser	Tyr	Met	Val	Asn	-	-	-	Ser	Lys	Tyr	Val	Asn	Pro	Lys
X1A	Met	Ser	Ser	Phe	Leu	Ile	-	-	-	Ser	Ser	Asn	Tyr	Val	Asp	Pro	Lys

Fig. 5 : Alignment of various homeo-protein N-terminal sequences. A hole of three aa appears in Hox-5.1 and Xhox1A (X1A) sequences (dashes). Position one is thought to define the Met initiation codon. Amino-acids found in all murine homeo-proteins so far described are boxed. Sequences are from references cited in Fig. 4, except Xhox1A (Harvey et al., 1986).

A third region of aa conservation is observed at the N-terminus of various homeo-gene products (for details, see 15, 21, 35, 37). As for the homeo-domain and the conserved pentapeptide, comparison of murine sequences so far available reveals high conservation of aa identity at certain positions (Fig. 5). For example, the Met-Ser-Ser triplet at positions 1 to 3 has been found in all murine homeo-proteins so far examined, as well as the Asn^{7} and Tyr$^{13/10}$ (Fig. 5). Beside these general similarities, it is striking that the Hox-1.3 sequence closely matches the Hox-2.1 (a Hox-1.3-like gene) sequence all along the first 30 aa. Conversely, Hox-1.1 and 2.3 are highly similar to each other but diverge from

1.3/2.1 at several positions. Some of the same substitutions in Hox-1.1/2.3 are also found in the Hox-3.1 homeo-protein (probably in all Hox-3.1-like genes), again suggesting that only some substitutions are tolerated and that the function(s) assigned to this protein domain might be subject to sub-specification (see for example amino-acids 9 to 12 in both Hox-2.3 and 3.1). This observation is reinforced by the analysis of Hox-1.4-like genes since similar variations are found in Hox-5.1, Hox-1.4 (B.G., unpublished results) and Hox-2.6 (Krumlauf, personal communication). In this case, most of these variations are also found in the *Xenopus* Xhox1A homeo-protein, a *Xenopus* member of the Hox-1.4-like family which also includes Dfd (see positions 8, 14-17 in Fig. 5 and Fig. 6).

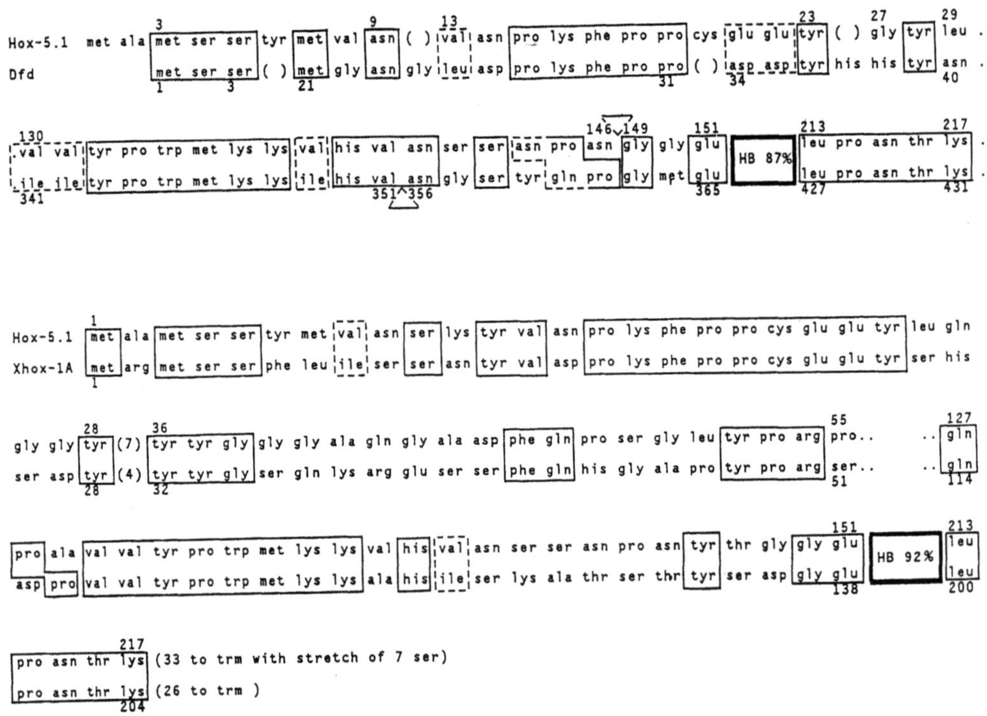

Fig. 6 : Comparisons of the mouse Hox-5.1 gene with the *Drosophila* Deformed (Dfd) gene (top) and the *Xenopus* Xhox1A gene (bottom). For more details see Featherstone et al. (1988). The figure is designed in the manner of Regulski et al. (1987). Sequences are from the two ref. cited above and Harvey et al. (1986).

In fact, the similarities between Hox-5.1, the *Xenopus* Xhox1A, and the *Drosophila* Deformed (Dfd) gene extend beyond these three sub-domains (the homeo-box, the conserved pentapeptide and the N-terminal part) (Fig. 6, see Regulski et al., 1987 and Featherstone et al., 1988 for details). Thus, sub-families of murine homeo-genes can be matched with genes from *Xenopus* and *Drosophila* to define an evolutionarily related set. Such conservation confirms the probable functional importance of the protein sub-domains used to group these genes in sub-families. These extended similarities between insects and vertebrate homeo-proteins also suggest that they may act at the protein and DNA levels in a similar manner, though the developmental processes they regulate may be different.

Conclusion

Murine homeo-genes are a gene superfamily which may encode a battery of regulatory proteins. It seems probable that the diversity within this superfamily was generated during evolution by a first step involving gene duplication followed by divergence and large scale duplication of clusters. Homeo-genes produced in this manner have kept several protein domains in common (the homeo-box, the conserved pentapeptide and the N-terminus), variations in which can be used as criteria to classify them into sub-families. This classification, based on protein sequences, is therefore correlated with the linear arrangement of these genes on chromosomes (at least for the Hox-1 and Hox-2 complexes). The high conservation of protein sub-domains strongly suggests that each sub-family may share functional features such as the specificity and stability of DNA-binding or of protein-protein interactions. On the other hand, transcription studies (see all the references cited above) indicate that spatial or temporal coexpression of genes from the same group is not required during normal development. More results are needed at the levels of gene structure and expression before models integrating functional and regulatory aspects of this gene network can be proposed.

Acknowledgements

We thank P. Chambon for continuous interest, C. Werlé and B. Boulay for the illustrations, and the secretarial staff for typing the manuscript. This work was supported by grants from the CNRS, INSERM, ARC and FRMF. M.S.F. is a post-doctoral fellow of the MRC of Canada.

References

1. Colberg-Poley, A.M., Voss, S.D., Chowdhury, K. and Gruss, P. (1985a) Structural analysis of murine genes containing homeo-box sequences and their expression in embryonal carcinoma cells. Nature 314: 713-716.
2. Colberg-Poley, A.M., Voss, S.D., Chowdhury, K., Stewart, C.L., Wagner, E.F. and Gruss, P. (1985b) Clustered homeo-boxes are differentially expressed during murine development. Cell 43: 39-45.
3. Odenwald, W., Taylor, C., Palmer-Hill, F., Tani, M. and Lazzarini, R. (1987b) Expression of the murine Mox6-3 homeo-box protein is regulated by cell-cell contact. ICSU Short Repts. 7: 126.
4. Duboule, D., Baron, A., Mähl, P. and Galliot, B. (1986) A new homeo-box is present in overlapping cosmid clones which define the mouse Hox-1 locus. EMBO J. 5: 1973-1980.
5. Wolgemuth, D.J., Engelmyer, E., Duggal, R.N., Gizang-Ginsberg, E., Mutter, G.L., Ponzetto, C., Viviano, C. and Zakeri, Z.F. (1986) Isolation of a mouse cDNA coding for a developmentally regulates testis-specific transcript containing homeo-box homology. EMBO J. 5: 1229-1235.
6. Rubin, M.R., Toth, L.E., Patel, M.D., d'Eustachio, P. and Chi Nguyen-Huu, M. (1986) A mouse homeo-box gene is expressed in spermatocytes and embryos. Science 233, 663-667.
7. McGinnis, W., Hart, C.P., Gehring, W.J. and Ruddle, F.H. (1984b) Molecular cloning and chromosome mapping of a mouse DNA sequence homologous to homeotic genes of Drosophila. Cell 38: 675-680.
8. Gaunt, S.J. (1987) Homeo-box gene Hox-1.5 expression in mouse embryos : earliest detected by *in situ* hybridization is during gastrulation. Development 101: 51-60.
9. Fainsod, A., Awgulewitsch, A. and Ruddle, F.H. (1987) Expression of the murine homeo-box gene Hox-1.5 during embryogenesis. Developmental Biology 124: 125-133.
10. Baron, A., Featherstone, M.S., Hill, R.E., Hall, A., Galliot, B. and Duboule, D. (1987) Hox-1.6 : a mouse homeo-box-containing gene member of the Hox-1 complex. EMBO J. 6: 2977-2986.
11. Rubin, M.R., Toth, L.E., Patel, M.D., d'Eustachio, P. and Chi Nguyen-Huu, M. (1986) A mouse homeo-box gene is expressed in spermatocytes and embryos. Science 233: 663-667.
12. Jackson, I.J., Schofield, P. and Hogan, B. (1985) A mouse homeo-box gene is expressed during embryogenesis and in adult kidney. Nature 317: 745-748.
13. Krumlauf, R., Holland, P.W.H., McVey, J.H. and Hogan, B.L.M. (1987) Developmental and spatial patterns of expression of the mouse homeo-box gene, Hox-2.1. Development 99: 603-617.

14. Hart, C.P., Awgulewitsch, A., Fainsod, A., McGinnis, W. and Ruddle, F.H. (1985) Homeo-box gene complex on mouse chromosome 11 : molecular cloning expression in embryogenesis, and homology to a human homeo-box locus. Cell 43: 9-18.
15. Meijlink, F., de Laaf, R., Verrijzer, P., Destrée, O., Kroezen, V. and Deschamps, J. (1987) A mouse homeo-box containing gene on chromosome 11 : sequence and tissue-specific expression. Nucl. Acids Res. 17: 6773-6786.
16. Deschamps, J., de Laaf, R., Joosen, L., Meijlink, F. and Destrée, O. (1987a) Abundant expression of homeo-box genes in mouse embryonal carcinoma cells correlates with chemically induced differentiation. Proc. Natl. Acad. Sci. USA 84: 1304-1308.
17. Lonai, P., Arman, E., Czosnek, H., Ruddle, F.H. and Blatt, C. New murine homeo-boxes; structure, chromosomal assignment and differential expression in the embryo and in adult erythropoiesis. Gene, in press.
18. Awgulewitsch, A., Utset, M.F., Hart, C.P., McGinnis, W. and Ruddle, F.H. (1986) Spatial restriction in expression of a mouse homeo-box locus within the central nervous system. Nature 320: 328-335.
19. Breier, G., Bucan, M., Francke, U., Colberg-Poley, A.M. and Gruss, P. (1986) Sequential expression of murine homeo-box genes during F9 EC cell differentiation. EMBO J. 5: 2209-2217.
20. Le Mouellic, H., Condamine, H. and Brûlet, P. Pattern of transcription of the homeo-gene Hox-3.1 in the mouse embryo, in press.
21. Featherstone, M.S., Baron, A., Gaunt, S.J., Mattei, M.G. and Duboule, D. (1988) Hox-5.1 defines a homeo-gene-locus on mouse chromosome 2, submitted for publication.
22. Sharpe, P.T., Miller, J.R., Evans, E.P., Burtenshaw, M.D. and Gaunt, S.J. Hox 6: a mouse homeo-box gene locus on chromosome 14, in press.
23. Dony, C. and Gruss, P. (1987) Specific expression of the Hox-1.3 homeo-box gene in murine embryonic structures originating from or induced by the mesoderm. EMBO J. 6: 2965-2975.
24. Gehring, W.J. (1985) The homeo-box : A key to the understanding of development ? Cell 40: 3-5.
25. Gehring, W.J. (1987) Homeo-boxes in the study of development. Science 236: 1245-1252.
26. Laughon, A. and Scott, M.P. (1984) Sequence of a Drosophila segmentation gene : protein structure homology with DNA-binding proteins. Nature 310: 25-31.
27. Martin, G.R., Boncinelli, E., Duboule, D., Gruss, P., Jackson, I., Krumlauf, R., Lonai, P., McGinnis, W., Ruddle, F. and Wolgemuth, D. (1987) Nomenclature for homeo-box-containing genes. Nature 325: 21-22.
28. Bucan, M., Yang-Feng, T., Colberg-Poley, A.M., Wolgemuth, D.J., Guenet, J.L., Francke, U. and Lehrach, H. (1986) Genetic and cytogenetic localisation of the homeo-box containing genes on mouse chromosome 6 and human chromosome 7. EMBO J. 5: 2899-2905.
29. Rabin, M., Hart, C.P., Ferguson-Smith, A., McGinnis, W., Levine, M. and Ruddle, F.H. (1985) Two homeo-box loci mapped in evolutionarily related mouse and human chromosomes. Nature 314: 175-178.
30. Mavilio, F., Simeone, A., Boncinelli, E. and Andrews, P.W. Activation of four homeo-box gene clusters in human embryonal carcinoma cells induced to differentiate by retinoic acid, in press.
31. Harvey, R.P., Tabin, C.J. and Melton, D.A. (1986) Embryonic expression and nuclear localization of Xenopus homeo-box gene products. EMBO J. 5: 1237-1244.
32. Regulski, M., Harding, K., Kostriken, R., Karch, F., Levine, N. and McGinnis, W. (1985) Homeo-box genes of the Antennapedia and Bithorax complexes of Drosophila. Cell 43: 71-80.

33. McGinnis, W., Garber, R.L., Wirz, J., Kuroiwa, A. and Gehring, W.J. (1984a) A homologous protein coding sequence in Drosophila homeotic genes and its conservation in other metazoans. Cell 38: 403-409.
34. Hart, C.P., Fainsod, A. and Ruddle, F.H. (1987) Sequence analysis of the murine Hox-2.2, Hox-2.3 and Hox-2.4 homeo-boxes : evolutionary and structural comparisons. Genomics 1: 182-195.
35. Regulski, M., McGinnis, N., Chadwick, R. and McGinnis, W. (1987) Developmental and molecular analysis of Deformed : a homeotic gene controlling Drosophila head development. EMBO J. 6: 767-777.
36. Hoey, T., Doyle, H.J., Harding, K., Wedeen, C. and Levine, M. (1986) Homeo-box gene expression in anterior and posterior regions of the Drosophila embryo. Proc. Natl. Acad. Sci. USA 83: 4809-4813.
37. Mavilio, F., Simeone, A., Giampaolo, A., Faiella, A., Zappavigna, V., Acampora, D., Poiana, G., Russo, G., Peschle, C. and Boncinelli, E. (1986) Differential and stage-related expression in embryonic tissues of a new human homeo-box gene. Nature 324: 664-668.
38. Beachy, P.A., Helfand, S.L. and Hogness, D.S. (1985) Segmental distribution of Bithorax complex proteins during Drosophila melanogaster development. Nature 313: 545-551.
39. Weinzierl, R., Axton, J.M., Ghysen, A. and Akam, M. (1987) Genes and Development 1: 386-397.
40. Kessel, M., Schulze, F., Fibi, M. and Gruss, P. (1987) Primary structure and nuclear localization of a murine homeo-domain protein. Proc. Natl. Acad. Sci. USA 84: 5306-5310.
41. Mlodzik, M., Fjose, A. and Gehring, W.J. (1985) Isolation of caudal, a Drosophila homeo-box-containing gene with maternal expression, whose transcripts form a concentration gradient at the preblastoderm stage. EMBO J. 4: 2961-2969.

CELL ADHESION IN MORPHOGENESIS

Jean Paul THIERY
Laboratoire de Physiopathologie du Développement, CNRS-Ecole Normale Supérieure, 46 rue d'Ulm - 75230 Paris Cedex 05 France.

Animal morphogenesis results from a complex set of autonomous and interactive cellular behaviors which include cell proliferation, adhesion, movement, differentiation and death. How these different primary processes are so precisely regulated during embryonic development is one of the major issues that faces the molecular embryologist.

In insects, a vast body of data has accumulated to show the importance of segmentation and homeotic genes in the establishement of the body plan. However, as described in several chapters of this book the corresponding genes in vertebrates have not yet been shown to be directly implicated in pattern formation.

To try and decipher the intricate network of interactions occuring during morphogenesis a variety of approaches have been designed to detect functionally important molecules. These approaches are particularly necessary in vertebrates since it is already apparent that morphoregulatory genes must operate in concert with epigenetic phenomena to achieve the body plan.

A major effort has been devoted to a molecular understanding of the mechanism of cell adhesion since this primary process is directly involved in each morphogenetic step including epithelial remodeling, epithelial-mesenchymal interconversion and individual cell migrations. These studies have allowed the identification of adhesion molecules which appear to play a critical role in morphogenesis. In this chapter the contribution of studies on cell adhesion to the understanding of morphogenesis is considered.

I. ADHESION MOLECULES

There are two major categories of adhesion molecules. The cell adhesion molecules (CAMs) mediate ligation between cells; these are generally large intrinsic cell surface glycoproteins. The substrate adhesion molecules (SAMs) are involved in the interactions between cells, basal laminae and extracellular matrices. SAMs include extracellular matrix molecules and their corresponding cell surface receptors.

The discovery that some of these adhesion molecules appear early during development and are functionally expressed later in different tissues (Edelman and Thiery, 1985) has led to the hypothesis illustrated in Figure 1.

NATO ASI Series, Vol. H26
Cell to Cell Signals in Mammalian Development
Edited by S. W. de Laat et al.

Figure 1

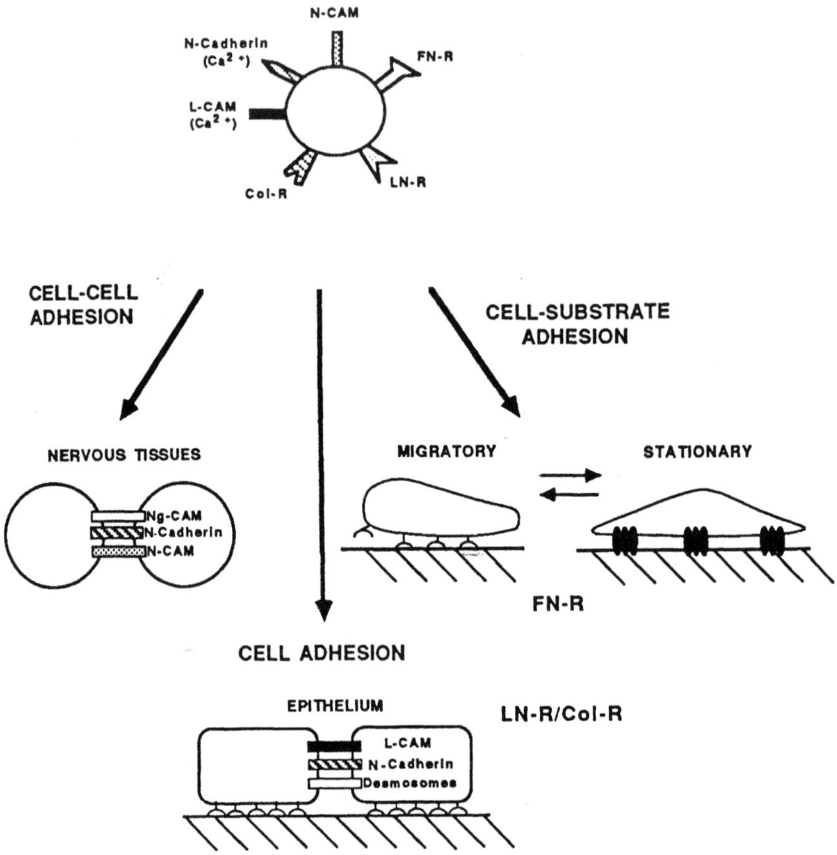

Each embryonic cell carries several distinct adhesion systems which at least in part are utilized in different morphogenetic processes. During histogenesis and differentiation other more cell type restricted adhesion molecules will complement the primary adhesion systems. This hypothesis contrasts with those in which a vast repertoire of recognition molecules was considered to be necessary to achieve the correct assembly of cells into functional organs. Altogether less than 20 different cell adhesion molecules and possibly no more than 30 substrate adhesive molecules have been identified so far. However, mechanisms such as alternative splicing of the primary transcript and post-translational modifications will probably increase the size of this repertoire.

Indeed this model must implicate finely regulated control mechanisms providing enough specificity but also allowing plasticity. Before discussing this issue in this article, I shall provide a brief description of the structure and functional properties of adhesion molecules in relation to certain morphogenetic events.

Cell adhesion molecules (CAMs)

Primary CAMs are those which are expressed early in development and show a widespread distribution in tissues derived from the three primary germ layer.

N-CAM: The neural cell adhesion molecule was the first discovered cell-adhesion molecule and is by now the most extensively studied (Thiery et al., 1977; Edelman and Thiery, 1985; see Goridis this volume).

N-CAM is encoded by a single gene of about 80kb which contains at least 19 exons. Three major forms of N-CAM polypeptides are found to be expressed at the cell surface (Figure 2). Each polypeptide shares the same amino terminal region carrying five immunoglobulin-like domains. The large size polypeptide (180kd) has a long cytoplasmic domain which can interact with cytoskeletal elements. The 140kd polypeptide differs only by its shortened cytoplasmic domain, while the surface-associated 120kd polypeptide is anchored to the plasma membrane by phosphatidyl-inositol (Cunningham et al, 1987). Other variants recently isolated in muscle cells differ at some sites close to the plasma membrane (Goridis, this volume). These different polypeptide chains are formed by an alternative splicing mechanism of the primary N-CAM transcript.

Figure 2

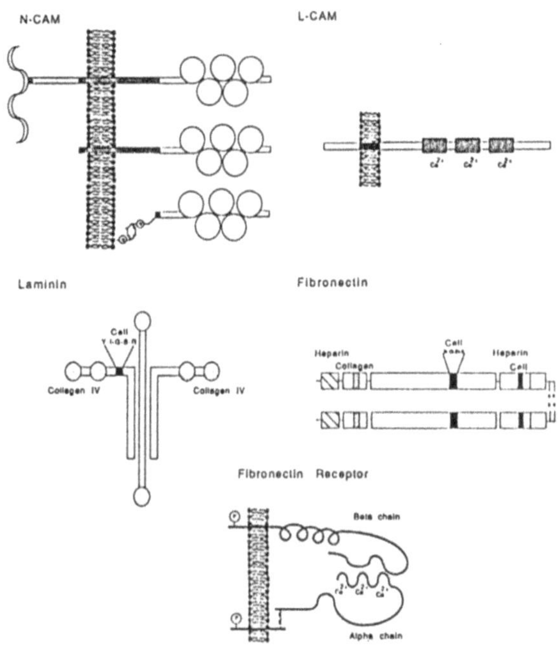

N-CAM is a member of the immunoglobulin superfamily; another adhesion molecule the myelin-associated glycoprotein (MAG), which shares significant partial identity sequence with N-CAM, also belongs to this family. Most members of this family contain immunoglobulin-like domains encoded by one single exon whereas the domains of N-CAM, and the lymphocyte antigen CD4 are encoded by two exons. A detailed comparative analysis of the sequences associated with these domains suggests that each domain is derived from an ancestral gene which was duplicated. N-CAM domains are closer to Ig C-domains in length but resemble V domains in terms of their sequence. Therefore the N-CAM gene may have diverged before V and C domains diverged from each other. It was postulated that the gene for a cell adhesion molecule precursor has given rise to a wide variety of surface molecules endowed with self-self and self-non-self recognition functions; growth factor receptor and other molecules such as Thy1 and alpha1 B glycoprotein, whose functions remain to be determined may have also emerged during evolution from an ancestral N-CAM gene (Edelman,1987).

One of the unique features of N-CAM is its very high carbohydrate content. Three alpha 2-8 polysialic acid chains are clustered on the fifth immunoglobulin-like domain and the amount of these unusual oligosaccharides diminishes during development. Two extreme forms of N-CAM designated embryonic and adult N-CAM have been identified; the embryonic forms contain as much as 30% carbohydrate mostly polysialic acid while the adult forms have than less a third of the amount found in embryonic N-CAM.

The aggregation of N-CAM-containing liposomes or N-CAM coated covaspheres must implicate a direct homophilic binding mechanism i.e. one N-CAM on one cell binds to the same N-CAM on the other cell. The same conclusion was drawn for the aggregation of L cells stably transfected with N-CAM (Edelman et al,1987).

Cell surface modulation of N-CAM binding activity can be controlled by a variety of mechanisms including alternative splicing of the primary transcript, chemical modification, prevalence at the cell surface, anisotropic distribution as well as cis and trans interactions.

Increase in relative amounts of the 180kd polypeptide chain through the control of alternative splicing may favor stronger binding since this polypeptide may be anchored to fodrin and becomes immobile in the plane of the membrane (Pollerberg et al,1986).

Studies on the kinetics of aggregation of N-CAM-containing liposomes have shown that a two-fold increase in surface density leads to a 30-fold increase in binding rate. Fifth order kinetics in the rate of aggregation strongly suggests that N-CAM oligomers may form at the cell surface (Hoffman et al., 1983). In addition a 4 to 5-fold increase in binding rate is observed when embryonic N-CAM is replaced by adult N-CAM, in the liposomes. The large partial specific volume and the negatively charged glyconjugates could induce conformational changes, particularly in reducing the flexibility of the hinge region and as a consequence cis and trans interactions could be weakened. These data show that the prevalence of N-CAM at the cell surface and their carbohydrate content are critical paramaters for the binding strength.

It was not a surprise to find a high proportion of embryonic N-CAM during the ontogeny of the neuronal network or during regeneration of peripheral nerves as well as in regions of constant regeneration such as the olfactory bulb of mammals and in staggerer, a mouse mutant with a defective cerebellum. The adult forms, in contrast, are associated with tissues which have acquired their definitive anatomical organization (Chuong et al,1984).

L-CAM and N-cadherin: In contrast to N-CAM, L-CAM and N-cadherin mediated adhesions require calcium. The liver cell adhesion molecule L-CAM originally isolated from chick embryo hepatocytes (Gallin et al, 1983) is found in mammals and has been referred to by different acronyms (uvomorulin, E-cadherin, Arc 1, gp 80/120).

L-CAM is also derived, at least in mammals, from one single gene but in contrast to N-CAM is expressed as a single transmembrane polypeptide chain of 124kd which does not exhibit any partial identity sequence with N-CAM. However L-CAM shares similarities with N-cadherin (Hatta et al., 1988). Several putative calcium binding sites have been mapped on the extracellular portion of L-CAM (Ringwald et al., 1987). Calcium may be necessary for the proper configuration of L-CAM and N-cadherin and it also confers resistance to proteolysis (Figure 2).

Transfection of mouse L-cells with a complete cDNA coding for L-CAM and N-cadherin strongly suggest that both molecules act through a direct homophilic interaction (Nagafuchi et al,1987; Edelman et al,1987; Hatta et al., 1988).

In summary, N-CAM, L-CAM and N-Cadherin constitute the three primary CAMs and are very likely to be found in all vertebrates. In the chick, the three CAMs are expressed early after fertilization on the same blastodermal cells (Edelman et al,1983; Hatta and Takeichi, 1986). However, L-CAM becomes predominant on some epithelia such as the ectoderm and the endoderm (Thiery et al,1984) while N-CAM and N-cadherin become enriched in the neural plate and the somitic mesoderm (Thiery et al,1982; Duband et al,1988a). Coexpression or sequential expression of N-CAM, N-cadherin and L-CAM are frequently observed during histogenesis such as in ectodermal placodes, in the mesonephric rudiments and at specific sites of endodermal budding. Later, the three CAMs are found in many different tissues.

Secondary CAMs have a restricted tissue distribution and appear later during histogenesis. An increasing number of secondary CAMs are being discovered; however, the exact function and the structure of most of these molecules remains to be established.

Ng-CAM: The chicken neuron-glia cell adhesion molecule (also called L1 or NILE in mammals) is certainly the most extensively characterized so far (Edelman,1986b). Ng-CAM appears on post-mitotic neurons particularly on neurites and on immature Schwann cells (Thiery et al,1985a). It is a transmembrane glycoprotein of 200kd which is cleaved to 135kd and 80kd chains. Ng-CAM mediates interactions between glial cells and neurons through an heterophilic binding mechanisms. Ng-CAM can also mediate direct interaction between neurites to form fascicules through an homophilic binding mechanism (Grumet and Edelman,1988).

Cell-CAM 105 was originally isolated from rat hepatocytes and shown to be involved in hepatocyte colony formation. In vivo it is restricted to villi of the bile caniculi; this cell adhesion molecule has been detected in several other tissues including the intestinal epithelium where it is present also on the villi (Obrink 1986).

P-cadherin: The recently described placental cadherin belongs to the family of calcium-dependent adhesion molecules. It shares similarities with L-CAM and N-cadherin but has a much more restricted tissue distribution (Nose et al., 1987).

Substrate Adhesion Molecules (SAMs).

Fibronectins (FNs) and their receptors (Figure 3) constitute the best characterized substrate adhesion system (see Yamada, 1983; Ruoshlati and Pierschbacher, 1986; Hynes, 1987; Dufour et al., 1988b; see also Zardi, this volume). FNs are large glycoproteins made up of two similar polypeptide chains (250 kd) held together by disulfide bonds. Primary sequence analysis in several species has revealed the existence of three different types of internal homologies. Type I and type II homologies are encoded by one exon, whereas 15 of the type III are encoded by two exons each. However two type III homologies (ED 1 and ED 2) are encoded by one exon; these two sequences are subjected to alternative splicing. Plasma FNs synthesized by hepatocytes are devoid of these two type III homologies, while fibroblasts, endothelial and epithelial cells make variants both with and without the EDs. FNs possess many specific binding domains that have been localized on the molecules. Heparin and fibrin binding domains have been mapped at the carboxy- and amino-terminal portions of the molecules. The collagen binding site resides within a 14 amino-acid region localized at the junction of a type II and a type I homology (Owens & Baralle, 1986). The short tetrapeptide sequence arginine-glycine-aspartate-serine (RGDS), well-exposed on a hydrophilic loop of a type III homology, is a primary cell binding site. However the affinity of small FN fragments containing RGDS is almost two orders of magnitude lower than intact FNs (Akiyama et al., 1985). Comparisons with larger-size peptides and more recently with fusion proteins has led to the localization of a new site which together with RGDS provide a binding affinity close to that of intact FNs (Obara et al., 1987; Obara et al., 1988). This region has been defined as the cell binding site I. A second cell binding site associated with sequences contained in the connecting segment to the last type III homology (IIICS) has been recently identified. This site is involved in adhesion of mouse melanoma cells and of some embryonic cells (Humphries et al., 1987; Dufour et al., 1988a). This region is of great interest since a differential splicing mechanism associated with the IIICS region leads to the synthesis of 3 different variants in rat FNs and 5 variants in human FNs whose abundancies vary among different tissues though in a less obvious way as compared to the EDs. Altogether 20 mRNAs can be formed from a single gene, possibly leading to as many as 400 different FNs. The exact biological function of the different segments (ED1, ED2 and IIICS) whose presence in the FN molecules is regulated at the splicing level remains to be defined. ED1 and ED2 sequences do not show any distinctive feature when compared with the other 15 units of the same class of internal homologies, which are constitutively present in all FN polypeptides. The existence of tissue-specificity in their expression, as well as of a remarkable sequence conservation among vertebrates which in the case of ED2 is much higher than for the rest of the molecule (Gutman & Kornblihtt, 1987; Schwarzbauer et al., 1987; Zardi, this volume), suggesting that these sequences may play key roles in modulating the different biological properties of FNs.

Several cell surface FN receptors have been characterized; the most advanced studies have been performed with the 140kd glycoprotein complex for which direct evidence was provided for an interaction with the RGDS adhesive signal on FNs. In mammals, the FN receptors are composed of one alpha and one beta polypeptide chain (150 and 120kd respectively) (Figure 3); many different cell types express receptors which are closely related to FN receptors.

Figure 3

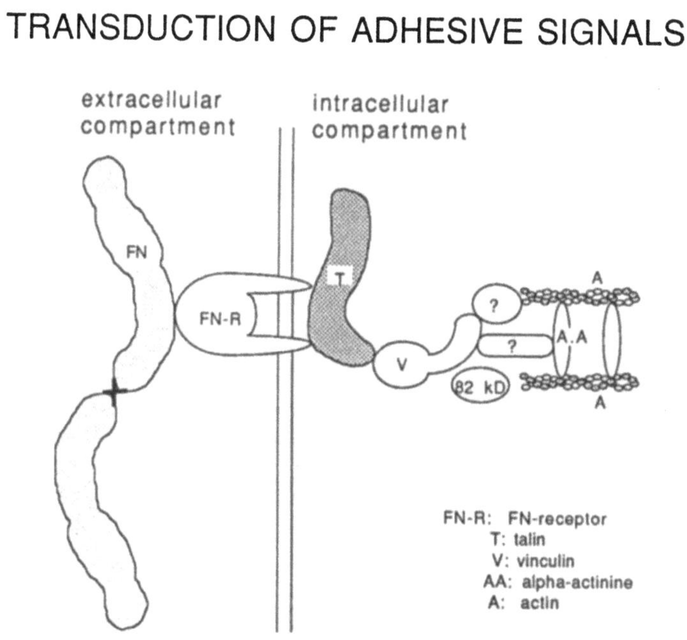

This large family includes IIb-IIIa on platelets, VLA antigens, the vitronectin receptor, LFA-1,

Mac-1, and p150,95 (Hynes,1987; Buck & Horwitz, 1987).

FNs and their receptors appear very early during embryogenesis and are expressed later in many different tissues (Duband et al., 1986). Immunofluorescent localization of the FN receptors on cultured cells has shown that they codistribute with extracellular FN fibrils and intracellular actin cables; they are also present at sites of focal contacts in association with talin, vinculin, and alpha-actinin. The current model is that FN receptors form a transmembrane linkage between FNs and the cytoskeleton (Buck & Horwitz, 1987) (Figure 4). One of the main features of the interaction between FNs and their receptors is low or moderate affinity (Kd $\approx 10^{-6}$ M). This rather weak interaction has been considered to be essential for the establishment of transient binding of migrating cells to the extracellular matrix. Of potential interest is the fact that FN receptors can be phosphorylated on tyrosine present in their cytoplasmic domains. It is postulated that the phosphorylated receptors do not interact with talin and remain in a diffuse state in the plasma membrane. Indeed pp60 sarc can phosphorylate these receptors (Hirst et al., 1986). The transformed cells retain as many FN receptors but lose the ability to form focal contacts and more generally are unable to assemble FN fibrils at their surface. The morphology of the transformed cells can be transiently modified by adding exogenous cellular FNs resulting in the cells becoming more spread, suggesting that not all receptors are defective (Yamada et al., 1976).

MODES OF ADHESION OF AVIAN
NEURAL CREST CELLS

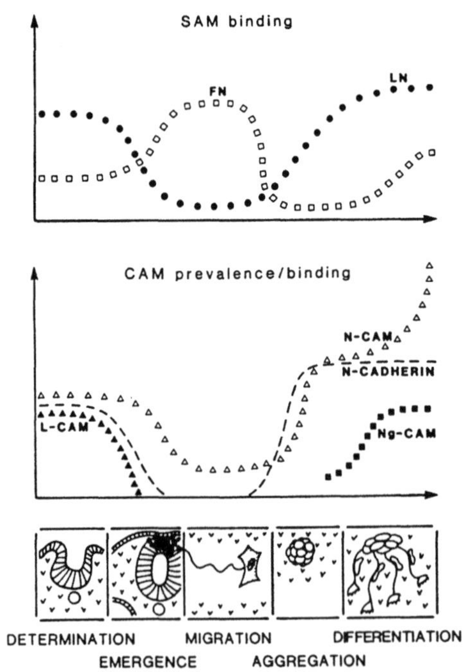

Laminin (LN) is a very large glycoprotein detected early in embryos which is subsequently, principally localized in the basal lamina of epithelial cells, on peripheral nerves, muscles and endothelial cells. LN is composed of one A chain (400kd) and two smaller chains B1 (230 kd) and B2 (220kd) (Martin & Timpl, 1987; Graf et al., 1987) and the primary structure of these three chains has just been determined. Similarly to FNs, LN is a multifunctional glycoprotein which interacts with cell surfaces and with other extracellular matrix molecules, particularly with type IV collagen, a major constituent of basal lamina (Figure 3). It also interacts with heparan sulfate proteoglycan, entactin and with itself within the basal lamina.

A 68kd polypeptide cell surface receptor for laminin has been isolated and partially cloned (Wewer et al., 1986). This receptor may interact with a pentapeptide sequence tyrosine-isoleucine-glycine-serine-arginine (YIGSR) recently identified on the B1 chain. However as in the case of FNs there may be several other cell binding sites recognized by different receptors, particularly in the case of neurons. Much attention has been paid to LN, since tumor cells must interact with and traverse basal laminae (Liotta, 1984). LN can promote the metastatic potential of melanoma cells, whereas LN fragments, antibodies to LN or the YIGSR peptides coinjected with melanoma cells inhibit pulmonary metastases (Iwamoto et al., 1987).

Cytotactin also called, tenascin, hexabrachion and myotendinous antigen is a recently characterized extracellular matrix glycoprotein containing disulfide-linked polypeptide chains. Sequence analyses showed the presence of regions sharing partial identities with β and γ fibrinogen chains and with FN type III homologies; cytotactin also contains four epidermal growth factor-like repeats (Jones et al., 1988). Cytotactin is made by a variety of mesenchymal cells and glial cells and binds to a chondroitin sulfate proteoglycan made by neurons (and numerous other cell types) as well as to FNs (Hoffman et al,1988). Cytotactin appears in discrete sites in the embryo in a cephalocaudal and in a mediolateral wave (Crossin et al,1986).

Cells can attach to cytotactin but do not spead. It has been suggested that cytotactin modulates the different binding properties of FNs allowing spreading and migration of cells (Tan et al,1987). Cytotactin is particularly abundant during chondrogenesis where it could be involved in the stabilization and condensation of cells forming the cartilage (Mackie et al., 1987).

II. MORPHOGENETIC PROCESSES

The distribution and functions of several CAMs and SAMs have been analyzed at different stages of development. As previously mentioned primary CAMs appear at the blastula stage along with some of the SAMs. The same primary and secondary CAMs and SAMs have been detected in all the vertebrates analyzed so far. In fact certain functional studies have been carried out in the vertebrate which offers the most suitable model system at one given stage of development (see Thiery et al. 1985b for a review).

Here I shall focus on two model systems involving epithelial-mesenchymal cell interconversion and cell-migrations.

Epithelial-mesenchymal cell conversion and migrations

Somitogenesis: In birds, gastrulating cells have been shown to lose their epithelial phenotype and migrate between the ectoderm and the primitive endoderm; they were found to lose very rapidly both N-CAM and L-CAM. Some of the newly formed mesodermal cells gave rise to somites along the neural tube. This process was studied in detail since it is a clear-cut example of epithelial-mesenchymal cell interconversion (Duband et al,1987). While mesenchymal cells regroup in the metamerized epithelial somites, N-CAM and N-cadherin increase at their surfaces, and N-cadherin molecules progressively accumulate at the level of intermediate junctions. Both adhesive mechanisms mediated by N-CAM and by the calcium-dependent N-cadherin operate in an in vitro aggregation assay. However, anti-N-cadherin antibodies can disrupt this transient epithelium much more rapidly, as if its calcium adhesion mode is predominant and more amenable to regulation in vivo. Later when somites dissociate to give rise to dermis, muscle and sclerotome, one finds also a correlation with the modulation of N-CAM and N-cadherin. In particular, N-cadherin disappears very rapidly from the surface of the newly formed mesenchymal cells (i.e., dermis and sclerotome) while it is maintained together with N-CAM at the surface of myoblasts.

Neural crest ontogeny: Another striking example of modulations of CAMs has been found throughout the development of neural crest cells. These cells are formed at the dorsal border of the neural epithelium; however during neural tube closure, neural crest cells lose their epithelial phenotype, disperse and migrate in different territories of the embryo to give rise to many different derivatives including mesodermal structures in the head, the peripheral nervous system and melanocytes (Le Douarin, 1982). Their departure from the neural tube is accompanied by the successive loss of L-CAM, N-Cadherin and N-CAM. In fact N-cadherin is very precisely modulated at the onset of migration (Duband et al., 1988a). Later when these cells reach their target sites, particularly those giving rise to the peripheral nervous system, a de novo expression of N-CAM and N-cadherin is noticed as they re-aggregate; however, L-CAM never reappears on these cells. During early differentiation into neurons and glia, the secondary CAM Ng-CAM, is also expressed mostly at the surface of neurites and on immature Schwann cells (Figure 5).

Figure 5

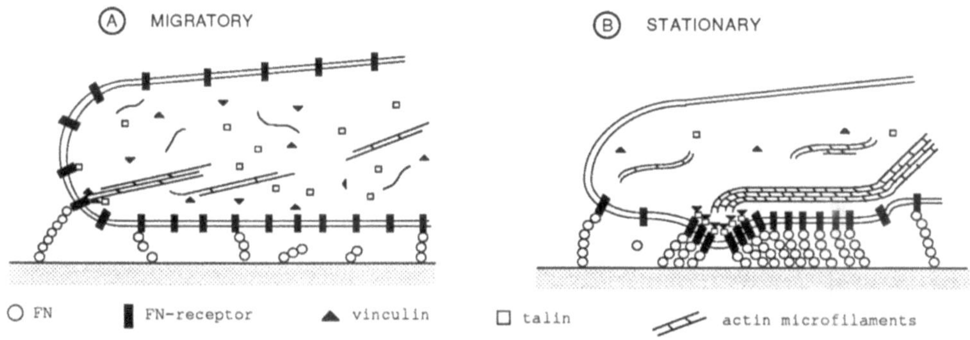

During their migration, neural crest cells and other migratory cells encounter a FN-rich extracellular matrix with also contains collagen type I (Duband & Thiery, 1987), while the epithelial basal lamina forming the edge of the pathways contains LN and type IV collagen. Crest cells which localize lateral to the neural tube to give rise to the dorsal root ganglion progressively accumulate in the anterior part of each somite at the moment when sclerotome starts to form. Cytotactin is at first found concentrated in the basement membrane of the posterior portion of each somite; subsequently it starts to appear in the anterior part of the dissociating somite at the time neural crest cells accumulate in this region. At this stage FNs, collagen I and the chondroitin sulfate proteoglycan are uniformely distributed in the somite. Thereafter, the chondroitin sulfate proteoglycan becomes restricted to the posterior half of the somites. The resultant pattern of alternating stripes of cytotactin (anterior portion) and proteoglycan (posterior portion) together with the final location of neural crest cells in the cytotactin rich region suggest that cytotactin together with its ligands FNs differentially change the migratory behavior of these cells (Tan et al,1987). Binding studies in vitro have revealed that neural crest cells adhere preferentially to FN-containing matrices and that the binding is specific for FNs. However, later on during differentiation into neural cells, these cells bind equally well to LN and to FNs (Rovasio et al., 1983).

Recent studies have shown the involvement of the 140kd receptor complex in in vitro and in vivo neural crest cell migration (Duband et al., 1986). In vitro, the cell binding site I can promote attachement and spreading of neural crest cells, while the cell binding site II only permits attachment. Both sites are implicated in cell motility (Dufour et al., 1988a). Therefore, crest cells display behavior intermediate between fibroblastic cells which recognize only the cell binding site I and that of the the malignant derivative which recognizes almost exclusively the IIICS domain which forms the second cell binding site.

The dynamics of the FN receptors was analyzed during migration and the acquisition of the stationary state with the FRAP (Fluorescence Recovery After Photobleaching) technique (Duband et al., 1988b). More than 80% of the FN receptors are mobile in the plane of the plasma membrane in migratory cells while most of these receptors become immobile in stationary cells. These receptors are regrouped into focal contacts and fibrillary streaks; the latter may be coupled directly to actin filaments while the former are associated with talin, itself associated with vinculin (Figure 6). We have been able to reduce migratory ability by increasing artificially the affinity of the interactions between FN receptors and their substrate. Conversely the disruption of the interaction between FN receptors and FNs in the focal contacts by specific antibodies leads to the redistribution of FN receptors in the plasma membrane. These studies are conceptually very important since they indicate that low affinity interactions between the FN receptor and FNs are favorable for transient attachment and migration, while this interaction becomes stronger when FN receptors cluster at focal adhesion sites to form a multimolecular complex with cytoskeletal elements.

Figure 6

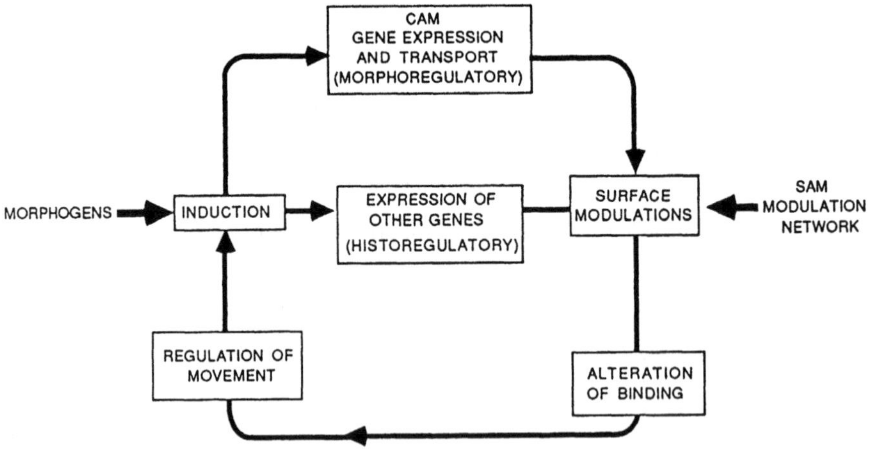

As shown above, neural crest cells undergo a very well-defined pattern of expression of different adhesive systems during their ontogeny. While the cell-cell adhesion systems are predominant before and after migration, the FN-FN receptor system becomes most important during migratory processes.

Evidence has also been provided for a well-defined pattern of expression of adhesion molecules during histogenesis of the central nervous system. Studies have particularly focused on the cerebellum which offers a remarkable model system to study cell migration.

Development of the cerebellum: The ontogeny of cerebellar layers involves all the primary processes in a very defined time sequence. The internal granule layer contains post-mitotic neurons which have migrated from the external proliferating granular layer along the radial Bergman glia. At the onset of migration in birds the post mitotic granule cells express N-CAM and Ng-CAM while cytotactin is associated with the radial glia. A striking inhibition of migration into the molecular layer was obtained with monovalent antibodies to Ng-CAM. Antibodies to cytotactin induced the arrest of the granule cells in the molecular layer. Even though antibodies to N-CAM had only a marginal effect, this cell adhesion molecule and most importantly the 180 kd polypeptide was present on the surface of the migrating neurons along with the cytotactin-binding proteoglycan. It is therefore very likely that at least two CAMs and two SAM are involved in the positioning of neurons and in the development of the neuronal network (Chuong et al,1987).

Secondary inductive events are operating throughout early embryogenesis and particularly in epithelial-mesenchymal cell interactions. The skin appendages are a particularly striking example of pattern formation which results from such reciprocal interactions.

Feather formation: A detailed analysis of feather development has revealed a remarkably coordinated expression of two primary CAMs: N-CAM and L-CAM (Chuong and Edelman,1985a,b). Cells from the dorsal portion of the somites dissociate from the epithelium to give rise to dermis while losing N-cadherin and N-CAM. These cells accumulate under the L-CAM positive epidermis while reexpressing N-CAM. Such condensations induce the formation of ectodermal placodes which become organized into an hexagonal pattern. From each placode expressing L-CAM and N-CAM a feather filament will form and develop into barbs and barbules. During the development of these structure barb ridges become only L-CAM positive while the valley between two barb ridges are N-CAM positive and L-CAM negative. A similar phenomenon occurs within each barb to form barbules. Finally all the L-CAM positive cells synthesize keratins while the N-CAM positive cells die.

Feather formation provides one striking example of a remarkably coordinated expression of both CAMs and SAMs and the other primary processes of cell division cell migration cell differentiation and cell death. During induction processes to form the placode, an anisotropic distribution of FN and collagens develops; while laminin and FNs also form a basement membrane separating the mesodermal pulp from the ectoderm in the feather rudiment.

Inductive events and subsequent morphogenesis can be perturbed if one of the critical components of the primary process of cell adhesion is altered. Dorsal skin of young chick embryos can be cultured in vitro to form a hexagonal pattern of the feather germ. Addition of monovalent antibodies to L-CAM at the onset of the culture leads to a disturbed pattern of the mesodermal condensates and of the ectodermal placodes (Gallin et al,1986). These experiments suggest that the inductive signals from the mesoderm direct the altered ectoderm into a different developmental pathway and or that the responding tissue sends a signal which modifies the growth pattern of mesodermal condensates.

III. ADHESION MOLECULES ARE PRIMARY REGULATORS OF MORPHOGENESIS

The different morphogenetic events described above illustrate the important role played by adhesion molecules. By their remarkably controlled dynamic spatio-temporal expression, adhesion molecules are now considered to form the mechanochemical link between genes and morphology.

Perturbation experiments made in a variety of developmental systems show that these molecules do not simply act as ligands between cells but rather contribute to the shaping of the embryo.

The morphoregulator hypothesis (Edelman,1984; Edelman, 1986a) considers that regulatory genes and structural genes determining the appearance and function of the adhesion molecules are the essential genetic components controlling morphogenesis. These genes may be under a separate control from those involved in histogenesis and cytodifferentiation.

The set of genes coding for the morphoregulatory molecules are responding to signals arising across borders between cell collectives with different previous histories. Induction of CAM cycles and the development of a complex SAM network provides the necessary conditions to alter growth pattern and movement of cells. The number of molecules and signals need not be large since the different modes of surface modulations of CAMs and the diversified effects of the interactive SAMs occuring in different amounts could lead to a vast repertoire of developmental pathways.

One of the most important features in vertebrate embryogenesis is the trigger of morphogenetic events and of the differentiation program by inductive signals mediated by cell collectives brought together by a series of complex movements. Indeed for many years, attempts were made to identify specific inducers. It is only very recently that for the first time two fairly ubiquitous growth factors: FGF and TGF ß2 have been shown to induce mesoderm formation in amphibian blastula in vitro (Weeks and Melton, 1987; Rosa et al., 1987). This discovery may provide a novel approach to address control mechanisms implicated in the trigger of specific morphogenetic pathways. TGF ß has already been shown to stimulate the production of FNs and their surface receptors (Ignotz and Massagué, 1987). NGF is known to increase the surface expression of N-CAM and Ng-CAM in cultured pheochromocytoma (Friedlander et al., 1986; Prentice et al., 1987). It is therefore intriguing to consider the possibility that some growth factors have morphogenetic activities (see Sporn 1988) and through their action on the control of CAM and SAM gene expression these factors will determine the type of interactions that each cell should have when assembling into tissues.

Acknowledgment:

The author's laboratory is supported by grants from CNRS, INSERM, The ligue Nationale Française contre le Cancer, the Association pour la Recherche sur le Cancer, The Fondation pour la Recherche Médicale and The March of Dimes-Birth Defects. The author thanks Dr. Giovanni Levi and Dr. Jelena Gavrilovic for critical reading of the manuscript.

REFERENCES

Akiyama, S.K., E., Hasegawa, T., Hasegawa, and K.M., Yamada. 1985. The interaction of fibronectin fragments with fibroblastic cells. J. Biol. Chem. 260: 13256-13260.

Buck, C.A., and A.F., Horwitz. 1987. Cell surface receptors for extracellular matrix molecules. Ann. Rev. Cell Biol. 3: 179-205.

Chuong, C.-M., and G.M., Edelman. 1984. Alterations in neural cell adhesion molecules during development of different regions of the nervous system. J. Neurosci. 4: 2354-2368.

Chuong, C.-M., and G.M., Edelman. 1985a. Expression of cell adhesion molecules in embryonic induction. II. Morphogenesis of adult feathers. J. Cell Biol. 101: 1027-1043.

Chuong, C.-M., and G.M., Edelman. 1985b. Expression of cell adhesion molecules in embryonic induction. I. Morphogenesis of nestling feathers. J. Cell Biol. 101: 1009-1026.

Chuong, C.-M., K.L., Crossin, and G.M., Edelman. 1987. Sequential expression and differential function of multiple adhesion molecules during the formation of cerebellar cortical layers. J. Cell Biol. 104: 331-342.

Crossin, K.L., S., Hoffman, M., Grumet, J.P., Thiery, and G.M., Edelman. 1986. Site-restricted expression of cytotactin during development of the chicken embryo. J. Cell Biol. 102: 1917-1930.

Cunningham, B.A., J.J., Hemperly, B.A., Murray, E.A., Prediger, R., Brackenbury, and G.M., Edelman. 1987. Neural cell adhesion molecule: structure, immunoglobulin-like domains, cell surface modulation, and alternative RNA splicing. Science 236: 753-884.

Duband, J.-L., S., Rocher, W.-T., Chen, K.M., Yamada, and J.P., Thiery. 1986. Cell adhesion and migration in the early vertebrate embryo: Location and possible role of the putative fibronectin-receptor complex. J. Cell Biol. 102: 160-178.

Duband, J.-L., and J.P., Thiery. 1987. Distribution of laminin and collagens during avian neural crest development. Development 101: 461-478.

Duband, J.-L., S., Dufour, K., Hatta, M., Takeichi, G.M., Edelman, and J.P., Thiery. 1987. Adhesion molecules during somitogenesis in the avian embryo. J. Cell Biol. 104: 1361-1374.

Duband, J.-L., G.H., Nuckolls, A., Ishihara, T., Hasegawa, K.M., Yamada, J.P., Thiery, and K., Jacobson. 1988. The fibronectin receptor exhibits high lateral mobility in embryonic locomoting cells but is immobile in focal contacts and fibrillar streaks in stationary cells. J. Cell Biol. in press.

Duband, J.-L., T., Volberg, I., Sabanay, J.P., Thiery, and B., Geiger. 1988. Spatial and temporal distribution of adherens-junction associated adhesion molecule A-CAM during avian embryogenesis. Development 103: 000-000.

Dufour, S., J.-L., Duband, M.J., Humphries, M., Obara, K.M., Yamada, and J.P., Thiery. 1988a. Attachment, spreading, and migration of avian neural crest cells are mediated by multiple adhesion sites on fibronectin molecules. EMBO J. submitted.

Dufour, S., J.-L., Duband, A.R., Kornblihtt and J.P., Thiery. 1988b. The role of fibronectins in embryonic cell migrations. TIG in press.

Edelman, G.M., W.J., Gallin, A., Delouvée, B.A., Cunningham, and J.P., Thiery. 1983. Early epochal maps of two different cell adhesion molecules. Proc. Natl. Acad. Sci. USA 80: 4334-4388.

Edelman, G.M. 1984. Cell adhesion and morphogenesis: The regulator hypothesis. Proc. Natl. Acad. Sci. USA 81: 1460-1464.

Edelman, G.M. and J.P., Thiery. 1985. The Cell in Contact: Adhesion and junctions as morphogenetic determinants. John Wiley and Sons, New York. pp. 507.

Edelman, G.M. 1986a. Cell adhesion molecules in the regulation of animal form and tissue pattern. Ann. Rev. Cell Biol. 2: 81-116.

Edelman, G.M. 1986b. Cell adhesion molecules in neural histogenesis. Ann. Rev. Physiol. 48: 417-430.

Edelman, G.M. 1987. CAMs and Igs: Cell adhesion and the evolutionary origins of immunity. Immunol. Rev. 100: 11-45.

Edelman, G.M., B.A., Murray, R.-M., Mege, B.A., Cunningham, and W.J., Gallin. 1987. Cellular expression of liver and neural cell adhesion molecules after transfection with their cDNAs results in specific cell-cell binding. Proc. Natl. Acad. Sci. USA 84: 8502-8506.

Friedlander, D.R., R. Brackenbury, and G.M. Edelman. (1985). Conversion of embryonic form to adult forms of N-CAM in vitro results from de novo synthesis of adult forms. J. Cell Biol. 101: 412-419.

Gallin, W.J., G.M., Edelman, and B.A., Cunningham. 1983. Characterization of L-CAM, a major cell adhesion molecule from embryonic liver cells. Proc. Natl. Acad. Sci. USA 90: 1038-1042.

Gallin, W.J., C.-M., Chuong, L.H., Finkel, and G.M., Edelman. 1986. Antibodies to liver cell adhesion molecule perturb inductive interactions and alter feather pattern and structure. Proc. Natl. Acad. Sci. USA 83: 8325-8329.

Graf, J., Y., Iwamoto, M., Sasaki, G.R., H.K., Kleinman, F.A., Robey, and Y., Yamada. 1987. Identification of an amino-acid sequence in laminin mediating cell attachment, chemotaxis and receptor binding. Cell 48: 989-996.

Grumet, M., and G.M., Edelman. 1988. Neuron-glia cell adhesion molecule interacts with neurons and astroglia via different binding mechanisms. J. Cell Biol. 106: 487-505.

Gutman, A., and A.R., Kornblihtt. 1987. Identification of a third region of cell-specific alternative splicing in human fibronectin mRNA. Proc. Natl. Acad. Sci. USA 84: 7179-7182.

Hatta, K., A., Nose, A., Nagafuchi and M., Takeichi. 1988. Cloning and expression of cDNA encoding a neural calcium-dependent cell adhesion molecule: Its identity in the cadherin gene family. J. Cell Biol. 106: 873-881.

Hatta, K., and M., Takeichi. 1986. Expression of N-cadherin adhesion molecules associated with early morphogenetic events in chick development. Nature 320: 447-449.

Hirst, R., A., Horwitz, C., Buck, and L., Rohrschneider. 1986. Phosphorylation of the fibronectin receptor complex in cells transformed by oncogenes that encode tyrosine kinases. Proc. Natl. Acad. Sci. USA 83: 6470-6474.

Hoffman, S., and G.M., Edelman. 1983. Kinetics of homophilic binding by embryonic and adult forms of the neural cell adhesion molecule. Proc. Natl. Acad. Sci. USA 80: 5762-5766.

Hoffman, S., K.L., Crossin, and G.M., Edelman. 1988. Molecular forms, binding functions, and developmental expression patterns of cytotactin and cytotactin-binding proteoglycan, an interactive pair of extracellular matrix molecules. J. Cell Biol. 106: 519-532.

Humphries, M.J., V.K. Komoriya, S.K., Akiyama, K., Olden, and K.M., Yamada. 1987. Identification of two distinct regions of the type III connecting segment of human plasma fibronectin that promote cell type-specific adhesion. J. Biol. Chem. 262: 6886-6892.

Hynes, R.O 1987. Integrins: A family of cell surface receptors. Cell 48: 549-554.

Ignotz, R.A. and J. Massagué.1987. Cell adhesion protein receptors as targets for transforming growth factor-ß action. Cell 52: 189-197.

Iwamoto, Y., F.A., Robey, J., Graf, M., Sasaki, H.K., Kleinman, Y., Yamada, and G.MR, Martin. 1987. YIGSR, a synthetic laminin pentapeptide, inhibits experimental metastasis formation. Science, 238: 1132-1134.

Jones, F.S., M.P. Burgoon, S. Hoffman, K.L., Crossin, B.A., Cunningham and G.M., Edelman. 1988. A cDNA clone for cytotactin contains sequences similar to epidermal growth factor-like repeats and segments of fibronectin and fibronogen. Proc. Natl. Acad. Sci. USA, 85: 2186-2190.

Le Douarin, N.M. 1982. The Neural Crest. Cambridge University Press, Cambridge.

Liotta, L.A. 1984. Tumor invasion and metastasis. Am. J. Path. 339-348.

Mackie, E.J., I., Thesleff, and R. Chiquet-Ehrismann. 1987. Tenascin is associated with chondrogenic and osteogenic differentiation in vivo and promotes chondrogenesis in vitro. J. Cell Biol. 105: 2569-2579.

Martin, G.R., and R., Timpl. 1987. Laminin and other basement membrane components. Ann. Rev. Cell Biol. 3: 57-85.

Nagafuchi, A., Y., Shirayoshi, K., Okazadi, K., Yasuda, and M., Takeichi. 1987. Transformation of cell adhesion properties by exogenously introduced E-cadherin cDNA. Nature 329: 341-343.

Nose, A., A., Nagafuchi, and M., Takeichi. 1987. Isolation of placental cadherin cDNA: Identification of a novel gene family of cell-cell adhesion molecules. EMBO J. 6: 3655-3661.

Obara, M., M.S., Kang, S., Rocher-Dufour, A., Kornblihtt, J.P., Thiery, and K.M., Yamada. 1987. Expression of the cell-binding domain of human fibronectin in E. coli: Identification of sequences promoting full to minimal adhesive function. FEBS Lett. 213: 261-264.

Obara, M., Kang, M.S., and Yamada, K.M. (1988). Site-directed mutagenesis of the cell-binding domain of human fibronectin: separable, synergistic sites mediate adhesive function. Cell, in press.

Obrink, B. (1986). Epithelial cell adhesion molecules. Exp. Cell Res., 163: 1-21.

Owens, R.J., and F.E., Baralle. 1986. Mapping the collagen-binding site of human fibronectin by expression in Escherichia coli. EMBO J. 5: 2825-2830.

Pollerberg, G.E., M., Schachner, and J., Davoust. 1986. Differentiation state-dependant surface mobilities of two forms of the neural cell adhesion molecule. Nature 324: 462-465.

Prentice, H.M., S.E. Moore, J.G., Dickson, P., Doherty and F.S. Walsh. 1987. Nerve growth factor-induced changes in neural cell adhesion molecules (N-CAM) in PC12 cells. EMBO J. 6: 1859-1863.

Rosa, F., Roberts, A.B., Danielpour, D., Dart, L.L., Sporn, M.B., and Dawid, I.B. (1988) Mesoderm induction in amphibians : The role of TGF-ß2-like factors. Science, 239:783-785.

Rovasio, R.A., A., Delouvée, K.M., Yamada, R., Timpl, and J.P., Thiery. 1983. Neural crest cell migration: Requirement for exogenous fibronectin and high cell density. J. Cell Biol. 96: 462-473.

Ruoslahti, E., and M.D., Pierschbacher. 1986. Arg-Gly-Asp: A versatile cell recognition signal cell. Cell 44: 517-518.

Schwarzbauer, J.E., R.S., Patel, D., Fonde, and R.O., Hynes. 1987. Multiple sites of alternative splicing of the rat fibronectin gene transcript. EMBO J. 6: 2573-2580.

Sporn, M.B., and A.B. Roberts. 1988. Peptide growth factors are multifunctional. Nature 332: 217-219.

Tan, S.S., K.L., Crossin, S., Hoffman, and G.M., Edelman. 1987. Asymmetric expression in somites of cytotactin and its proteoglycan ligand is correlated with neural crest cell distribution. Proc. Natl. Acad. Sci. USA 84: 7977-7981.

Thiery, J.P., R., Brackenbury, U., Rutishauser, and G.M., Edelman. 1977. Adhesion among neural cells of the chick embryo. II. Purification and characterization of a cell adhesion molecule from neural retina. J. Biol. Chem. 252: 6835-6845.

Thiery, J.P., J.-L., Duband, U., Rutishauser, and G.M., Edelman. 1982. Cell adhesion molecules in early chicken embryogenesis. Proc. Natl. Acad. Sci. USA 79: 6737-6741.

Thiery, J.P., A., Delouvée, W., Gallin, B.A., Cunningham, and G.M., Edelman. 1984. Ontogenetic expression of cell adhesion molecules: L-CAM is found in epithelia derived from the three primary germ layers. Dev. Biol. 102: 61-78.

Thiery, J.P., A., Delouvée, M., Grumet, and G.M., Edelman. 1985a. Appearance and regional distribution of the neuron-glia cell adhesion molecule (Ng-CAM) in the chick embryo. J. Cell Biol. 100: 442-456.

Thiery, J.P., J.-L., Duband, and G., Tucker. 1985. Cell migration in the vertabrate embryo: Role of cell adhesion and tissue environment in pattern formation. Ann. Rev. Cell Biol. 1: 91-113.

Volk, T., O., Cohen, and B., Geiger. 1987. Formation of heterotypic adherens-type junctions between L-CAM-containing liver cells and A-CAM-containing lens cells. Cell 50: 987-994.

Weeks, O.L., and D.A., Melton. 1987. A maternal mRNA localized to the vegetal hemisphere in Xenopus eggs codes for a growth factor related to TGFß. Cell 51: 861-867.

Wewer, U.M., L.A., Liotta, M., Jaye, G.A., Ricca, W.N., Drohan, A.P., Claysmith, Rao, C.N., Wirth, P., Coligan, J.E., Albrechtsen, R., Mudryj, M. and M.E., Sobel. 1986. Altered levels of laminin receptor mRNA in various human carcinoma cells that have different abilities to bind laminin. Proc. Natl. Acad. Sci. USA 83: 7137-7141.

Yamada, K.M., S.S., Yamada, and I., Pastan. 1976. Cell surface protein partially restores morphology, adhesiveness, and contact inhibition of movement to transformed fibroblasts. Proc. Natl. Acad. Sci. USA 73: 1217-1221.

Yamada, K.M. 1983. Cell surface interactions with extracellular materials. Ann. Rev. Biochem. 52: 761-799.

GENERATION OF DIVERSITY IN THE NCAM ADHESION PROTEIN OF RODENTS

C. Goridis[1], M.-J. Santoni[1], M.-R. Hirsch[1], J. Barbas[1], D. Barthels[2] and W. Wille[2].

[1] Centre d'Immunologie INSERM-CNRS de Marseille-Luminy
Case 906
F-13288 Marseille Cedex 9
France

[2] Institut für Genetik der Universität Köln
Weyertal 121
D-5000 Köln 41
FRG

ABSTRACT

NCAM (neural cell adhesion molecule) is the name given to a family of closely related cell surface glycoproteins that serve as ligands in the formation of cell-cell contacts and are thought to play a central role in a number of developmental processes. Three size classes of NCAM polypeptides (NCAM-180, -140 and -120 according to their apparent Mr) are differentially expressed during development and differentiation. Their size differences are due to membrane-associated and cytoplasmic domains of different length : NCAM-140 and -180 are transmembrane proteins whereas NCAM-120 is attached to the membrane by covalently bound phospholipid. The three isoforms share five N-terminal domains that are homologous to immunoglobulin domains. They are specified by separate mRNAs which are derived from a single gene by a combination of differential splicing and alternative polyadenylation. Additional diversity in NCAM polypeptides appears to be generated by the inclusion of small exons which encode extracellular segments and can combine with 3' exons coding for transmembrane or lipid-linked proteins. The patterns of processing of the NCAM gene are different depending on cell type and developmental stage and change during differentiation of clonal cell lines suggesting that they are regulated by _trans_-acting factors.

NATO ASI Series, Vol. H26
Cell to Cell Signals in Mammalian Development
Edited by S. W. de Laat et al.

INTRODUCTION

Since the early studies of Holtfreter (1939), cell adhesion has been recognized as a key mechanism in animal morphogenesis. At the molecular level, cell adhesion has been shown to depend on the presence of specialized contact receptors known as cell adhesion molecules (Edelman, 1983, 1986 ; Takeichi et al, 1985 ; Obrink, 1986 ; Hynes, 1987 ; Thiéry et al this volume). They are operationally defined, in the first place by the inhibitory effect of antibodies directed against them in a suitable adhesion assay. For some, but not all cell adhesion molecules thus defined has one been able to demonstrate either an inhibitory effect of the purified molecules themselves or their direct binding to a cell surface receptor. If only antibody inhibition has been obtained, one may say that the antigen is involved in interactions leading to cell-cell or cell-matrix contacts. If inhibition by the purified molecules or their binding to the cell surface has been demonstrated, one may conclude that they are ligands in the formation of cell-cell or cell-matrix contacts. The function of cell adhesion molecules may be limited to the formation of physical cell-cell or cell-matrix contacts. However, there is evidence that cell adhesion can also generate signals which lead to alterations in gene expression. For example, direct contact between glial cells and neurons in vitro promotes biochemical changes in glial cells (Linser and Moscona, 1983), and the maturation of dopaminergic neurones appears to be accelerated when co-cultured with their target neurones, an effect which could be mimicked by adding membrane preparations (Prochiantz et al, 1981).

Adhesion molecules may be classified on the basis of the type of adhesion they mediate. CAMs (cell adhesion molecules) mediate cell-cell contacts without being concentrated in morphologically recognizable contact structures. JAMs (junctional adhesion molecules) also mediate cell-cell contacts but are assembled in junctional complexes such as tight junctions, the various types of adherens junctions or gap junctions. SAMs (substrate adhesion molecules) mediate cell-extracellular matrix contacts. NCAM, the first adhesion molecule to be identified (Thiéry et al, 1977), functions mostly as a CAM, although recent results indicate that some forms of it may also mediate cell-matrix interactions (Cole and Glaser, 1986).

NCAM is one of the best known cell adhesion molecules and may serve as a model for understanding the molecular basis of cell adhesion and its regulation. The name NCAM for neural cell adhesion molecule was chosen since NCAM was first identified in the nervous system. However, NCAM is not specific to nervous tissue : during embryonic development, NCAM is present on derivatives of all three germ layers. In the perinatal period, prominent expression of NCAM is also seen in skeletal muscle and in the kidney. Nevertheless, in the normal adult animal, NCAM is found mainly on neurones and glial cells of the central and peripheral nervous system (see for reviews, Edelman, 1986 ; Rutishauser, 1986 ; Ekblom et al, 1986).

NCAM was the first adhesion molecule for which direct binding of the purified protein to the cell surface (Edelman et al, 1983 ; Goridis et al, 1983 ; Rutishauser, 1983) and to itself (Hoffman and Edelman, 1983) has been demonstrated. NCAM is a large cell surface glycoprotein. Its binding site has been attributed to the N-terminal part but may extend into the middle portion of the molecule which is carbohydrate-rich and bears unusual sugars, unbranched polymers of sialic acid (Cunningham et al, 1983 ; Finne et al, 1983 ; Gennarini et al, 1984 ; Frelinger et al, 1986 ; Cole et al, 1986 ; Cunnigham et al, 1987). The C-terminal part anchors the molecule into the membrane and may promote interactions with the cytoskeleton.

NCAM is not a single molecule, but rather a group of closely related proteins, that differ at the carbohydrate or the protein level (see for review, Rutishauser and Goridis, 1986). In the following, we will focus on the primary structure and polypeptide diversity of rodent NCAM proteins and on how the different isoforms are generated.

THE PRIMARY STRUCTURE OF MOUSE NCAM PROTEINS.

Three size classes of NCAM proteins have been identified in various species. Their desialo-forms migrate in SDS-polyacrylamide gels with apparent Mr's of 180, 140 and 120 x 10^3 (although the true Mr's of the polypeptides are considerably smaller, i.e. around 120,000, 90,000 and 80,000 for the 180, 140 and 120 kD species of mouse NCAM, respectively). In accord with other authors (Pollerberg et al, 1986 ; Sunshine et al, 1987), we call them NCAM-180, -140 and -120. They are schematically depicted in Fig. 1, which summarizes results from biochemical and cDNA sequencing data (see Rutishauser and Goridis, 1986 ; Barthels et al, 1987 ; Cunningham et al, 1987). Their size differences are due to membrane-associated and cytoplasmic domains of different sizes. The N-terminal region of the NCAM proteins is organized into 5 domains which are homologous to each other and to immunoglobulin (Ig) domains. The polysialic acid is borne by two, perhaps three, sites on the fifth domain. The site for NCAM-NCAM binding has not been precisely mapped yet; the available data indicate that it is within the 5 Ig-like domains.

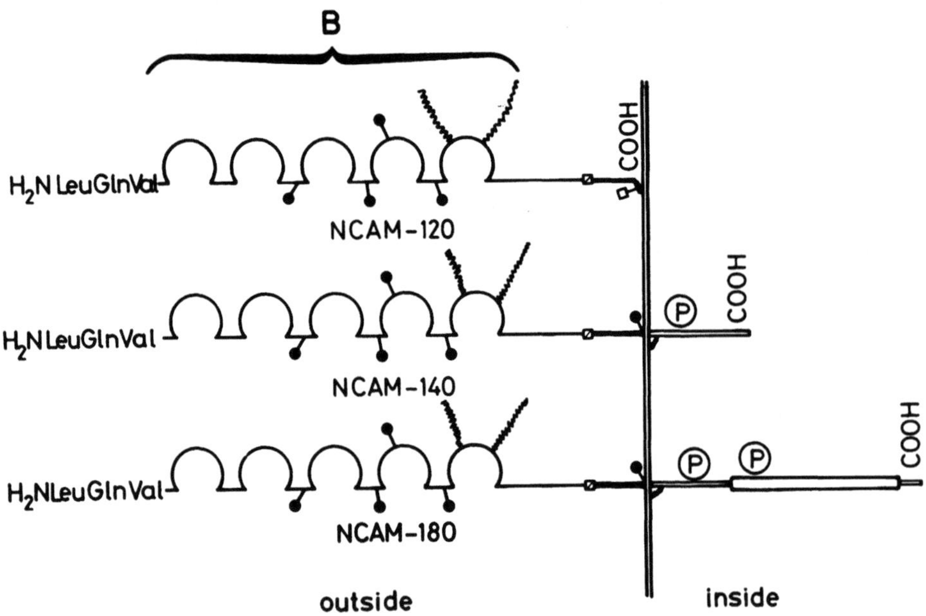

Figure 1. Schematic representation of the 3 size classes of NCAM proteins. B, binding site, , sites for potential Asn-linked glycosylation, , polysialic acid, extra-sequence in the membrane proximal region, , glycophospholipid anchor, , fatty acid residues, P , phosphoserine residues.

The suggestion that the N-terminal NCAM domains are structurally similar to Ig or Ig-like domains is fully borne out by secondary structure predictions indicating that each domain is folded like Ig constant domains into 7 strands that form 2 beta-sheets (Santoni et al, 1988). This Ig-like folding pattern may be important for NCAM-NCAM binding. In fact, all Ig-like domains tend to form dimers which are engaged in inter-chain associations. Dimer formation between two Ig domains may be involved in trans binding of NCAM molecules on apposing cells, but it may also lead to pairwise association of NCAM chains on the same cell thus creating a site for trans binding or changing the valency of interaction. This of course remains speculation as long as NCAM has not been crystallized. Several CAMs have now been found to contain Ig homology units. Apart from NCAM, these are myelin-associated glycoprotein (Arquint et al, 1987 ; Salzer et al, 1987), L1, another cell adhesion molecule of the nervous system (M. Schachner, pers. comm.), and the CD2-LFA-3 couple (Seed, 1987) and I-CAM (Simmons et al, 1988), which are involved in adhesion of lymphocytes.

Rodent NCAM cDNA clones coding for NCAM-140 and -120 have identical 5' regions which code for the common extracellular domains of the proteins (although this does not exclude the presence of variant forms in this region, see below). As expected, they differ in their 3' parts (Barthels et al, 1987 ; Santoni et al, 1987 ; Small et al, 1987). The coding region of NCAM-120 ends with a short stretch of moderately hydrophobic, uncharged amino acids not contained in the clones that code for NCAM-140. Clones coding for NCAM-140 have a longer open reading frame that predicts a transmembrane and a cytoplasmic domain. It has been shown first for chicken NCAM by Murray et al (1986) that the larger cytoplasmic region of NCAM-180 is due to an additional domain inserted close to the C-terminus. We have now isolated and sequenced this domain at the cDNA and the genomic level (Barbas et al, 1988 ; Barthels et al, 1988) ; it is inserted at the same point as it is in chicken NCAM.

The C-terminus of NCAM-120 is reminiscent of proteins which are anchored in the membrane not by a transmembrane segment but by phospholipid. In the case of the Trypanosome surface coat protein, it has been shown that such a hydrophobic extension is cleaved immediately after biosynthesis followed by transfer of the new C-terminus to a glycophospholipid which contains ethanolamine, several sugars and phosphatidylinositol as membrane anchor (see for review Low et al, 1986).

These proteins are released from the membrane by phosphatidylinositol-specific phospholipase C, and this we have shown to be the case for NCAM-120 (He et al, 1986). Pulse-chase experiments demonstrate that also NCAM-120 acquires its lipid anchor within minutes after completion of polypeptide biosynthesis, and that cultured cells spontaneously release this NCAM form into the extracellular milieu (He et al, 1987).

Do three polypeptides with different modes of membrane-attachment and with cytoplasmic domains of different size represent the entire structural heterogeneity of the NCAM polypeptides ? Previous immunochemical data (Williams et al, 1985) suggested that there may be additional sources of heterogeneity, although it is not clear whether this diversity was associated with the polypeptide, the carbohydrate or some other structural alteration. We have approached this question by S1 nuclease protection experiments done with brain mRNA of different developmental stages. Major splice sites are of course easily revealed at the points where the transcripts that code for NCAM-120, -140 and -180 diverge from each other. In addition, these experiments indicated the existence of a major splice site in the membrane-proximal, extracellular region (Barthels et al, 1987). Precisely at this point, Frank Walsh and his collaborators (Dickson et al, 1987) have found an extra 37 amino acid sequence in a human muscle NCAM cDNA that codes for NCAM-120. We have now isolated mouse brain cDNAs for NCAM-120 and -140 which contain 7 additional amino acids inserted at the same place, the first five of which match with the corresponding human muscle sequence. Not only muscle NCAM may thus contain extra sequences in the membrane-proximal region and these sequences occur in both, lipid-anchored and membrane-spanning NCAM proteins.

DISTINCT NCAM TRANSSCRIPTS AND PROTEINS ARE GENERATED BY DIFFERENTIAL SPLICING AND ALTERNATIVE POLYADENYLATION.

We have cloned the transcriptional unit of the mouse NCAM gene, which covers around 80 kb, in 6 overlapping cosmids and one lambda clone. When fragments free of repetitive sequences cut out from the cosmids are used as probes in Southern blots on genomic DNA cut with the same restriction enzymes, single bands of corresponding size are recognized. A representative experiment of this type is shown in Fig. 2. Identical results are obtained under low stringency conditions. Hence, it can be concluded that there is only one NCAM gene in the haploïd mouse genome and that closely related genes are also not present in the mouse.

The exon-intron organization of the gene, in so far as we have determined it, and its relationship with the mRNAs and the proteins is schematically depicted in Fig. 3. Exon-containing fragments from the 5' part of the gene hybridize in Northern blots with the four prominent size classes of NCAM mRNAs present in the adult mouse brain (2.9, 5.2, 6.7 and 7.4 kb in size, Gennarini et al, 1986). Exons contained in these regions are thus constitutively expressed in the four mRNA species ; they code for the shared extracellular parts of the protein isoforms. The 3' part of the gene contains 5 exons, inclusion or exclusion of which together with the alternative use of two poly(A) addition signals accounts for the size differences of the mRNAs (Barbas et al, 1988 and Fig. 3). A large exon encodes the diversity region of the two smaller transcripts. Its very 5' part codes for the hydrophobic extension of NCAM-120. A first polyadenylation signal is used by the 2.9 kb mRNA. The 5.2 kb mRNA uses a second signal 2.3 kb further downstream. Three 3' exons are shared by the 6.7 and 7.4 kb transcripts and code for the transmembrane and cytoplasmic domains common to NCAM-140 and -180 ; in between lies an exon specific for the 7.4 kb mRNA that encodes the additional cytoplasmic domain of NCAM-180. Obviously, the 7.4 and 6.7 kb mRNA species code for NCAM-180 and -140, respectively, and both, the 5.2 and 2.9 kb transcripts code for NCAM-120. A small extra exon specifies the sequence inserted in the membrane-proximal region ; our cDNA data shows that it can freely combine with the 3' exons coding for NCAM-140 and -120 ; we do not know yet whether it occurs together with the NCAM-180-specific exon.

While it is clear that splice site selection dictates inclusion or exclusion of the exon specific for the 7.4 kb mRNA and of the one encoding the membrane-proximal extra-sequence, several different mechanisms could control the generation of the mRNAs that code for NCAM-120. The transscripts could terminate near one of the three poly(A) addition sites thus creating different precursors for the 7.4/6.7, the 5.2 and the 2.9 kb mRNAs. The two other possibilities are that specific RNA production is determined by selective cleavage and poly(A) addition at different sites on the same primary transcript or that inclusion of the exon specific for the two small transcripts precedes the selection of the polyadenylation signals it contains.

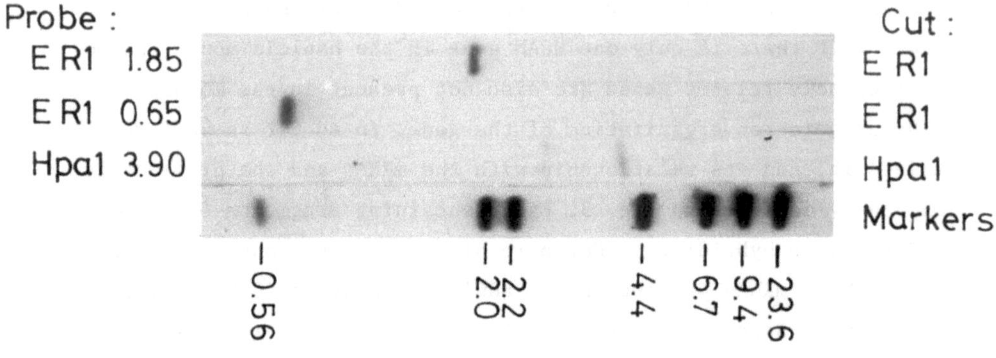

Figure 2. Hybridization of cosmid fragments from the mouse NCAM gene to genomic DNA from mouse liver. Sizes are indicated in kb. ER1, probes or genomic DNA cut with Eco RI; Hpa 1, probes or genomic DNA cut with Hpa I.

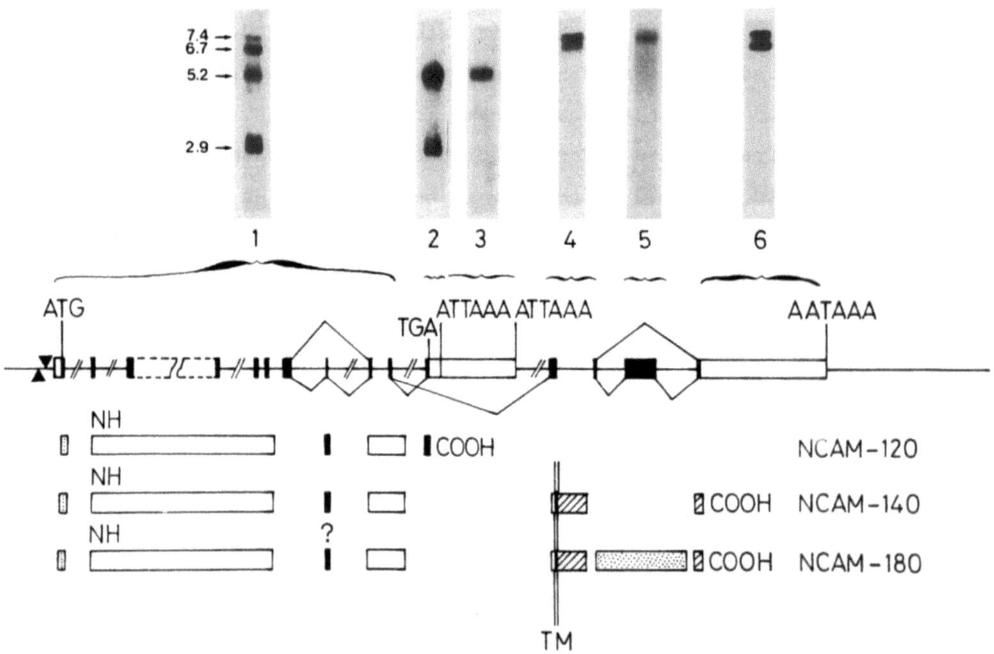

Figure 3. Schematic representation of mouse NCAM mRNAs, gene, and proteins.

The exon-intron organization of most of the chicken NCAM gene has been reported (Owens et al, 1987). The mouse gene appears to be organized very similarly. Most striking is the conservation of intron positions within the translational reading frame at every splice junction. However, a chicken homologue for the small exon specifying the membrane-proximal extra sequence has not been reported and the NCAM-120-specific exon appears to be organized differently in the mouse.

REGULATION OF EXPRESSION OF THE NCAM GENE.

Studies from a number of laboratories have shown that different NCAM proteins are expressed by different cell types and at different stages of development (see for reviews, Hirn et al, 1985 ; Cunningham, 1986 ; Nybroe et al, 1988 ; Walsh, 1988). This is due to regulation of the expression of the different mRNA species ; in other words what is controlled appears to be the processing of the gene, although we cannot exclude at present that mRNA stability is a contributing factor. For example, during brain development, there is a shift from the predominant expression of the two larger NCAM mRNAs in the embryonic brain towards a prevalence of the two smaller transscripts in the adult. There are also large changes in the overall expression of the NCAM gene in the brain with a peak around birth and an around 6 fold drop during postnatal development (Gennarini et al, 1986). Similar changes have been noted during muscle development (Covault et al, 1986). Of particular interest is the finding that the change in the NCAM mRNA pattern that occurs during muscle development can be reproduced in clonal muscle cell lines (Covault et al, 1986 ; Moore et al, 1987). Formal proof that these changes are controlled at the level of pre-mRNA processing is still lacking ; if so, we must postulate the existence of trans-acting factors that regulate the splicing or the choice of the poly(A) addition signal.

The studies on NCAM expression during muscle development also show that the overall expression of the gene is controlled independently of its processing, probably at the level of transcription. As a first step towards identifying the cis and trans elements involved, we have cloned and partially sequenced the 5' flanking region of the NCAM gene. The region contains several classical control elements : a CAAT box, two TATA boxes, a SpI factor binding sequence. However, by S1 nuclease protection and primer extension experiments, we have located two major transcription initiation sites which are too distant to be controlled by the CAAT and the TATA boxes. The region does contain, though, a strong promoter active in NCAM-expressing mouse neuroblastoma but not in NCAM-negative L cells. When the chloramphenicol acteylase gene was coupled to the genomic region comprising the identified transcriptional start sites, high levels of enzyme expression were detected in transfected N2A neuroblastoma cells which were as high as those achieved in L cells using the Rous sarcoma virus promoter (Table I).

Table I
Promoter activity of the 5' flanking region of the mouse NCAM gene.

Plasmid/ cell type	CAT activity (cpm)	Transfection efficiency	Relative promoter activity
BX1-CAT			
NCAM			
N2A cells	670,000	1.0	100
L cells	5,000	0.2	3.7
BL-CAT3			
promoter-less			
N2A cells	1,000	0.9	0.2
L cells	2,000	0.1	4.9
RSV-CAT			
N2A cells	4,000	0.8	0.8
L cells	27,000	0.1	49

A 852 bp fragment (BX-1) from the 5' flanking region of the mouse NCAM gene was cloned into pBLCAT3 (Luckow and Schütz, 1987). Transfection efficiency was determined and CAT assays done as described (Kimura et al, 1986). pBLCAT3 and RSV-CAT were kind gifts of G. Schütz and A. Israël, respectively.

CONCLUSION

The immediate consequences of the regulated processing of the NCAM gene are clear : different NCAM isoforms are expressed by different cell types and during differentiation. The question remains, however, as to how this diversity affects the function of NCAM as an adhesion protein. The most obvious types of protein diversity concern the transmembrane or cytoplasmic domains which cannot directly participate in the binding. They may do so indirectly ; for example, NCAM-180 accumulates at sites of cell-cell contacts (Pollerberg et al, 1985), has binding affinity for brain spectrin (Pollerberg et al, 1987) and has a lower mobility in the plane of the membrane than NCAM-140 (Pollerberg et al, 1986). Clearly, these functions can be attributed to its cytoplasmic extra domain. Since the binding strength of NCAM interactions is greatly influenced by valency (Hoffman and Edelman, 1983), the lateral mobility of NCAM chains may be an important parameter in determining adhesivity.

The most striking difference between NCAM-120 and the other two proteins is the fact that the small isoform is attached to the membrane by phospholipid. Recent results show that NCAM-120 can be spontaneously released into the extracellular milieu (He et al, 1987) and it has been shown previously that it exists in part in soluble form (Gennarini et al, 1984b ; Nybroe et al, 1985). Hence, NCAM-120 may serve dual roles both as a cell-surface molecule and as an extracellular protein. The higher mobility of lipid-linked as compared to transmembrane proteins (Noda et al, 1987) may be biologically important. Once released from the cells, NCAM-120 may also act as a soluble competitor of cell adhesion or become incorporated into extracellular matrix.

The function of the small membrane-proximal additional sequences is still obscure. When visualized by electron microscopy (Hall and Rutishauser, 1985), NCAM appears as a rod with a hinge or bend, the location of which may correspond to the extra sequences. One may thus speculate that even short sequences inserted at this point may serve to render the hinge more flexible or to change its angle. Another question which remains to be clarified is whether or not NCAM has some signalling function. Many of these questions may now be approached by expressing different NCAM isoforms via transfection, and such studies have already been initiated (Edelman et al, 1987).

Finally, selective adhesion results not only from variations in structure and expression of a single species but also from the interplay of different kinds of adhesion molecules and it may very well be that contacts initiated by one change the expression of another adhesion protein. Soluble factors, as exemplified in other chapters, undoubtedly contribute to the complexity of such systems and we have still a long way to go to be able to understand the circuits that control developmental programs.

LITERATURE REFERENCES

Arquint M, Roder J, Chia LS, Down J, Wickinson D, Bayley H, Braun P, Dunn R (1987) Molecular cloning and primary structure of myelin-associated glycoprotein. Proc Natl Acad Sci 84:600-604

Barbas JA, Chaix JC, Steinmetz M, Goridis C (1988) Differential splicing and alternative polyadenylation generates distinct NCAM transcripts proteins in the mouse. EMBO J 7:625-632

Barthels D, Santoni MJ, Wille W, Ruppert C, Chaix JC, Hirsch MR, Fontecilla-Camps JC, Goridis C (1987) Isolation and nucleotide sequence of mouse NCAM cDNA that codes for a Mr 79,000 polypeptide without a membrane-spanning region. EMBO J 6:907-914

Barthels D, Vopper G, Wille W (1988) NCAM-180, the large isoform of the neural cell adhesion molecule of the mouse, is encoded by an alternatively spliced transcript. Submitted for publication

Cole GJ, Glaser L (1986) A heparin-binding domain from NCAM is involved in neural cell-substratum adhesion. J Cell Biol 102:403-412

Cole GJ, Loewy A, Cross NV, Akeson R, Glaser L (1986) Topographic localization of the heparin-binding domain of the neural cell adhesion molecule NCAM. J Cell Biol 103:1739-1744

Covault J, Merlie JP, Goridis C, Sanes JR (1986) Molecular forms of NCAM and its RNA in developing and dernervated skeletal muscle. J Cell Biol 102, 731-739

Cunningham BA, Hoffman S, Rutishauser U, Hemperly JJ, Edelman GM (1983) Molecular topography of the neural cell adhesion molecule N-CAM : surface orientation and location of sialic acid-rich and binding regions. Proc Natl Acad Sci USA 80:3116-3120

Cunningham BA (1986) Cell adhesion molecules : a new perspective on molecular embryology. Trends Biochem Sci 11:423-426

Cunningham BA, Hemperley JJ, Murray BA, Prediger EA, Brackenbury R Edelman GM (1987) Neural cell adhesion molecules : structure, immunoglobulin-like domains, cell surface modulation and alternative RNA splicing. Science 236:799-806

Dickson G, Gower HJ, Barton HC, Prentice HM, Elsom VL, Moore SE, Cox RD, Quinn C, Putt W, Walsh FS (1987) Human muscle neural cell adhesion molecule (NCAM) : Identification of a muscle-specific sequence in the extracellular domain. Cell 50:1119-1130

Edelman GM (1983) Cell adhesion molecules. Science 219:450-457

Edelman GM, Hoffman S, Chuong CM, Thiery JP, Brackenbury R, Gallin WJ, Grumet M, Greenberg ME, Hemperly JJ, Cohen C, Cunningham BA (1983) Structure and modulation of neural cell adhesion molecules in early and late embryogenesis. Cold Spring Harbor Symposia on Quantitative Biology 48:515-526

Edelman GM (1986) Cell adhesion molecules in the regulation of animal form and tissue pattern. Annu Rev Cell Biol 2:81-116

Edelman GM, Murray BA, Mege RM, Cunningham BA, Gallin WJ (1987) Cellular expression of liver and neural cell adhesion molecules after transfection with their cDNAs results in specific cell-cell binding Proc Natl Acad Sci USA 84:8502-8506

Ekblom P, Vestweber D, Kemler R (1986) Cell-matrix interactions and cell adhesion during development. Annu Rev Cell Biol 2:27-47

Finne J, Finne U, Deagostini-Bazin H, Goridis C (1983) Occurrence of α2-8 linked polysialosyl units in a neural cell adhesion molecule. Biochim Biophys Res Comm 112:482-487

Frelinger AL III, Rutishauser U (1986) Topography of NCAM structural and functional determinants. II. Placement of monoclonal antibody epitopes. J Cell Biol 103:1729-1737

Gennarini G, Rougon G, Deagostini-Bazin H, Hirn M, Goridis C (1984a) Studies on the transmembrane disposition of the neural cell adhesion molecule NCAM. A monoclonal antibody recognizing a cytoplasmic domain and evidence for the presence of phosphoserine residues. Europ J Biochem 142:57-64

Gennarini G, Hirn M, Deagostini-Bazin H, Goridis C (1984b) Studies on the transmembrane disposition of the neural cell adhesion molecule NCAM. The use of liposome-inserted radioiodinated NCAM to study its transbilayer orientation. Eur J Biochem 142:65-73

Gennarini G, Hirsch MR, He HT, Hirn M, Finne J, Goridis C (1986) Differential expression of mouse neural cell adhesion molecule mRNA species during brain development and in neural cell lines. J Neuroscience 6:1983-1990

Goridis C, Deagostini-Bazin H, Hirn M, Hirsch MR, Rougon G, Sadoul R, Langley OK, Gombos G, Finne J (1983) Neural surface antigens during nervous system development. Cold Spring Harbor Symposia on Quantitative Biology 48:527-537

Hall AK, Rutishauser U (1985) Phylogeny of a neural cell-adhesion molecule. Dev Biol 110:39-46

He HT, Barbet J, Chaix JC, Goridis C (1986) Phosphatidylinositol is involved in the membrane attachment of NCAM-120, the smallest component of the neural cell adhesion molecule. EMBO J 5:2489-2498

He HT, Finne J, Goridis C (1987) Biosynthesis, membrane association and release of NCAM-120, a phosphatidylinositol-linked form of the neural cell adhesion molecule. J Cell Biol 105:2489-2500

Hirn M, Deagostini-Bazin H, Gennarini G, Santoni MJ, He HT, Hirsch MR, Goridis C (1985) Structural and functional studies on NCAM neural cell adhesion molecules J Physiol 80:247-254

Hoffman S, Edelman GM (1983) Kinetics of homophilic binding by embryonic and adult forms of the neural cell adhesion molecule. Proc Natl Acad Sci USA 80:5762-5766

Holtfreter J (1939) Gewebeaffinität, ein Mittel der embryonalen Formbildung. Arch Exp Zellforsch 23:169-209

Hynes RO (1987) Integrins, a family of cell surface receptors. Cell 48:549-554

Kimura A, Israël A, Le Bail O, Kourilsky P (1986) Detailed analysis of the mouse $H\text{-}2K^b$ promoter : enhancer like sequences and their role in the regulation of class I gene expression. Cell 44:261-272

Linser PJ, Moscona AA (1983) Hormonal induction of glutamine synthetase in cultures of embryonic retina cells : requirement for neurone-glia contact interactions. Dev Biol 96:529-534

Low MG, Ferguson MAJ, Futerman AH, Silman I (1986) Covalently attached phosphatidylinositol as a hydrophobic anchor for membrane proteins. Trends Biochem Sci 11:212-215

Luckow B, Schütz G (1987) CAT constructions with multiple unique restriction sites for the functional analysis of eukaryotic promoters and regulatory elements. Nucl Acids Res 15:5490

Moore SE, Thompson J, Kirkness V, Dickson JG, Walsh FS (1987) Skeletal muscle neural cell adhesion molecule (NCAM) : Changes in protein and mRNA species during myogenesis of muscle cell lines. J Cell Biol 105:1377-1386

Murray BA, Hemperly JJ, Prediger EA, Edelman GM, Cunningham BA (1986) Alternatively spliced mRNAs code for different polypeptide chains of the chicken neural cell adhesion molecule (N-CAM). J Cell Biol 102:189-193

Noda M, Yoon K, Rodan G, Koppel DE (1987) High lateral mobility of endogenous and transfected alkaline phospatase : a phosphatidylinositol-anchored membrane protein. J Cell Biol 105:1671-1677

Nybroe O, Albrechtsen M, Dahlin J, Linneman D, Lyles JM, Møller CJ, Bock E (1985) Biosynthesis of the neural cell adhesion molecule. Characterization of polypeptide C. J Cell Biol 101:2310-2315

Nybroe O, Linneman D, Bock E (1988) NCAM biosynthesis in brain. Neurochem Int, in press

Obrink B (1986) Epithelial cell adhesion molecules. Exp Cell Res 163:1-21

Owens GC, Edelman GM, Cunningham BA (1987) Organization of the neural cell adhesion molecule (NCAM) gene : Alternative exon usage as the basis for different membrane-associated domains. Proc Natl Acad Sci USA 84:294-298

Pollerberg EG, Sadoul R, Goridis C, Schachner M (1985) Selective expression of the 180 kD component of the neural cell adhesion molecule N-CAM during development. J Cell Biol 101:1921-1929

Pollerberg EG, Schachner M, Davoust J (1986) Differentiation state-dependent surface mobilities of two forms of the neural cell adhesion molecule. Nature 324:462-465

Pollerberg E, Burridge K, Krebs KE, Goodman SR, Schachner M (1987) The 180-kD component of the neural cell adhesion molecule NCAM is involved in cell-cell contacts and cytoskeleton-membrane interactions. Cell Tissue Res 250:227-236

Prochiantz A, Daguet MC, Herbet A, Glowinski J (1981) Specific stimulation of in vitro maturation of mesencephalic dopaminergic neurones by striatal membranes. Nature 293:570-572

Rutishauser U (1983) Molecular and biological properties of a neural cell adhesion molecule. Cold Spring Harbor Symposia on Quantitative Biology 48:501-514

Rutishauser U (1986) Differential cell adhesion through spatial and temporal variations of NCAM. Trends Neurosci 9:374-378

Rutishauser U, Goridis C (1986) NCAM : The molecule and its genetics. Trends Genet 2:72-76

Salzer JL, Homes WP, Colman DR (1987) The amino acid sequences of the myelin-associated glycoproteins : Homology to the immunoglobulin gene superfamily. J. Cell Biol 104:957-965

Santoni MJ, Barthels D, Barbas JA, Hirsch MR, Steinmetz M, Goridis C, Wille W (1987) Analysis of cDNA clones that code for the transmembrane forms of the mouse neural cell adhesion molecule (NCAM) and are generated by alternative RNA splicing. Nucleic Acids Res 15:8621-8641

Santoni MJ, Goridis C, Fontecilla-Camps JC (1988) Molecular modelling of the immunoglobulin-like domains of NCAM ; implications for the positioning of functionally important sugar side chains. J Neuroscience Res in press

Seed B (1987) An LFA-3 cDNA encodes a phospholipid-linked membrane protein homologous to its receptor CD2. Nature 329:840-842

Simmons D, Makgoba MW, Seed B (1988) ICAM, an adhesion ligand of LFA-1 is homologous to the neural cell adhesion molecule NCAM. Nature 331:624-627

Small SJ, Shull GE, Santoni MF, Akeson R (1987) Identification of a cDNA clone that contains the complete coding sequence for a 140-kD rat NCAM polypeptide. J Cell Biol 105:2335-2345

Sunshine J, Balak K, Rutishauser U, Jacobson M (1987) Changes in neural cell adhesion molecule (NCAM) structure during vertebrate neural development. Proc Natl Acad Sci USA 84:5986-5990

Takeichi M, Noro CY, Shirayoshi Y, Hatta K (1985) Calcium-dependent cell-cell adhesion system : its molecular nature, cell type specificity and morphogenetic role, In the Cell in Contact (Edelman GM and Thiéry JP eds) New York John Wiley and Sons, pp 219-232

Thiery JP, Brackenbury R, Rutishauser U, Edelman GM (1977) Adhesion among neural cells of the chick embryo. II. Purification and characterization of a cell adhesion molecule from neural retina. J Biol Chem 252:6841-6845

Walsh FS (1988) The NCAM gene is a complex transcriptional unit. Neurochem Int in press

Williams RK, Goridis C, Akeson R (1985) Individual neural cell types express immunologically distinct N-CAM forms. J Cell Biol 101:36-42

Ultrastructural Localization of the Cell Adhesion Molecule Uvomorulin using Site-directed Antibodies

R. Kemler and H. Schwarz

Max-Planck-Institute für Immunobiologie, Stübeweg 51, D-7800 Freiburg, F.R.G.

Max-Planck-Institute für Biologie, Correnstrasse 38, D-7400 Tübingen, F.R.G.

INTRODUCTION

Uvomorulin belongs to the class of Ca^{++}-dependent cells adhesion molecules (CAMs). Members of this class have several features in common; i) they express their adhesive function only in the presence of Ca^{++}, ii) they are protected by Ca^{++} from proteolytic degradation and iii) they are integral membrane glycoproteins with a similar molecular weight of 120 kDa.

Uvomorulin expresses its adhesive function very early in mouse development during the compaction process of preimplantation embryos (1). Compaction is a prerequisite for the generation of cell diversity in the preimplantation embryo and leads to the generation of the trophectoderm and inner cells mass lineages (2). It is this compaction process which is blocked by the monoclonal anti-uvomorulin antibody, DECMA-1 (3). Studies in the synthesis and on cell surface localization in early embryos demonstrate that the protein is already present on the cell surface of unfertilized eggs but that uvomorulin synthesis starts in late two-cell embryos with the activation of the embryonal genome (4). However, during preimplantation development uvomorulin is specifically redistributed on the cells surface of blastomeres. This redistribution is a developmentally regulated process since it occurs only in cells committed for epithelial cell differentiation. The redistribution leads to a concentration of uvomorulin in the baso-lateral membrane and finally into the assembly of the protein in the zonula adherens of trophectodermal cells. These results suggest that during preimplantation development the redistri-

NATO ASI Series, Vol. H26
Cell to Cell Signals in Mammalian Development
Edited by S. W. de Laat et al.

bution of uvomorulin is an initial event that might trigger compaction which is a prerequisite for the generation of trophectoderm (5).

A similar cell surface localization of uvomorulin was also found in adult intestinal epithelial cells where it is highly concentrated in the intermediated junctions of the junctional complex (6). Monoclonal antibody DECMA-1 which blocked the compaction process of early embryos disrupts also the epithelial layer of cultured cells indicating that uvomorulin is involved in cell-cell contacts of embryonal and epithelial cells (3).

Attempts were made towards a more detailed structural analysis of uvomorulin and a cDNA clone was isolated from a bacteriophage lambda gt11 cDNA library with immunological techniques (7). Subsequently the complete primary structure of the mature proteins was established (8). This allowed important insights into the structural features of uvomorulin which might be of functional significance. The extracellular part of the molecule exhibits three internally repeated domains which are most likely generated by gene duplication. The gene duplication might have been generated during evolution under a strong selection for the uvomorulin function and it is assumed that the molecular mechanism of cell adhesion is mediated by these domains.

Analysis of the hydrophobicity of the amino acid sequence suggested that uvomorulin contains a single membrane-spanning region. The membrane intergration of uvomorulin was confirmed in immunofluorescence tests where antibodies directed against different parts of the molecule were analyzed (8).

In this report we have examined the staining pattern of antibodies against the external and the cytoplasmic part of uvomorulin on an electronmicroscopical level.

MATERIALS AND METHODS

Anti-Uvomorulin-antibodies

Rabbit antisera were raised against different parts of the uvomorulin molecule. Antibodies against the external part of the protein (anti-84 kDa) were induced against the 84 kDa fragment obtained by trypsin digestion in the presence of Ca^{++} and were affinity purified on a 84 kDa Sepharose column (Pharmacia). Antibodies against the cytoplasmic domain (anti-UMc) were generated against the 3' sequences of uvomorulin cDNA expressed as fusion protein (for details see Ref. 8).

Plastic embedding

Mouse small intestine was fixed in 2% freshly prepared formaldehyde and 0.2% glutaraldehyde in 0.1M phosphate buffer (pH 7.4) for 2h at 4°C. Tissues were cut into small pieces, dehydrated, and low-temperature embedded in Lowicryl K4M as described (9). Ultra thin sections (50-70 nm) were cut with a diamond knife on a LKB ultratome III. To minimize unspecific labelling, sections were washed in a solution of 0.5% BSA, 0.2% gelatine in PBS, pH 3-4. The same solution was used to dilute antibodies and colloidal gold particles.

Immuno-gold labelling

For the immunoelectron analysis of uvomorulin localization affinity purified anti-84 kDa and anti-UMc were used at 10μg/ml and 25μg/ml respectively. Bound antibodies were detected with protein A-labeled colloidal gold (7 nm). After immonolabeling sections were stained with 1% neutral uranyl-acetate and 0.4% lead citrate and examined in a Philips EM 201 at 60kV. Controls included rabbit antibodies with unrelated specificity and protein A-labelled colloidal gold.

RESULTS AND DISCUSSION

Previously it has been shown that uvomorulin is preferentially localized in the intermediate junctions of intestinal epithelial cells (6). The junctional complex of epithelial cells is composed of tight junctions, intermediate junctions and desmosomes and separates the apical and base—lateral membrane domains of the cells.

This polarization of the epithelial cell suface is a prerequisite of proper epithelial functions (10). Tight junctions have been considered as a major component for the maintenance of the polarized membrane compartments since the boundary of the distribution of membrane proteins between apical and baso—lateral domains coincides with tight junctions (11). Although this view is supported by some experimental results more data have not given additional evidence. A monoclonal antibody (anti—Z0.1) against a tight junction protein seems not to recognize an integral membrane protein or interfere with cell—cell contact (12). Moreover, the use of a transepithelial resistance assay to study the role of tight junctions in the formation of epithelia led to the discovery of the canine uvomorulin (19). This and other observations discussed below support the idea that uvomorulin is a major junctional element for the preservation of the integrity of the epithelial cell layer. Uvomorulin is concentrated in the intermediate junctions of the junctional complex (Fig. 1A). It is also found in less amount on the baso—lateral membrane domains. At present there seems to be no molecular differences between uvomorulin localized in the junctions and at the baso—lateral membrane.

From a detailed structural analysis the association of uvomorulin with the plasma membrane become transparent (8). Uvomorulin is an integral membrane protein with a cytoplasmic domain of around 17kDa. This is confirmed by immunoelectron analysis where the staining pattern of antibodies against the extracellular and cytoplasmic part of the protein was compared (Fig. 1A and B).

Antibodies against the extracellular part of the protein were almost completely localized in the intercellular space of the intermediate junctions (Fig. 1A). In contrast, antibodies against the cytoplasmic domain of uvomorulin concentrated below the plasma membrane in the cytoplasm (Fig. 1B). By using highly purified antibody preparations virtually no back—ground labeling was observed.

To verify the differential localization of antibodies against the extracellular and cytoplasmic part of the molecule the distribution of gold particles obtained with

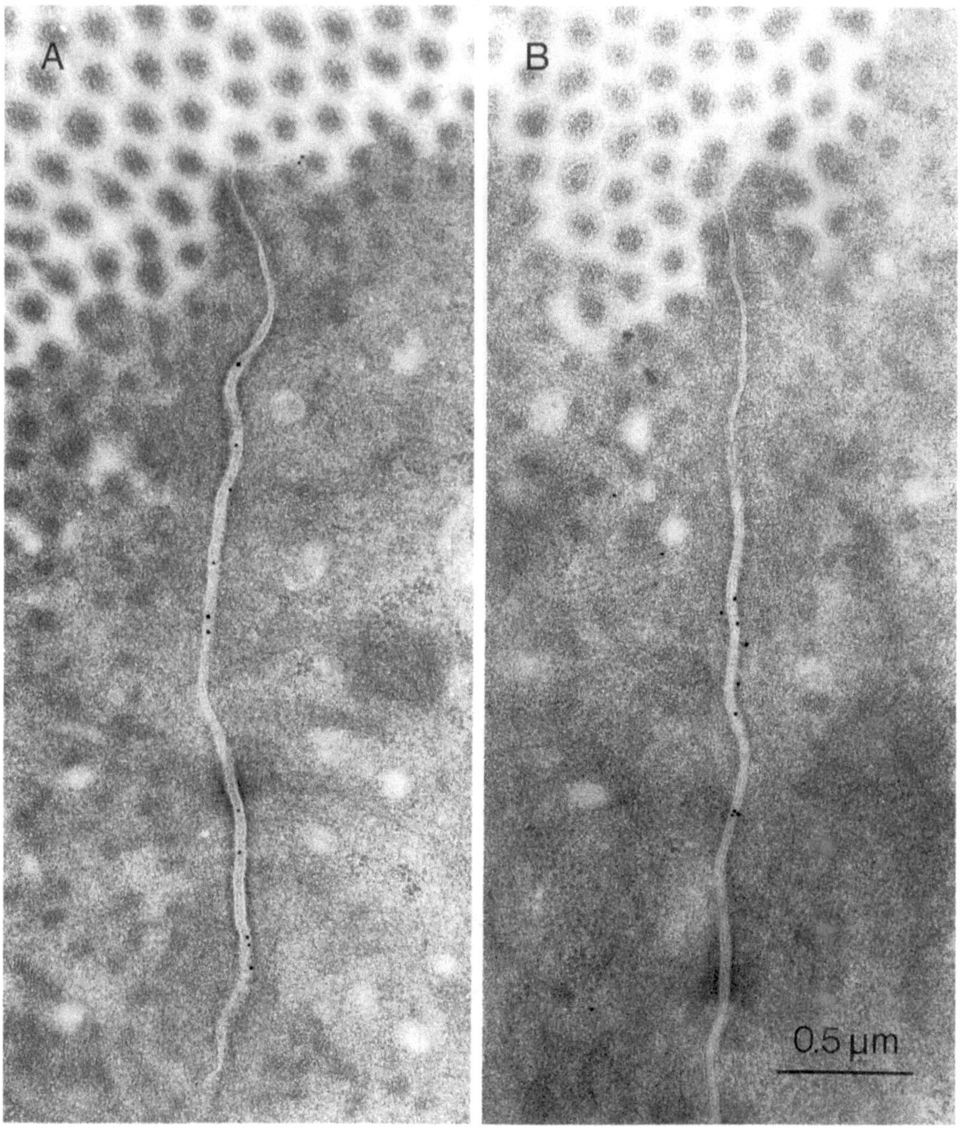

Figure 1. Electron micrographs showing immunolocalization of uvomorulin in the intermediate junctions of mouse intestinal epithelial cells from Lowicryl-embedded tissue. Antibodies against the extracellular part of uvomorulin are preferentially localized in the intercellular space (A). Antibodies against the cytoplasmic domain of uvomorulin concentrate below the plasma membrane in the cytoplasm (B).

each labelling was measured. For this only intermediate junctions cut perpendicular to the section plane were used (Fig. 1). In these sections the boundary between the cytoplasm and the plasma membrane is clearly recognizable as an electrondense line.

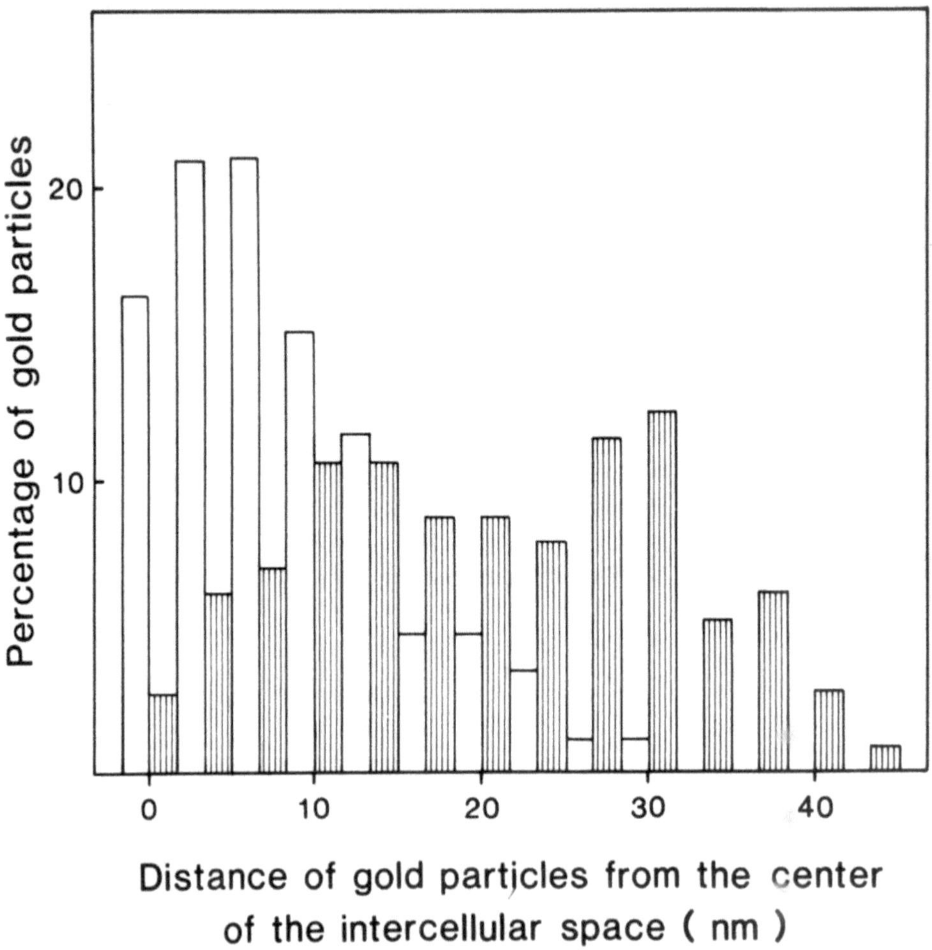

Figure 2. Measurement of the distribution of gold particles in the intermediate junctions detecting antibodies against the external (white columns) and cytoplasmic (striated columns) part of uvomorulin. For details see text.

The distribution of gold particles were measured in relative distance to an imagined midline halfway between the two membrane/cytoplasm boundaries. The distribution profile of 200 gold particles is summarized in Fig. 2. Antibodies aginst the extracellular part of uvomorulin concentrate near the midline of the intercellular space (white columns), whereas antibodies against the cytoplasmic part (striated columns) are preferentially localized close to the plasma membrane.

This result provides further evidence for a possible connection of uvomorulin to cytoplasmic components. In intestinal epithelial cells a contractile ring of microfilaments is localized at the level of the intermediate junctions (14). In the view of uvomorulin as an adhesive factor preserving the integrity of an epithelial sheet it would make sense that uvomorulin is in some way connected to cytoskeletal structures. Preliminary data obtained from the analysis of uvomorulin protein has given first support for such a view. Anti-uvomorulin antibodies detect in immunoprecipitation experiments – but not in immunoblots – additional polypeptides (4). Whereas in immunoblots only the 120 kDa uvomorulin molecule is recognized, a set of additional proteins of 102, 92 and 88 kDa is constantly seen in immunoprecipitation. The 102 kDa protein has been analyzed to some extent (unpubl.observ.). The result favours the idea that the 102 kDa protein is a structually unrelated cytoplasmic component which forms most likely a complex with the cytoplasmic domain of uvomorulin. Complete tryptic peptide analysis reveals structural differences between uvomorulin and the 102 kDa protein. Immunoprecipitation experiments with microsomal vesicles demonstrate that the 102 kDa protein is not protected from Proteinase K digestion. The 102 kDa protein is not glycosylated and therefore most likely not an integral membrane protein. These obeservations together suggest that the 102 kDa protein is associated with the cytoplasmic domain and that it could represent a first molecular component for a possible cytoplasmic anchorage of uvomorulin. To define more precisely the cytoplasmic connection of uvomorulin will be an important step towards a more general understanding of the function of uvomorulin. It should be recalled that the cytoplasmic domains of uvomorulin and other Ca^{++} dependent CAMs are highly homologous (8). The evolutionary conservation of this domain underlines its functional importance one of which might be to mediate the cytoplasmic connection of uvomorulin.

ACKNOWLEDGEMENTS

We thank Ms. Rosemary Brugger for typing the manuscript. The work was supported by the Deutsche Forschungsgemeinschaft (Ke 246/6-1).

LITERATURE

1. Kemler R, Babinet C, Eisen H, Jacob F (1977) Surface antigen in early differentiation. Proc Natl Acad Sci USA 74:4449-4452
2. Ekblom P, Vestweber D, Kemler R (1986) Cell-matrix interactions and cell adhesion during development. Ann Rev Cell Biol 2:27-47
3. Vestweber D, Kemler R (1985) Identification of a putative cell adhesion domain of uvomorulin. EMBO J 13A:3393-3398
4. Vestweber D, Gossler A, Boller K, Kemler R (1987) Expression and distribution of cell adhesion molecule uvomorulin in mouse preimplantation embryos. Devel Biol 124:451-456
5. Kemler R, Gossler A, Mansour A, Vestweber D (1988) The cell adhesion molecule uvomorulin. Cell in Contact II, in press
6. Boller K, Vestweber D, Kemler R (1985) Cell adhesion molecule uvomorulin is localized in the intermediate junctions of adult intestinal epithelial cells. J Cell Biol 100:327-332
7. Schuh R, Vestweber D, Riede I, Ringwald M, Rosenberg UB, Jäckle H, Kemler R (1986) Molecular cloning of mouse cell adhesion molecule uvomorulin: cDNA contains a B1 related sequence. PNAS 83:1364-1368
8. Ringwald M, Schuh R, Vestweber D, Eistetter H, Lottspeich F, Engel J, Dölz R, Jähnig F, Epplen J, Mayer S, Müller C, Kemler R (1987) The structure of cell adhesion. EBMO J 6:3647-3653
9. Roth J, Bendayan M, Carlemalm E, Villiger W, Gravito RM (1981) Enhancement of structural preservation and immunocytochemical staining on low temperature. J Histochem Cytochem 29:663-671
10. Simons K, Fuller SD (1985) Cell surface polarity in epithelia. Ann Rev Cell Biol 1:243-288
11. Hubbard AL, Bartles JR, Braiterman LT (1985) Identification of rat hepatocyte plasma membrane proteins using monoclonal antibodies. J Cell Biol 100:1115-1125
12. Stevenson BR, Siliciano JD, Mooseker MS, Goodenough DA (1986) Identification of ZO-1: a high molecular weight polypeptide associated with the tight junction (zonula occludens) in a variety of epithelia. J Cell Biol 103:755-766
13. Gumbiner B, Simons K (1986) A functional assay for proteins involved in establishing an epithelial occluding barrier: identification of a uvomorulin-like polypeptide. J Cell Biol 102:457-468
14. Burgess DR (1982) Reactivation of intestinal epithelial cell brush border motility: ATP-dependent contraction via a terminal web contractile ring. J Cell Biol 95:853-863

MUTATIONAL ANALYSIS OF THE EGF RECEPTOR PROTO-ONCOGENE

J. Schlessinger[1,*], A.M. Honegger [1] and A. Ullrich[2].

[1]Rorer Biotechnology, Inc.
680 Allendale Road
King of Prussia, PA 19046.

Growth factors and their specific cell surface receptors play an important role in the control of cell growth and differentation. Ligand-receptor interaction on the cell surface is translated into activation of intercellular pathways, initiating a sequence of events which eventually results in a specific cellular response. Several growth factor receptors posses protein tyrosine kinase activity that is intrinsic to the receptor molecules. Signalling by tyrosine kinase activation is shared by at least five known hormones and growth factors and their corresponding receptors. These receptor tyrosine kinases (RTKs) constitute a family of structually-related receptor polypeptides with a rapidly increasing number of members (Yarden and Ullrich, 1988). In recent years it has become clear that the study of receptor function and receptor-activated signalling pathways will provide a better understanding of basic biological problems in the areas of cell biology, endocrinology and development. As a first step toward the detailed analysis of growth factor activated cellular signalling mechanism, structural characterization of the RTK family resulted in complete primary structures of all known RTKs. One surprising result of these efforts further reinforced the pivotal role of RTK molecules in signal transduction: structurally modified RTK genes were found to be part of the genomes of acutely oncogenic retroviruses (Heldin and Westermark, 1984;

[2]Department of Developmental Biology, Genentec, Inc. 460 Point San Bruno Boulevard, South San Francisco, CA 94080. *On leave from the Weizmann Institute, Rehovot, Israel.

Sherr, 1987). Moreover, secretion of various ligands for RTKs was found to be

NATO ASI Series, Vol. H26
Cell to Cell Signals in Mammalian Development
Edited by S. W. de Laat et al.

widely associated with the transformed phenotype (Sporn and Roberts, 1985; Ross et al., 1986) and may therefore play a crucial role in oncogenesis. More recently, the availability of complementary DNA (cDNA) clones encoding RTKs enabled construction of structurally altered receptors, which provided insights into the molecular mechanism of signal transduction.

Epidermal growth factor (EGF) is a polypeptide growth factor of 53 animo acid residues (Carpenter and Cohen, 1979), which mediates its biological responses by binding to and activating a specific cell surface receptor termed the EGF-receptor. The EGF-receptor is a glycoprotein of molecular weight 170,000 and it is a member of the RTK gene family. Following the purification of EGF-receptor (Yarden et al., 1985) and its partial sequencing (Downward et al., 1984) the complete primary structure of the EGF-receptor was deducted from the sequence of cDNA clones (Ullrich et al., 1984). The mature EGF-receptor is composed of three major regions. A large glycosylated extracellular EGF binding region which is anchored to the plasma membrane by a single membrane spanning region of 23 hydrophobic amino acids. A cytoplasmic portion containing the kinase domain which contains consensus residues typical to the tyrosine kinase gene family (reviewed in Hunter and Cooper, 1985; Yarden and Ullrich, 1987). The Lys721 residue and consensus sequence Gly-X-Gly-X-Phe-Gly-X-Val, located 15 residues upstream to the lysine residue, probably function as part of the ATP binding site in the kinase domain. The binding of EGF to the receptor induces activation of the protein tyrosine kinase (Ushiro and Cohen, 1980) leading to self-phosphorylation and phosphorylation of various cellular substrates. In intact cells, autophosphorylation occurs mainly on Tyr1173 while at least two additional tyrosine residues located at the C-terminal end of the EGF-receptor are also phosphorylated when EGF is added to solubilized membranes or to pure EGF-receptor (Downward et al., 1984a). It has been suggested that autophosphorylation of EGF–receptor releases a negative constraint leading to enhanced phosphorylation of exogenous substrates by the EGF–receptor (Betrics and Gill, 1985).

Quantitative binding experiments with radiolabeled EGF indicate that the stoichiometry of ligand to the EGF–receptor is 1:1 (Weber et al., 1984). Nevertheless, analyses of binding experiments of radiolabeled–EGF to intact cells according to the method of Scatchard, reveal non–linear plots which are interpreted as an indication of the presence of different receptor classes with distinct affinities toward EGF. Hence, high affinity EGF–recptors with apparent Kd of 0.1–0.3 nM comprise 5–10% of the total receptor while the remaining low affinity receptors have an apparent Kd of 2–5 nM (King and Cuatrecasas, 1982). The treatment of cells expressing EGF–receptor with the tumor promoter phorbol myristate acetate (PMA) or with growth factors such as PDGF or bombesin, each binding to its own distinct receptor, abolishes the high affinity state of the EGF–receptor and also reduces the tyrosine kinase activity of the EGF–receptor (Shoyab et al., 1979; Cochet et al., 1984; Iwashita and Fox, 1984; Wrann et al., 1980). This process termed "receptor transmodulation" is probably mediated by the Ca^{+2} sensitive kinase–C, which has been shown to phosphorylate the EGF–receptor on several sites (reviewed in Schlessinger, 1986; Hunter and Cooper, 1985). One of these sites is Thr654 (Hunter et al., 1984; Davis and Czech, 1985) which is located 10 animo acids from the plasma membrane region. It has been proposed that the phosphorylation of Thr654 allosterically regulates ligand binding affinity and protein tyrosine kinase activity of the EGF–receptor (Hunter and Cooper, 1985).

EGF–receptors are randomly distributed on the cell surface (Schlessinger, et al., 1978; Haigler et al., 1987; 1979) and they undergo rapid lateral (Hillman and Schlessinger, 1982) and rotational (Zidovezki et al., 1981; 1985) diffusion.

Following ligand binding however, the occupied receptors cluster in coated-pits and after internalization both EGF and the receptor are degraded by lysomal enzymes (Carpenter and Cohen, 1979). In the absence of EGF the half-life of the EGF-receptor is 10-12 hours. EGF binding decreases EGF-receptor half-life to approximately 1 hour (Stoscheck et al., 1985; Honegger et al.,1987a). Two reports provide evidence suggesting that receptor degradation may occur after several rounds of receptor recycling (Dunn et al., 1986; Murthy et al., 1986).

Receptor oligomerization induced by EGF has been demonstrated using various independent methods including morphological (Haigler et al., 1979), biophysical (Hillman and Schlessinger, 1982; Zidovezki et al., 1981; 1985) and biochemical (Yarden et al., 1987a; 1987b; Cochet et al., 1988) approaches. The capacity of EGF-receptors to undergo oligomerization is an intrinsic property of the EGF-receptor since pure EGF-receptor undergoes rapid ligand-induced, temperature sensitive, reversible receptor dimerization (Yarden et al., 1987b). On the basis of these and other results an allosteric aggregation model was formulated for the activation of the kinase function of the EGF-receptor by EGF (Schlessinger, 1986; Yarden et al., 1987b). According to this model, monomeric EGF-receptors are in equilibrium with oligomeric receptors. The binding of EGF stabilizes the oligomeric state which in turn, brings together adjacent cytoplasmic kinase domains leading to their activation by receptor-receptor interactions. Such a mechanism bypasses the requirement for an energetically unfavorable conformational change to be transmitted through the transmembrane region connecting the two functional portions; as essential feature of every model which assumes an intramolecular mechanism for the activation of the tyrosine kinase by the binding of EGF to the extracellular domain (Weber et al., 1984; Betrics and Gill, 1985; Staros et al., 1985).

A versatile approach for the analysis of the various domains of the EGF-receptor and their potential role in signal transduction is to use site-directed in vitro mutagenesis combined with transfection experiments into cultured animal cells. The approach was used to generate various cell lines which express a functional human EGF-receptor and various receptor mutants. In the remaining portion of this article we will describe recent studies aimed to elucidate the role of the protein tyrosine kinase of the EGF-receptor, and the role of EGF-receptor in oncogenesis.

The role of the protein tyrosine kinase function of the EGF-receptor.

Several different EGF-receptor mutants were generated in order to explore the role of the protein tyrosine kinase function of the EGF receptor. These include: A point mutant in which Lys721 was replaced by an alanine residue (K721A, Honegger et al., 1977,a,b); an EGF-receptor containing two mutations in which Lys721 was replaced by a methionine residue and Thr654 was replaced by an alanine residue (Chen et al., 1987) and an insertional mutant containing four additional amino acids in the kinase domain after residue 708 (iM-708, Prywes et al., 1986; Livneh et al., 1987). The most complete analysis was performed with the point mutant (K721A), and therefore the biological properties of this mutant will be described in detail.

The wild type human EGF-receptor was expressed in various cell lines including NIH-3T3 cells that contain undetectable amounts of endogenous EGF-receptors. The expressed wild type EGF-receptor behaves like the native EGF-receptor and is able to stimulate all the known responses to EGF so far tested. Hence, these transfected cells provide a reliable system for exploring the role of the various domains of the EGF-receptor.

Unlike wild type EGF-receptor which exhibits EGF-dependent protein kinase activity the mutant receptor expressed in K721A lacks protein tyrosine kinase activity in vitro (Honegger et al., 1987a), and in vivo (Honegger et al., 1987b). Despite this deficiency, the mutant receptor is normally processed and expressed on the cell surface where it exhibits both typical high and low affinity binding sites. Moreover, as found with the wild type receptor, phorbol ester treatment abolished the high affinity binding sites and induced phosphorylation of both wild type and mutant EGF-receptors on serine and threonine residues (Honegger et al., 1987b). It is concluded therefore, that the kinase activity of the EGF-receptor is not required for processing and expression of EGF-receptor on the cell surface, for the display of high and low affinity EGF-receptors or for transmodulation mediated by phorbol ester. Inspite of having normal ligand binding characteristics, the kinase defective mutant in unable to stimulate various responses of EGF suggesting that the tyrosine kinase activity is essential for EGF-receptor signal transduction.

The kinase negative mutant is unable to stimulate both early responses such as $[Na^+]/[H^+]$ exchange, Ca^{+2} influx, inositol phosphate formation (Moolenaar et al., 1987), proto-oncogene c-fos and c-myc expression, S-6 ribosome phosphorylation, and delayed responses such as DNA synthesis and "foci-formation" (Honegger et al., 1987b). Hence, ligand binding alone provides a necessary, but insufficient step while protein tyrosine kinase activity induced by EGF is essential for signal transduction.

The inability of the kinase-defective receptor mutant to stimulate inositol lipid hydrolysis is of special interest. This result suggest a possible linkage between the protein tyrosine kinase function, and the phosphoinositol signalling pathway. Regulatory proteins in this pathway, such as G-proteins or phospholipase-C may serve as substrates and thus, be regulated by protein tyrosine kinases.

The kinase activity is also essential for normal receptor cellular routing (Honegger et al., 1987b). Although, the rate of internalization and degradation of

125I-EGF was similar for both wild type and mutant EGF-receptors, the mutant receptors were not down-regulated and degraded in response to EGF, but were instead recycled to the cell surface for reutilization. It was concluded therefore, that the protein tyrosine kinase activity of EGF-receptor is not essential for ligand internalization and degradation. However, the degradation of EGF-receptor after endocytosis is dependent on the kinase activity of the receptor (Honegger et al., 1987a). On the basis of these results, we propose a model for the cellular trafficking of EGF-receptors. It is proposed that EGF-receptors are internalized and recycled continuously under normal conditions and that activated receptor is trafficked to lysosome for degradation following several rounds of recycling so that efficiency of receptor degradation is approximately 30%. As a consequence, in every three rounds of recycling, on average, activated receptor is trafficked to lysosome where it is degraded. However, in the kinase defective mutant the efficiency of the degradative pathway is greatly reduced, and therefore, the mutant receptors undergo continuous recycling delivering the ligand molecules to lysosomes for degradation. The kinase activity serves the active role of specifically targeting the EGF-receptor for degradation probably by phosphorylation of specific, as yet, unidentified substrates involved in receptor trafficking.

Hence, both signal transduction and normal receptor trafficking depend on a functional protein tyrosine kinase suggesting that both processes are regulated by tyrosine phosphorylation of cellular substrates.

<u>The role of the EGF-receptor in oncogenesis</u>

The v-erbB oncogene of avian erythroblastosis virus encodes a truncated EGF-receptor devoid of most of the extracellular ligand binding region and also 34 amino acids at the C-terminous (Downward et al., 1983; Ullrich et al., 1984;

Yamamoto et al., 1983; Nilsen et al., 1985). Hence, we have proposed that the v-erbB oncogene transforms by acting as a constitutively activated EGF receptor (Downward et al., 1984; Ullrich et al., 1984; Kris et al., 1985; Lax et al., 1985). However, Riedel et al. have shown (1987) that a chimeric receptor composed of the extracellular and transmembrane regions of the human EGF-receptor attached to the cytoplasmic region of avian v-erbB was still able to transform culture rat-1 cells. It appears that the v-erbB protein in the chimeric receptor retains its transforming activity. Moreover, the binding of EGF to the chimeric receptor augments its transforming activity. This suggests that in addition to the extracellular and C-terminal deletions in v-erbB, structural differences between v-erbB and the cytoplasmic region of the EGF-receptor may be major factors which infuence the transforming potential of v-erbB (Riedel et al., 1987).

The EGF receptor gene was found to be amplified in several types of human tumors including gliomas and epidermoid carcinomas (Libermann et al., 1984; Hendler and Ozanne, 1984; Libermann et al., 1985; Yamamoto et al., 1986). Structural alterations were also observed (Libermann et al., 1985) but the most frequent lesion found was gene amplification with concomitant overexpression of the EGF-receptor. Overexpression of EGF-receptor may be important in oncogenesis since EGF is able to induce transformation in transfected NIH-3T3 cells which overexpress the EGF-receptor (Riedel et al., 1988; DiFiore et al., 1988. The EGF receptor may play a role in oncogenesis through both autocrine or paracrine mechanisms since various animal and human tumor cells produce TGF-a to EGF or TGF-a and also possess elevated levels of EGF-receptors.

Acknowledgements

The work supported in part by grants from NIH CA-25820, from U.S. Israel Binational Science Foundation and from D.K.F.Z.

References

Betrics PJ and Gill GN (1985) J Biol Chem 260:14642–14647
Carpenter G and Cohen S (1979) Ann Rev Biochem 48:193–216
Chen SW, Lazar SC, Poenie M, Tsien RY, Gill GN and Rosenfeld GM (1987) Nature 328:820–823.
Cochet C, Gill GN, Meisenheider J, Cooper JA and Hunter T (1984) J Biol Chem 259:2553–2558
Cochet C, Kashles O, Chambaz E, Borrello I, King CR and Schlessinger J (1988) J Biol Chem 7:3290–3295
Davis RJ and Czech M (1985) Proc Natl Acad Sci USA 82:1974–1978
Downward J, Parket P and Waterfield MD (1984a) Nature 311:483–485
Downward J, Yarden Y, Mayes E, Scrace G, Totty N, Stockwell P, Ullrich A, Schlessinger J and Waterfield MD (1984b) Nature 307:521–527
Dunn WA, Connolly TP, Hubbard AL (1986) J Cell Biol 102:24–36
Haigler HT, Ash JF, Singer SJ and Cohen C (1978) Proc Natl Acad Sci USA 75:5025–5029
Haigler HT, McKanna JA and Cohen C (1979) J Cell Biol 81:382–395
Heldin C-H and Westermark B (1974) Cell 37:9–20
Hendler FJ and Ozanne BW (1984) J Clin Invest 74:647–651
Hillman GM and Schlessinger J (1982) Biochemistry 21:1667–1672
Honegger AM, Dull TJ, Felder S, Van Obberghen E, Bellot F, Szapary D, Schmidt A, Ullrich A and Schlessinger J (1987a) Cell 51:199–209
Honegger AM, Szapary D, Schmidt A, Lyall R, Van Obberghen E, Dull TJ, Ullrich A and Schlessinger J (1987b) Mol Cell Biol F:4568–4571
Hunter T, Ling N and Cooper JA (1984) Nature 314:480–483
Hunter T and Cooper JA (1985) Ann Rev Biochem 54:897–930
Iwashita S and Fox CF (1984) J Biol Chem 259:2559–2567
King AC and Cuatrecasas P (1982) J Biol Chem 257:3053–3060
Lax I, Kris R, Sasson I, Ullrich A, Hayman MJ, Beug H and Schlessinger J (1985) EMBO J 4:3179–3182
Libermann TA, Razon N, Bartal AD, Yarden Y, Schlessinger J and Soreq H (1984) Cancer Res 44:735–760
Libermann TA, Nusbaum HR, Razon N, Kris R, Lax I, Soreq H, Whitle M, Waterfield M, Ullrich A and Schlessinger J. Nature 313:144–147
Lin CR, Chen SW, Lasar CS, Carpenter CD, Gill GN, Evans RM and Rosenfeld MD (1986) Cell 44:839–848
Livneh E, Prywes R, Kashles O, Reiss N, Sasson I, Ullrich A and Schlessinger J (1987) EMBO J 6:2669–2676
Livneh E, Prywes R, Dull TJ, Ullrich A and Schlessinger J (1988) Mel Cell Bio in Press
Moolenaar WH, Bierman AJ, Tilly BC, Verlaan I, Honegger AM, Ullrich A and Schlessinger J (1988) EMBO J 7:707–710
Murthy V, Basu M, Sen-Majundar A and Das M (1986) J Cell Biol 103:333–342
Riedel H, Schlessinger J and Ullrich A (1985) Science 236:197–200
Ross R, Raines EW and Bowen-Pope D (1986) Cell 46:155–169
Schlessinger J (1986) J Cell Biol 103:2067–2072
Schlessinger J, Shechter Y, Willingham MC and Pastan I (1978) Proc Natl Acad Sci USA 75:2659–2663
Schlessinger J, Schreiber AB, Levi A, Lax I, Libermann T and Yarden Y (1983) CRC Crit Rev Biochem 14:93–111
Sherr CJ, Rettenmier CW, Sacca R, Roussel MF, Look AT and Stanley R (1985) Cell 41:665–676
Shoyab M, DeLarco JE and Todaro GJ (1979) Nature 279:387–391
Sporn MB and Roberts AB (1985) Nature 313:745–747

Staros JV, Cohen S and Russo MW (1985) in Molecular Mechanisms of Transmembrane Signalling (Cohen P and935 Hously MD Eds) pp 253–278, Amsterdam/NY, Elsevier.
Stoscheck CM, Soderquist AM and Carpenter G (1985) Endocrinology 116:528–535
Ushiro H and Cohen S (1980) J Biol Chem 255:8363–8365
Weber W, Betrics PJ and Gill GN (1984) J Biol Chem 259:14631–14636
Wrann M, Fox CF and Ross R (1980) Science 210:1363–1364
Yamamoto T, Kamata N, Kawano H, Schimizu S, Kuroki T, Toyoshima K, Rikimaru K, Normara N, Ishizaki R, Pastian I, Gamou S and Schimizu N (1986) Cancer Res 46:414–416
Yamamoto T, Nishida T, Miyajimi N, Kawai S, Ooi T and Toyoshima K (1983) Cell 35:71–78
Yarden Y, Harari I and Schlessinger J (1985) J Biol Chem 260:315–319
Yarden Y and Schlessinger J (1987a) Biochemistry 26:1434–1442
Yarden Y and Schlessinger J (1987b) Biochemistry 26:1443–1451
Yarden Y and Ullrich A (1988) Ann Rev Biochem, in press
Zidovetzki R, Yarden Y, Schlessinger J and Jovin TM (1981) Proc Natl Acad Sci USA 78:6981–6985
Zidovetzki R, Yarden Y, Schlessinger J and Jovin TM (1986) EMBO J 5:247–250

ACTIVATION OF THE HUMAN EPIDERMAL GROWTH FACTOR RECEPTOR

L.H.K. Defize, J. Boonstra[1], P.M.P. van Bergen en Henegouwen[1], A.J. Verkley[1] and S.W. de Laat

Hubrecht Laboratory
Netherlands Institute for Developmental Biology
Uppsalalaan 8
3584 CT UTRECHT
The Netherlands

INTRODUCTION

Polypeptide growth factors have been recognized as important determinants in the regulation of cellular proliferation and differentiation (for a review: Sporn and Roberts, 1988). They exert their mitogenic or differentiating activity at nanomolar concentrations by binding to specific high-affinity receptor molecules localized in the plasma membrane. The ultimate biological action is dependent on the nature of the target cell. The number of identified growth factors is still rapidly growing. Several of these factors have been purified to homogeneity and their primary amino acid sequence has been determined completely or in part.

The study of growth factors and their receptors has been intensified after the discovery that al least some of these molecules show a striking homology with certain oncogene products: e.g. the sis oncogene encodes for a product homologous to PDGF (Doolittle et al. 1983) and the erb-B oncogene product is homologous to the intracellular domain of the EGF receptor (EGFR) (Downward et al. 1984a).

Furthermore, there is evidence that other oncogene products and also tumour promotors interfere with certain growth factor receptor-mediated cellular responses (Berridge and Irvine, 1984) in exerting their transforming or promoting activity. Thus, knowledge of the mode of action of growth factors seems indispensable for understanding the mechanisms of carcinogenesis.

[1] Dept. of Molecular Celbiology, University of Utrecht, Padualaan 8, 3584 CH UTRECHT, The Netherlands.

NATO ASI Series, Vol. H26
Cell to Cell Signals in Mammalian Development
Edited by S. W. de Laat et al.

Growth factors may serve as diffusible signal molecules in the interaction between cells and tissues. Such interactions could play a role in the regulation of differential growth of embryonic and somatic tissues, but also in the outgrowth of tumours (Sporn and Roberts, 1988). A number of polypeptide growth factors has been isolated from tumour cells, but also from cells equivalent to early embryonic cells (Heath and Isacke, 1984; van Veggel et al. 1987), which are capable, by themselves or in combination, to induce mitogenic stimulation as well as a transformed phenotype in normal cells. These so-called transforming growth factors (TGFs) could control the growth behaviour of neighbouring cells (paracrine stimulation) but also of the producing cell itself (autocrine stimulation), provided that the appropriate receptor molecules are expressed at the cell surface (De Larco et al. 1978).

How growth factor receptor interaction leads to the stimulation of cell proliferation and/or to new patterns of gene expression is still largely obscure, despite the great attention given to this important question. Among the various growth factors EGF and its receptor have probably been studied in most detail.

MOLECULAR ASPECTS OF EGF AND ITS RECEPTOR

EGF is a polypeptide of 53 amino acids, with a compact globular structure due to the presence of three intrachain disulfide bridges (Savage et al. 1973).

The first evidence that EGF binds in a specific, saturable manner to a cellular receptor came from binding studies in human fibroblasts (Hollenberg and Cuatracasas, 1973). The finding of Fabricant et al. (1977) that the human epidermoid carcinoma line A431 expresses an extraordinary amount of EGFR (2.10^6/cell) at its cell surface and thus provides an excellent source to study receptor structure and function, resulted in numerous attemps to characterize the EGFR molecule. The vast amount of data from these studies yielded the following picture (see also Fig. 1).

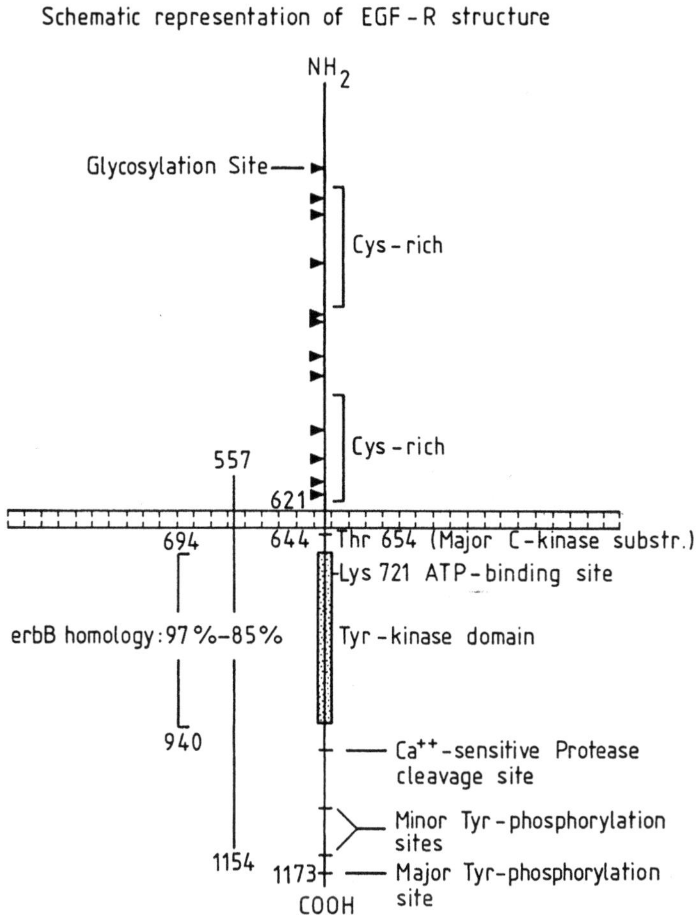

Fig. 1. Highly schematic linear representation of the EGF—receptor structure. The horizontal hatched bar represents the plasma membrane. The numbers indicate individual amino acids according to the numbering system proposed by Ullrich et al. (1984).

The protein part of the mature receptor is a single polypeptide chain of 1186 amino acid residues, as deduced form the nucleotide sequence of cDNA clones (Ullrich et al. 1984). The receptor can be divided into three domains, i.e. an extracellular, N—terminal domain of 621 amino acid residues containing the EGF binding site, an intramembraneous domain of 23 amino acids, and an intracellular, C—terminal domain of 542 amino acids. The extracellular domain is heavily glycosylated and characterized by a high amount of cysteine residues, localized in two major domains of each 160 amino acid residues long. The amino acids comprising the intramembraneous stretch have a strong hydrophobic nature, typical for the transmembrane segments of many intrinsic membrane proteins.

Finally, the cytoplasmic domain of the receptor contains protein tyrosine kinase activity. Binding of EGF to the receptor induces activation of this kinase (Ushiro and Cohen, 1980), leading to the phosphorylation of the receptor itself and other cellular proteins. This part of the receptor displays extensive homology with the v-erbB oncogene of Avian Erythroblastosis Virus (AEV). In the kinase domain, this homology is as high as 97%, which has led to the hypothesis that the transforming activity of the v-erbB product is due to the expression of a truncated EGFR-like molecule in the plasma membrane. In this view, the extracellular EGF binding domain is thought to act as a regulator for the intracellular tyrosine kinase activity. Because the erbB product lacks this control, it is constantly activated, and therefore transforms the cell in which it is expressed (Downward et al. 1984b). Phosphorylation of proteins on tyrosine residues is generally considered to be an important signal in growth regulating mechanisms. A large number of viral and cellular oncogene products have been shown to be protein-tyrosine kinases (for a review, see Hunter and Cooper, 1985). The EGFR belongs to a family of structurally related growth factor receptors with tyrosine kinase activity e.g. the platelet derived growth factor receptor (PDGFR), the receptor for insulin and insulin-like growth factors, and the receptor for colony stimulating factor I (CSF-1) (Yarden et al. 1986).

Several other features of the intracellular part of the EGFR are also depicted in Fig. 1: Lysine residue 721 lies within the ATP binding site of the EGFR (Russo et al. 1985). The indicated threonine residue on position 654 is a substrate for protein kinase C (see below). Phosphorylation of this residue decreases the EGF affinity and lowers the tyrosine kinase activity (Lee and Weinstein, 1978; Cochet et al. 1984), which could function as a regulatory feedback mechanism in the growth factor signalling pathway. The most extensively autophosphorylated tyrosine residue (number 1173) is found in the C-terminal part of the EGFR (Downward et al. 1984b).

OTHER SIGNAL TRANSDUCTION MECHANISMS

A common feature of many agonists of membrane receptors is their ability to induce the breakdown of phosphoinositides in the plasma membrane (for a review, see Berridge and Irvine, 1984). Two phosphorylated forms of phosphatidylinositol (PtdIns) are found in the plasma membrane, i.e. phosphatidylinositol-4-phosphate (PtdIns(4)P) and phosphatidylinositol-4,5-biphosphate (PtdIns(4,5)P_2). Agonist activation of a number of membrane receptors, especially those for the so-called Ca^{2+} mobilizing neurotransmitters, stimulates the hydrolysis of phosphoinositides by phospholipase C (PLC), which results in the production of the second messengers inositol-1,4,5-trisphosphate ($InsP_3$) and 1,2-diacyglycerol (DAG). $InsP_3$ is an effective mobilizer of cytoplasmic Ca^{2+} from intracellular stores, (Berridge and Irvine, 1984). As a consequence, a rapid, transient increase in the intracellular Ca^{2+} concentration ($[Ca^{2+}]_i$) is observed after receptor activation. DAG is a potent activator of the Ca^{2+}- and phospholipid-dependent protein kinase C (PKC) which phosphorylates proteins on serine and threonine residues. The evidence implicating phosphoinositide breakdown in the action of growth factors is rapidly increasing. Sawyer and Cohen (1981) have shown that EGF stimulates phosphatidylinositol (PtdIns) turnover and the production of DAG. A number of recent reports confirm this notion (Pike and Eakes, 1987; Hepler et al. 1987). This is also the case for PDGF and its receptor (Berridge et al. 1984).

An alternative line of evidence for the involvement of phosphoinositide breakdown in growth factor signalling comes from studies on the generation of ionic signals by growth factors. Such signals are among the earliest detectable responses to growth factor addition in quiescent cells. Generally growth factors, including EGF, activate an electroneutral, amiloride sensitive Na^+/H^+ exchanger, leading to a rapid alkalinization of the cytoplasm (Moolenaar, et al. 1983). Additional experiments have implicated protein kinase C in this action (Moolenaar et al. 1984a): TPA as well as the DAG derivative 1-oleoyl-2-acetyl-glycerol (OAG) can mimic growth factors in activating Na^+/H^+ exchange and raising pH_i. The simplest model to explain these observations is the activation of PKC by the growth factor, resulting in Na^+/H^+ exchange activation. Evidence that EGF indeed activates PKC comes from the fact that EGF bindings leads to phosphorylation of the EGFR on threonine 654 (Whiteley and Glaser, 1986).

Also a rapid rise in $[Ca^{2+}]_i$ appears to be involved in the signal transduction by growth factors. Using the intracellularly trapped fluorescent Ca^{2+}-indicator quin-2, it was found that EGF and PDGF induce a rapid but transient increase in $[Ca^{2+}]_i$ (Moolenaar et al. 1984b). This two to three fold increase in $[Ca^{2+}]_i$ represents one of the fastest cellular responses to mitogenic stimulation; it is detectable within a few seconds and reaches its maximum within 30-60 seconds after growth factor addition. While the Ca^{2+} signal in the case of PDGF appears to be almost entirely due to release from intracellular stores (Moolenaar et al. 1984b), the EGF induced Ca^{2+} signal is for the largest part mediated by Ca^{2+} influx (Moolenaar et al. 1986). Nevertheless, also in the action of EGF, at least a small part of the Ca^{2+} signal is the result of mobilization from intracellular stores, again indicating that phosphoinositide breakdown is a part of the EGF signal transduction pathway.

Last but not least, a number of recent studies suggest an intimate relationship between EGF induced cell proliferation and early signal transduction: when cells which are normally devoid of EGFR are induced to express EGFR by transfection with plasmids carrying the EGFR cDNA sequence, they become fully responsive to EGF, i.e.

not only do they respond mitogenically to the growth factor, but also the full repertoire of early responses occurs within these cells (Honegger et al. 1987a,b; Chen et al. 1987; Moolenaar et al. 1988). These studies also stress the importance of an intact protein-tyrosine kinase activity for other signals to occur. When the same cells are transfected with mutated EGFR which lack kinase activity, but are otherwise perfectly normal, there is a normal expression of receptors at the cell surface, but the cells are completely refractory to EGF (Honegger et al. 1987a,b; Chen et al. 1987; Moolenaar et al. 1988).

EGF-EGFR BINDING CHARACTERISTICS

The kinetics of EGF binding to its receptor are of a complicated nature, although the factor binds to the receptor with a 1:1 stoichiometry (Weber et al. 1984). Usually, receptor binding characteristics are established by determination of the relationship between specific ligand binding and ligand concentration under steady state conditions, followed by analysis according to the method of Scatchard. This analysis yields the apparent dissociation constant K_D, and the maximal binding capacity of the cells from which the total receptor number can be calculated. Curvilinearity of the Scatchard graph is usually interpreted as indicative for the presence of different classes of binding sites with different K_D's. EGF binding characteristics in a wide variety of cell lines, analyzed by the Scatchard method, reveal such a curvilinear relationship, and therefore indicate the presence of at least two classes of EGF binding sites, i.e. a high-affinity, low capacity class and a low-affinity, high capacity class (King and Cuatrecasas, 1982; Boonstra et al. 1985a,b; Honegger et al. 1987a). However, conclusions about the presence of two classes of EGF binding sites based upon Scatchard analysis, should be considered with caution as discussed in detail previously (Boonstra et al. 1985a; Carpenter, 1987). Nevertheless, several lines of evidence favor the presence of high and low affinity binding sites in EGF-receptor containing cells. Firstly, as mentioned above, the tumour promotor phorbol myristate acetate (PMA) has been demonstrated to specifically influence high-affinity EGF-binding in a variety of cell lines. Secondly, treatment of cells with other growth factors such as platelet derived growth factor (PDGF) or bombesin, each binding to its own distinct receptor, results in loss of high-affinity binding (Olashaw et al. 1986; Zachary et al, 1986), while an increase in high-affinity EGF receptors is observed after exposure of cells to glucocorticoids (Fanger et al. 1984) or cyclic AMP (Boonstra et al. 1987). The most compelling evidence in favor of the existence of two classes of binding sites is obtained by studies using antibodies directed against the EGF-receptor (Gregoriou and Rees, 1984; Defize et al. 1986, 1989a). In particular the monoclonal anti-EGF receptor antibody 2E9 (Defize et al. 1986) has been shown to exhibit unique properties in this respect. This 2E9 antibody is directed against the extracellular protein core of the receptor of a variety of human cells (Defize et al. 1988, 1989a). The antibody binds to the EGF receptor in a 1:1 ratio and has no direct effect on a variety of rapid responses evoked by EGF in intact cells (Defize et al. 1989a,b,c). Most importantly however, the antibody is demonstrated to block specifically EGF-binding to the low-affinity receptor population in a variety of cell lines, such as A431 and Hela cells (Fig. 2A,B) and a variety of transformed human keratinocyte cell lines (Boonstra and Ponec, 1988), while leaving EGF binding to the high-affinity population undisturbed (Defize et al. 1988, 1989a). Furthermore, in 3T3 cells transfected with a mutated EGF receptor that exhibited no high-affinity EGF binding site, a complete inhibition of EGF binding by 2E9 was observed (Fig. 2C).

These results clearly demonstrate the existence of two classes of EGF binding sites in EGF-responsive cells.

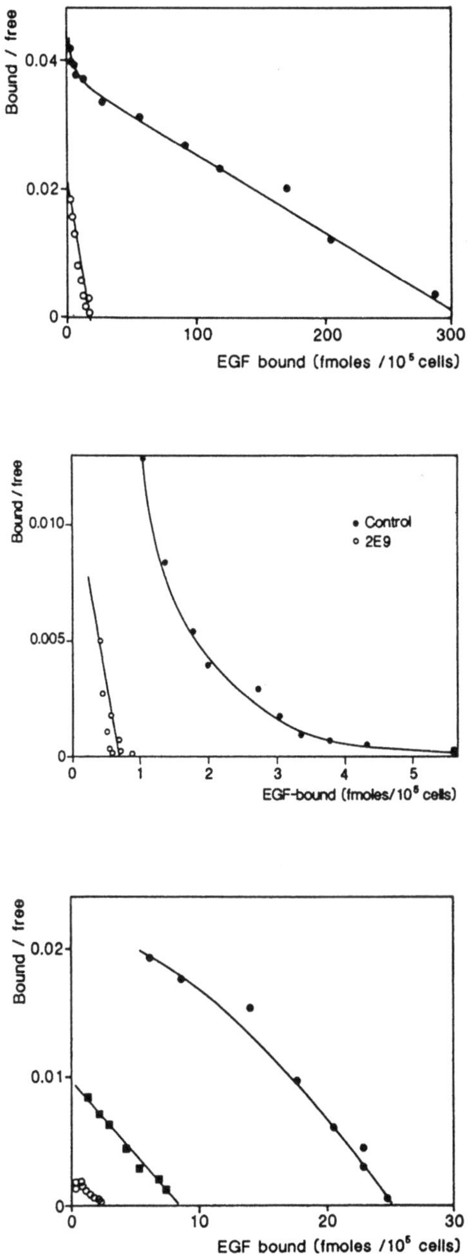

Fig. 2. Scatchard analysis of EGF binding to A431 cells (2A), HeLa cells (2B), and 3T3–IV cells (2C). Cells were either untreated (closed circles) or pretreated with 300 nM 2E9 for 30 minutes at room temperature. The closed squares in 2C represent cells pretreated with antibody 2E9 at a concentration of 30 nM. 3T3–IV cells are cells transfected with a truncated EGFR sequence that display only low–affinity binding. These cells were kindly provided by J. Schlessinger.

EGFR PROTEIN-TYROSINE KINASE ACTIVATION

Much uncertainty still exists about the mechanisms of activation of the EGFR. How can binding of EGF to the extracellular domain lead to activation of the intracellular kinase when the two domains are only connected via one short stretch of amino acids? There are two more or less opposing viewpoints in this issue. One model, called the intramolecular activation model, states that the activation occurs within a single EGFR molecule. In this model, EGF binding leads to a conformational change in the EGFR, which is somehow transferred to the intracellular domain, possibly via a "push-pull" mechanism. This in turn leads to autophosphorylation of the C-terminal part, which in the inactive situation functions as a negative regulator. Due to this autophosphorylation, the C-terminal folds back, thus exposing the kinase to other interacellular substrates (Bertics et al. 1985). Evidence in favor of this model was presented by Bertics and Gill (1985), who found that self-phosphorylation enhances the protein-tyrosine kinase activity of the EGFR. However, Downward et al. (1985) concluded that tyrosine phosphorylation of peptide substrates was independent of EGFR phosphorylation. Consistent with the latter result is the fact that deletion of the C-terminal part of the EGFR by site-directed mutagenesis results in a molecule that is still sensitive to EGF induced kinase activation (Livneh et al. 1986).

The other model, called the intermolecular activation model, states that activation occurs through the EGF driven mutual interaction between two receptor molecules. Is is this interaction which subsequently activates the kinase. Schlessinger, who strongly advocates this model (Schlessinger, 1986), suggests that the driving force behind the EGF induced EGFR dimerization is the higher affinity of receptor dimers for the growth factor. Thus, EGF binding shifts a hypothetical equilibrium between monomeric, inactive low affinity receptors and dimeric, active high affinity receptors to the latter form. Evidence for the model comes from the fact that several bivalent EGFR reagents are capable of activating the EGFR protein-tyrosine kinase, while immobilizing the EGFR on a solid substrate, thus preventing receptor-receptor interactions, leads to non-EGF stimulatability (Yarden and Schlessinger, 1987a, 1987b). Furthermore, addition of EGF to purified EGFR preparations as well as to intact cells leads to the rapid formation of receptor dimers (Yarden and Schlessinger, 1987b; Cochet et al. 1988).

Böni-Schnetzler and Pilch (1987) reported that EGFR dimers isolated from purified receptor preparations by sucrose density gradient centrifugation, display a four-fold higher autophosphorylation activity than monomers, while EGF addition shifts the monomer-dimer equilibrium to the dimeric form. Other data on purified EGFR preparations, however, argue against the intermolecular activation mechanism. Biswas et al. (1985), using a similar approach as Böni-Schnetzler and Pilch (1987), reported that receptor dimers are inactive, while EGF binding leads to monomerization and kinase activation. Northwood and Davis (1988) showed that in the process of EGFR activation by EGF, dimerization indeed occurs, but that sphingosine was equally potent as EGF in stimulating the kinase without the induction of dimerization. Lastly, Koland and Cerione (1988) showed that under certain conditions (in the presence of ammonium sulfate), EGF is capable of stimulating the kinase without significant EGFR aggregation.

The unique properties of 2E9 described above, enable us to test the prediction of the intermolecular activation model that high affinity EGFR are dimers and furthermore to measure the relative contribution of this receptor subclass in tyrosine kinase activation. After a preincubation of A431 with a saturating amount of 2E9 to fully block low affinity receptions, the cells were incubated with ^{125}I-EGF followed by chemical cross-linking. Polyacrylamide gel electrophoresis and autoradiography revealed the presence of monomers and dimers in control cells and cells treated with 2E9 (Fig. 3). Importantly, quantitation of the autoradiograms demonstrated no differences in the relative amount of dimers in control and 2E9

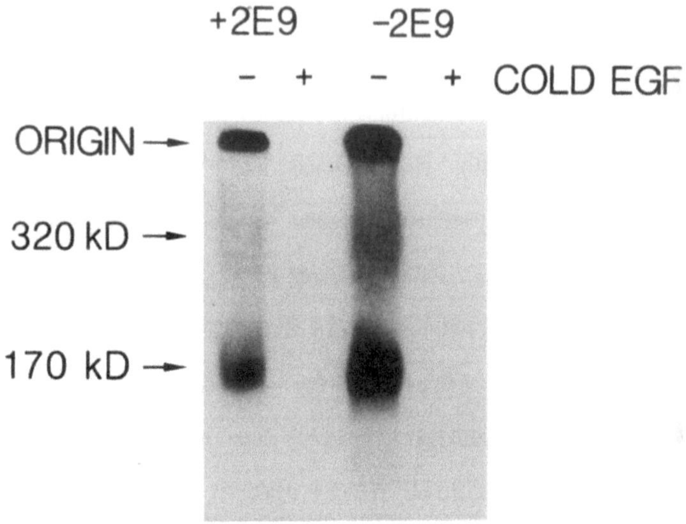

Fig. 3. Cross-linking of ^{125}I-labeled EGF to untreated (two rightmost lanes) or 2E9 pretreated (two leftmost lanes) A431 cells. Cells were treated as described in the text. To some cells, an excess of cold EGF was added. This is indicated in the figure by a +. The cross-linked complexes were separated on a 7-9% linear polyacrylamide gel. Autoradiography was with Kodak XAR-5 film. Exposure time was 50 hours at -70°C.

treated cells (Defize et al. 1989a). It was therefore concluded that there is no strict correlation between oligomerisation and affinity of the EGFR.

ASSOCIATION OF THE EGFR TO THE CYTOSKELETON

Since the state of oligomerisation is apparently not directly correlated with the regulation of EGFR affinity, some other feature must be involved. Association with (an) other cellular constituent(s) could be one explanation. A good candidate for this seems to be the cytoskeleton.

Numerous membrane proteins have been demonstrated to be associated to the cytoskeletal network of a variety of cells, including receptors for nerve growth factor (NGF) (Schechter and Bothwell, 1981), N-formylated peptide (Painter et al. 1987), cyclic AMP (Ludérus and van Driel, 1988), and acetylcholine (Bloch, 1986).

The EGF-receptor appears to be no exception. A variety of electronmicroscopical methods in combination with immunogold labeling strongly suggested a structural association between the EGF-receptor and the cytoskeleton of A431 cells. (Wiegant et al. 1986; Boonstra et al. 1989).

An interesting aspect of the interaction between the EGF-receptor and the cytoskeleton involves the nature of the cytoskeletal component. It has been suggested, as judged by the diameter of elements observed in the electron microscope, that actin is involved in the association (Wiegant et al. 1986). This suggestion has been supported by a co-localization of EGF-receptors and actin filaments by immunofluorescence microscopy (Bergen en Hengouwen, unpublished observations). Determination of the binding characteristics of the cytoskeleton-associated receptors of A431 cells by Scatchard analysis, revealed that these receptors are predominantly of the high-affinity class (Wiegant et al. 1986).

THE CONTRIBUTION OF HIGH AFFINITY EGFR IN INTRACELLULAR SIGNAL TRANSDUCTION

a) Tyrosine kinase activation

When intact, ^{32}P labeled A431 cells are preincubated with antibody 2E9 in order to block all low affinity EGF binding, followed by the addition of a saturating amount of EGF, activation of the EGFR tyrosine kinase as measured by receptor (auto) phosphorylation still occurs efficiently (Fig.4.).
This shows that EGF binding to the small subpopulation of high affinity receptors is sufficient to cause almost full tyrosine kinase activation and receptor (auto) phosphorylation.

Furthermore, combination of the total amount of EGFR exposed of the cell surface with the total amount of receptors phosphorylated on tyrosine residues after EGF binding, shows that cross-phosphorylation and possibly activation of low by high affinity receptors must occur (Defize et al. 1989b), even if the former are occupied by 2E9. This implies that after EGF binding, an interaction of high and low affinity receptors must occur. Although this assumption is not very easily testable, we have obtained some additional evidence for its validity.

Firstly, incubation of A431 cells with EGF for 2 hours at 4°C, resulted in a dramatic increase in the number of EGF-receptors associated to the cytoskeleton (Van Bergen en Hengouwen et al. 1988). Scatchard analysis revealed that these newly cytoskeleton-associated receptors were all of the low-affinity subclass.

Secondly, incubation of intact A431 cells with a saturating amount of ^{125}I-labeled 2E9 (300 nM), followed by EGF for various periods of time also revealed an increase in 2E9-labeled receptors associated to the cytoskeleton (Defize et al. 1989b). Since 2E9 prevents binding of EGF to low-affinity receptors, this provides evidence that binding of EGF to the high-affinity receptors alone is sufficient to

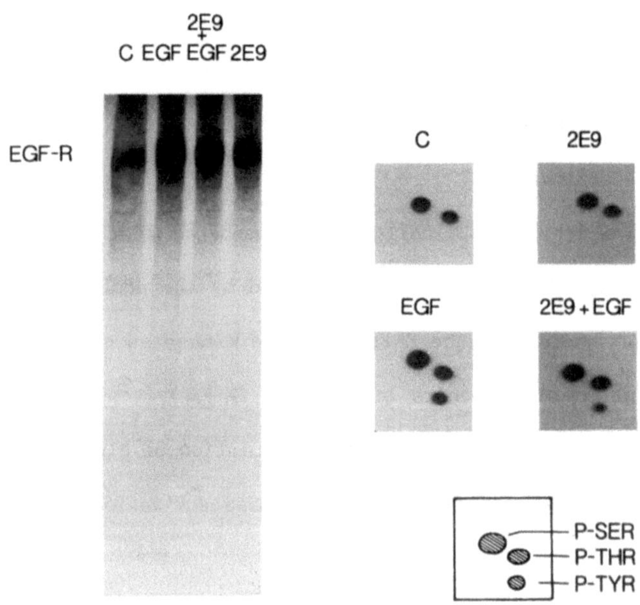

Fig. 4. Intact, ^{32}P labeled A431 cells were treated with 50 ng/EGF for 5 minutes at 37°C before or after 2E9 pretreatment. Subsequently EGFR was immunoprecipitated from the cells and subjected to phosphoamino acid analysis. The left panel shows an autoradiograph of the immunoprecipitates analysed by SDS-gel electrophoresis. The right panel shows phosphoamino acid analysis of the EGFR bands cut from the gel. P-SER=phosphoserine, P-THR=phosphothreonine, P-TYR=phosphotyrosine.

cause low-affinity receptors to become associated to the cytoskeleton, and hence to high affinity receptors.

These results strongly suggest that the cytoskeleton is involved in receptor oligomerization and kinase activation in intact cells, but further proof is needed to establish this phenomenon in more detail.

b) Other early intracellular signals

The obvious question whether exclusive occupation of high affinity EGFR with EGF would also be sufficient for the induction of other intracellular signal transduction mechanism was investigated using 2E9 as well. A number of EGF induced early signals were measured before and after 2E9 preincubation. Among the signals that were unaffected by 2E9 pretreatment were: i) induction of inositol lipid breakdown, ii) rise in pHi, iii) release of Ca^{2+} from intracellular stores, iv) phosphorylation of the receptor on threonine residue no 654 and v) the induction of c-fos expression (Defize et al. 1989c). Thus, also with respect to the occurence of the majority of the early signals, high affinity EGFR occupation by EGF seems sufficient. An interesting exception to the rule however, is the influx of Ca^{2+} from extracellular stores (Defize et al. 1989c). This is completely blocked by 2E9 pretreatment, suggesting low affinity receptors are involved in/necessary for the opening of plasma membrane Ca^{2+} channels. This interesting observation may provide a good starting point to investigate the nature of these channels.

In conclusion, we have shown that EGF assumption of high affinity EGFR alone is sufficient to fully activate the protein tyrosine kinase and also to evoke the majority of the early responses. Furthermore, the data provide evidence for the involvement of the cytoskeleton in this process and determining EGFR affinity. Last but not least, our data indicate that the EGF induced oligomerization observed in intact cells possibly occurs between high and low affinity receptors, resulting in cross phosphorylation/activation of the latter by the former.

ACKNOWLEDGEMENT

We thank Ms. S.N. Rusch and Mrs. M.C. Mulder-de Wildt for preparing the manuscript.

REFERENCES

Berridge MJ, Heslop JP, Irvine RF and Brown KD (1984) Inositol trisphosphate formation and calcium mobilization in Swiss 3T3 cells in response to platelet-derived growth factor. Biochem J 222:195–201

Berridge MJ and Irvine RF (1984) Inositol trisphosphate, a novel second messenger in cellular signal transduction. Nature 312:315–321

Bertics PJ, Weber W, Cochet C and Gill GN (1985) Regulation of the epidermal growth factor receptor by phosphorylation. J Cell Biochem 29:195–208

Bertics PJ and Gill GN (1985) Self-phosphorylation enhances the protein-tyrosine kinase activity of the epidermal growth factor receptor. J Biol Chem 260:14642–14647

Biswas R, Basu M, Sen-Majumdar A and Das M (1985) Intrapeptide autophosphorylation of the epidermal growth factor receptor: regulation of kinase catalytic function by receptor dimerization. Biochem 24:3795–3802

Bloch RJ (1986) Actin at receptor-rich domains of isolated acetylcholine receptor clusters. J Cell Biol 102:1447–1458

Böni-Schnetzler M and Pilch PF (1987) Mechanism of epidermal growth factor receptor antophosphorylation and high affinity binding. Proc Natl Acad Sci USA 84:7832–7836

Boonstra J and Ponec M (1988) High-affinity EGF receptors present in transformed keratinocytes cultured under low Ca^{2+}, proliferating conditions, submitted

Boonstra J, Mummery CL, van der Saag PT and de Laat SW (1985a) Two receptor classes for epidermal growth factor on pheochromocytoma cells, distinguishable by temperature, lectins and tumour promoters. J Cell Physiol 123:347–352

Boonstra J, de Laat SW and Ponec M (1985b) Epidermal growth factor receptor expression related to differentiation capacity in normal and transformed keratinocytes. Exp Cell Res 161:421–433

Boonstra J, Mummery CL, Feyen A, de Hoog WJ, van der Saag PT and de Laat SW (1987) Epidermal growth factor expression during morphological differentiation of pheochromocytoma cells, induced by nerve growth factor or dibutyryl cyclic AMP. J Cell Physiol 131:409–417

Boonstra J, van Belzen N, van Bergen en Henegouwen PMP, Hage WJ, van Maurik P, Wiegant FAC and Verkleij AJ. Verkleij AJ and Leunissen JLM (eds) (1989) The epidermal growth factor receptor. CRC Press

Carpenter G (1987) Receptors for epidermal growth factor and other polypeptide mitogens. Ann Rev Biochem 56:881–914

Chen WS, Lazar CS, Poenie M, Tsien RY, Gill GN and Rosenfeld MG (1987) Requirement for intrinsic protein tyrosine kinase in the immediate and late actions of the EGF receptor. Nature 328:820–823

Cochet C, Gill GN, Meisenhelder J, Cooper JA and Hunter T (1984) C-kinase phosphorylates the epidermal growth factor receptor and reduces its epidermal growth factor-stimulated protein-kinase activity. J Biol Chem 259:2553–2558

Cochet C, Kashles O, Chambaz EM, Borrello I, King CR and Schlessinger J (1988) Demonstration of epidermal growth factor induced receptor dimerization in living cells using a chemical covalent cross-linking agent. J Biol Chem 263:3290–3295

Defize LHK, Moolenaar WH, van der Saag PT and de Laat SW (1986) Dissociation of cellular responses to epidermal growth factor using anti-receptor monoclonal antibodies. EMBO J 5:1187–1192

Defize LHK, Arndt-Jovin DJ, Jovin TM, Boonstra J, Meisenhelder J, Hunter T, de Hey HT and de Laat SW (1988) A431 cell variants lacking the blood group A antogen display increased high-affinity EGF-receptor number, protein tyrosine kinase activity and receptor turnover. J Cell Biol 107:939–949

Defize LHK, Boonstra J and de Laat SW (1989a) A monoclonal antibody allows specific occupation of high affinity epidermal growth factor receptors with EGF: structural aspects of high-affinity EGF-binding in intact cells, submitted

Defize LHK, Boonstra J, Meisenhelder J, Hunter T, van Bergen en Henegouwen PMP and de Laat SW (1989b) High-affinity epidermal growth factor receptors (EGFR) play a major role in the EGF-induced protein-tyrosine kinase activity in intact A431 cells, submitted

Defize LHK, Meisenhelder J, Kruyer W, Tertoolen LGJ, Tilly BC, Hunter T, Boonstra J, Moolenaar WH and de Laat SW (1989c) Early signal transduction by epidermal growth factor occurs through binding to a subclass of high affinity receptors, submitted

de Larco JE and Todaro GJ (1978) Growth factors from murine sarcoma virus transformed cells. Proc Natl Acad Sci USA 75:4001–4005

Doolittle RF, Hunkapiller HW, Hood LE, Devare SG, Robbins KC, Aaronson SA and Antoniades HN (1983) Simian sarcoma virus onc gene V-sis, is derived from the gene (or genes) encoding a platelet-derived growth factor. Science 221:275–277

Downward J, Yarden Y, Mayes E, Scrace G, Totty N, Stockwell P, Ullrich A, Schlessinger J and Waterfield MD (1984a) Close similarity of epidermal growth factor receptor and verbB oncogene protein sequences. Nature 307:521–527

Downward J, Parker P and Waterfield M (1984b) Antophosphorylation sites on the epidermal growth factor receptor. Nature 311:483–485

Downward J, Waterfield M and Parker PJ (1985) Antophosphorylation and protein kinase C phosphorylation of the epidermal growth factor receptor. Effect on tyrosine kinase activity and ligand binding activity. J Biol Chem 260:14538–14546

Fabricant RN, De Larco JE and Todaro GJ (1977) Nerve growth factor receptors on human melanoma cells in culture. Proc Natl Acad Sci USA 74:565–569

Fanger BO, Viceps-Madore D and Cidlowski JA (1984) Regulation of high- and low-affinity epidermal growth factor receptor by glucocorticoids. Arch Biochem Biophys 235:141–149

Gregoriou M and Rees AR (1984) Properties of a monoclonal antibody to epidermal growth factor receptor with implications for the mechanisms of action of EGF. EMBO J 3:929–937

Heath JK and Isacke CM (1984) PC13 embryonal carcinoma-derived growth factor. EMBO J 3:2957–2962

Hepler JR, Nakahata N, Lovenberg TW, Di Giuseppi J, Herman B, Earp HS and Harden TK (1987) Epidermal growth factor stimulates the rapid accumulation of inositol (1,4,5)-trisphosphate and a rise in cytosolic calcium mobilized from intracellular stores in A431 cells. J Biol Chem 262:2951–2956

Hollenberg MD and Cuatracasas P (1973) Epidermal growth factor receptors in human fibroblasts and modulation of action by choleratoxin. Proc Natl Acad Sci USA 70:2964–2968

Honegger AM, Szapary D, Schmidt A, Lyall R, van Obberghen E, Dull TJ, Ullrich A and Schlessinger J (1987a) A mutant epidermal growth factor receptor with defective protein tyrosine kinase is unable to stimulate proto-oncogene expression and DNA synthesis. Mol Cell Biol 7:4568–4571

Honegger AM, Dull TJ, Felder S, van Obberghen E, Bellot F, Szapary D, Schmidt A, Ullrich A and Schlessinger J (1987b) Point mutation at the ATP binding site of EGF receptor abolishes protein tyrosine kinase activity and alters cellular routing. Cell 51: 199–209

Hunter T and Cooper JA (1985) Protein-tyrosine kinases. Ann Rev Biochem 54:897–930

King AC and Cuatrecasas P (1982) Resolution of high and low affinity epidermal growth factor receptors: inhibition of high-affinity component by low temperature, cycloheximide, and phorbol ester. J Biol Chem 257:3053–3060

Koland JG and Cerione RA (1988) Growth factor control of epidermal growth factor receptor kinase activity via an intramolecular mechanism. J Biol Chem 263:2230–2237

Lee LS and Weinstein IB (1978) Tumor promoting phorbol esters inhibit binding of epidermal growth factor to cellular receptors. Science 202:313–315

Livneh E, Prywes R, Kashles O, Reiss N, Sasson I, Mory Y, Ullrich A and Schlessinger J (1986) Reconstitution of human epidermal growth factor receptors and its deletion mutants in cultured hamster cells. J Biol Chem 261:12490–12497

Ludérus MEE and van Driel R (188) Interaction between the chemotactic cAMP receptor and a detergent-insoluble membrane residue of Dictyostelium discoideum. Modulation by guanine nucleotides. J Biol Chem 263:8326–8331

Moolenaar WH, Tsien RY, van der Saag PT and de Laat SW (1983) Na^+/H^+ exchange and cytoplasmic pH in the action of growth factors in human fibroblasts. Nature 304:645–648

Moolenaar WH, Tertoolen LGJ and de Laat (1984a) Phorbol ester and diacylglycerol mimic growth factors in raising cytoplasmic pH. Nature 312:371–374

Moolenaar WH, Tertoolen LGJ and de Laat SW (1984b) Growth factors immediately raise cytoplasmic free Ca^{++} in human fibroblasts. J Biol Chem 259:8066–8069

Moolenaar WH, Bierman AJ, Tilly BC, Verlaan I, Defize LHK, Honegger AM, Ullrich A and Schlessinger J (1988) A point mutation at the ATP-binding site of the EGF-receptor abolishes signal transduction. EMBO J 7:707–710

Moolenaar WH, Aerts RJ, Tertoolen LGJ and de Laat SW (1986) The epidermal growth factor induced calcium signal in A431 cells. J Biol Chem 261:279–284

Northwood IC and Davis RJ (1988) Activation of the epidermal growth factor receptor tyrosine protein kinase in the absence of receptor oligomerization. J Biol Chem 263:7450–7453

Olashaw NE, O'Keefe EJ and Pledger WJ (1986) Platelet-derived growth factor modulates epidermal growth factor receptors by a mechanisms distinct from that of phorbol esters. Proc Natl Acad Sci USA 83:3834–3838

Painter RG, Zahler-Bentz K and Dukes RE (1987) Regulation of the affinity state of the N-formylated peptide receptor of neutrophils: role of guanine nucleotide-binding proteins and the cytoskeleton. J Cell Biol 105:2959–2971

Pike LJ and Eakes AT (1987) Epidermal growth factor stimulates the production of phosphatidyl inositol monophosphate and the breakdown of polyphosphoinositides in A431 cells. J Biol Chem 262:1644–1651

Russo MW, Lukas TJ, Cohen S and Staros JV (1985) Identification of residues in the nucleotide binding site of the epidermal growth factor receptor kinase. J Biol Chem 260:5205–5208

Savage Jr. CR, Hash JH and Cohen S (1973) Epidermal growth factor: location of disulfide bonds. J Biol Chem 248:7669–7672

Sawyer ST and Cohen S (1981) Enhancement of calcium uptake and phosphatidyl turnover by epidermal growth factor in A431 cells. Biochemistry 20:6280–6286

Schlechter AL and Bothwell MA (1981) Nerve growth factor receptors on PC12 cells: Evidence for two receptor classes with differing cytoskeletal association. Cell 27:867–874

Schlessinger J (1986) Allosteric regulation of the epidermal growth factor receptor kinase. J Cell Biol 103:2067–2072

Sporn MB and Roberts AB (1988) Peptide growth factors are multifunctional. Nature 332:217–219

Ullrich A, Coussens L, Hayflick JS, Dull TJ, Gray A, Tam AW, Lee J, Yarden Y, Libermann TA, Schlessinger J, Downward J, Mayes ELV, Whittle N, Waterfield MD and Seeburg PH (1984) Human epidermal growth factor receptor cDNA sequence and aberrant expression of the amplified gene in A431 epidermoid carcinoma cells. Nature 309:418–425

Ushiro H and Cohen S (1980) Identification of phosphotyrosine as a product of epidermal growth factor–activated protein kinase in A431 cell membranes. J Biol Chem 255:8363–8365

van Bergen en Henegouwen PMP, de Kroon J, van Damme H, Verkley AJ and Boonstra J (1988) Ligand induced association of epidermal growth factor receptor tot the cytoskeleton of A431 cells. J Cell Biochem submitted

van Veggel JH, van Oostwaard TMJ, de Laat SW and van Zoelen EJJ (1987) PC13 embryonal carcinoma cells produce a heparin–binding growth factor. Exp Cell Res 169:280–286

Weber W, Bertics PJ and Gill GN (1984) Immunoaffinity purification of the epidermal growth factor receptor. Stoichimetry of binding and kinetics of self phosphorylation. J Biol Chem 259:14631–14636

Whiteley B and Glaser L (1986) Epidermal growth factor (EGF) promotes phosphorylation at threonine–654 of the EGF receptor:possible role of protein kinase C in homologous regulation of the EGF receptor. J Cell Biol 103:1355–1362

Wiegant FAC, Blok FJ, Defize LHK, Linnemans WAM, Verkleij AJ and Boonstra J (1986) Epidermal growth factor receptors associated to cytoskeletal elements of epidermoid carcinoma (A431) cells. J Cell Biol 103:87–94

Yarden Y, Escobedo JA, Kuang WJ, Yang–Feng TL, Daniel TO, Tremble PM, Chen EY, Ando ME, Harkins RN, Francke U, Fried VA, Ullrich A and Williams LT (1986) Structure of the receptor for platelet–derived growth factor helps define a family of closely related growth factor receptors. Nature 323:226–232

Yarden Y and Schlessinger J (1987a) Self–phosphorylation of epidermal growth factor receptors. Evidence for a model of intermolecular allosteric activation. Biochemistry 26:1434–1442

Yarden Y and Schlessinger J (1987b) Epidermal growth factor induces rapid, reversible aggregation of the purified epidermal growth factor receptor. Biochemistry 26:1443–1451

Zachary I, Sinnett–Smith JW and Rozengurt E (1986) Early events elicited by bombesin and structurally related peptides in quiescent Swiss 3T3 cells. I. Activation of protein kinase C and inhibition of epidermal growth factor binding. J Cell Biol 102:2211–2222

Insulin receptor tyrosine kinase and insulin action

E. Van Obberghen, S. Gammeltoft, Y. Le Marchand-Brustel, R.-Ballotti
INSERM U 145
Faculté de Médecine
Nice, France

INTRODUCTION

Regulation of cellular metabolism and growth by insulin is the result of a series of events initiated by the interaction of the hormone with its cell surface receptors. The insulin receptor structure has been elucidated through a variety of techniques (Van Obberghen, 1984), and more recently, the aminoacid sequence of the human insulin receptor precursor was unravelled by recombinant DNA technology (Ullrich et al., 1985; Ebina et al., 1985). Despite this progress the molecular mechanism of insulin action is still not entirely comprehended regarding the events following receptor binding and leading to the ultimate cellular responses. However, over the years considerable evidence has been gathered indicating that reversible phosphorylation contributes to the mechanism of insulin action (Denton, 1986). Further, a promising discovery was made by the demonstration that the insulin receptor is an insulin-sensitive protein kinase (Kasuga et al., 1982; Van Obberghen and Kowalski, 1982; Petruzzelli et al., 1982). This observation is of particular interest for our understanding of insulin-regulated processes, since it is now recognized that phosphorylation-dephosphorylation of proteins is a mechanism whereby many cellular functions are regulated by hormones and neurotransmitters. Furthermore, protein kinases are also constituents of receptors for several growth factors, implying that receptor kinase activity may represent a general mechanism in transmembrane sigalling of hormones and growth factors.

NATO ASI Series, Vol. H26
Cell to Cell Signals in Mammalian Development
Edited by S. W. de Laat et al.

INSULIN RECEPTOR PHOSPHORYLATION

In intact cells, insulin stimulates the phosphorylation of its receptor β-subunit (Van Obberghen and Kowalski, 1982; Kasuga et al., 1982). In these experiments, cells were preincubated with $[^{32}P]P_i$ to label cellular ATP; solubilized, and the glycoproteins purified on WGA-agarose. Immunoprecipitation by antibodies to insulin receptors followed by SDS/PAGE under reducing conditions and autoradiography revealed a labeled band (M_r 95 Kd), the phosphorylation of which was stimulated by insulin. Its identity with insulin receptor β-subunit was established based on its appropriate electrophoretic mobility and on the fact that it was not precipitated with nonimmune serum. In intact cells phosphoaminoacid analysis of the insulin receptor β-subunit showed phosphorylation of serine, threonine, and tyrosine under basal conditions. Insulin induced a rapid, several-fold increase in ^{32}P incorporation in tyrosine, followed by a slower rise in labeling of phosphoserine (Kasuga et al., 1982; Gazzano et al., 1983; White et al., 1985; Ballotti et al., 1987).

Subsequently, insulin stimulated phosphorylation of the insulin receptor β-subunit was demonstrated in cell-free systems using $[\gamma-^{32}P]$ATP and purified receptors (Van Obberghen and Kowalski, 1982; Kasuga et al., 1982; Van Obberghen et al., 1983; Shia and Pilch, 1983; Petruzzelli et al., 1982; 1984). These purified receptor preparations exhibited insulin-stimulated protein kinase activity which catalysed phosphorylation of both the β-subunit and exogenous substrates. With highly purified receptor, the phosphorylation occurred exclusively on tyrosine residues under basal conditions, and insulin stimulatory action was accounted for by a several-fold increase in phosphotyrosine. Thus, the tyrosine kinase appeared as a constituent of the insulin receptor. Further, the β-subunit contains an ATP-binding site as demonstrated by covalent affinity labeling (Van Obberghen et al., 1983; Shia and Pilch, 1983; Roth and Cassell, 1983). The simultaneous presence of phosphorylation sites and an ATP-binding site on the β-subunit indicates that the insulin receptor acts as its own tyrosine kinase. Further proof is the demonstration that the insulin-binding activity and insulin-dependent tyrosine-kinase activity copurified to homogeneity at a constant stoichiometric ratio (Petruzzelli et al., 1984). In addition, insulin binds to and promotes phosphorylation of the insulin receptor precursor (M_r 210 Kd) (Rees-Jones et al., 1984).

BIOCHEMISTRY OF THE INSULIN RECEPTOR KINASE

Following the identification of the insulin receptor protein kinase activity its biochemical properties have been investigated in detail. For the original references see in Gammeltoft and Van Obberghen (1986).

ROLE OF INSULIN RECEPTOR TYROSINE KINASE IN HORMONE ACTION

1. Characteristics of insulin receptor tyrosine kinase

Since the discovery that the insulin receptor is an insulin-dependent protein tyrosine kinase, it was anticipated that this receptor enzymic function was involved in insulin action (Gazzano et al., 1983). For the validity of this contention at least the following five criteria were expected to be fulfilled.

First, the insulin dose-response relationship of the kinase should be within the physiological range and correlate with that of binding to receptor. Several authors found that the kinase activation was half-maximal at an insulin concentration of 2-5 nM (ED_{50}), which corresponded to the apparent K_d of the receptor-insulin complex of the same solubilized receptor preparations (Kasuga et al., 1982; Shia and Pilch, 1983; Petruzzelli et al., 1984; Sadoul et al., 1985). In contrast, a dissociation between dose-respose curves of insulin binding and kinase activation was observed with soluble receptors from rat liver and human erythrocytes (Grigorescu et al., 1983). Here, the apparent K_d exceeded the ED_{50} by a factor of 3-10, suggesting that the phenomenon of 'spare receptors' observed for other insulin actions is also applicable for kinase activation. It is not clear whether these findings are explained by differences in tissues, purification procedures, or assay methods. In conclusion, in most instances the receptor kinase is activated by insulin concentrations within a physiological range corresponding to receptor binding.

Secondly, the receptor kinase should be capable of phosphorylating cellular substrates other than the receptor itself, in order to propagate the insulin response. The insulin receptor kinase can phosphorylate a number of substrates on tyrosine in vitro, although none of the proteins tested are proven to be physiologically relevant substrates. The two first "putative" substrates described were a 110-120 Kd protein, and a 185 Kd protein. In 1985 two laboratories, independently, identified a cellular protein "substrate" of Mr 110 Kd for the insulin receptor kinase in purified glycoproteins from rat liver and rabbit brown adipose tissue (Sadoul et al., 1985; Rees-Jones and Taylor, 1985). This glycoprotein appears as a monomeric structure, and is not part of the insulin receptor itself. Phosphorylation of the Mr 110 Kd protein and of the receptor β-subunit were stimulated by insulin in a remarkably similar dose-dependent fashion ($ED_{50} \simeq 1$ nM). Further, kinetic studies suggested that phosphorylation of the Mr 110 Kd protein occurred after activation and phosphorylation of the insulin receptor kinase. The nature and function of this endogenous substrate is as yet unknown. At the same period a different putative substrate for the insulin receptor kinase has been identified in a hepatoma cell line, Fao (White et al., 1985). This substrate does not contain carbohydrate moieties, and has a Mr 185 Kd. Since the reports on these two "putative" substrates a number of other phosphoproteins have been described including a 15 protein, that may play a role in insulin-mediated glucose transport (Bernier et al., 1987). However, at present it remains to be shown whether any of these proteins have a physiological significance. One is forced to admit that despite intensive efforts extremely little is known concerning "putative" substrates which is likely due to the fact that they are rare and labile. Further, as expected from the large array of biological responses induced by insulin a whole series of nonabundant regulatory proteins likely exist to account for the metabolic and growth promoting effects of insulin.

The third criterion is reversibility of insulin receptor phosphorylation. To exert a regulatory function, the phosphorylated and activated receptor kinase should return to basal activity through a dephosphorylation reaction. Lectin purified receptor preparations contained phosphatase activity, which slowly reduced the ^{32}P content of phosphorylated receptor, and which was insulin independent (Kowalski et al., 1983). Exposure of phosphorylated insulin receptor to alkaline phosphatase resulted in removal of about 50% of the β-subunit phosphotyrosine, and about 65% reduction in kinase activity (Yu and Czech, 1985). Thus, the insulin receptor kinase can by deactivated through dephosphorylation of tyrosine residues.

The fourth criterion regards the specificity of insulin effect on its receptor kinase. Several insulin analogues stimulated receptor phosphorylation with potencies relative to porcine insulin, which were identical with their relative binding affinities and potencies in other assay systems (Kasuga et al., 1982; Grigorescu et al., 1983). Finally, polyclonal antisera to insulin receptor, which exert insulin-like effects in several cell types, were also able to stimulate the receptor tyrosine kinase (Gammeltoft and Van Obberghen, 1986; Gherzi et al., 1987). In conclusion, the insulin effect on receptor phosphorylation has the affinity and specificity of a typical insulin receptor mediated event.

Taken together, the kinase activity of the insulin receptor seems to be a fundamental receptor property, since whenever insulin receptors are present, insulin-stimulated autophosphorylation occurs (Gammeltoft et al., 1984; Gammeltoft et al., 1985; Gazzano et al., 1985; Grigorescu et al., 1983; Kasuga et al., 1982; Petruzzelli et al., 1984; Shia and Pilch, 1983; Tanti et al., 1986; Van Obberghen and Kowalski, 1982)).

2. Insulin receptor tyrosine kinase and hormone action

A general consensus has been reached for a role of insulin receptor kinase in hormone action. A first series of suggestive observations was provided by studies showing that alterations in insulin action are associated with parallel alterations in insulin receptor tyrosine kinase activity. Thus, the receptor kinase is impaired in various insulin-resistant states including the syndrome of extreme insulin resistance type A (Grunberger et al., 1984), melanoma cell cultures (Häring et al., 1984), gold-thioglucose obese mice (Le Marchand-Brustel et al., 1985), and streptozotocin diabetic rats (Kadowaki et al., 1984). Conversely, insulin receptor kinase is hyperactive in insulin hyperresponsive adipocytes of young obese Zucker rats (Debant et al., 1987). Further, insulinomimetic agents (vanadate, lectins, trypsin) increased receptor autophosphorylation (Tamura et al., 1983; Roth et al., 1983). Introduction into mammalian cells of a monoclonal antibody, which inhibits insulin receptor kinase, blocks the rapid effects of insulin (Morgan and Roth, 1987). Conversely, microinjection of antiphosphotyrosine antibodies, which stimulate the insulin receptor kinase, induces enhanced insulin-evoked glucose transport and aminoacid uptake (Ballotti et al., submitted).

The most convincing and elegant evidence for the idea that insulin action depends on receptor protein tyrosine kinase activity comes from mutagenesis experiments involving the receptor. Thus, insulin receptors mutated on the ATP-binding site lack protein tyrosine kinase activity, and fail to mediate insulin post-receptor effects including glucose transport, glycogen synthesis, S6 kinase activity and thymidine uptake (Chou et al., 1987).

SERYL (AND THREONYL) PHOSPHORYLATION OF INSULIN RECEPTORS

In intact cells, the rapid insulin-stimulated phosphorylation of its receptor on tyrosine is followed by a slower serine phosphorylation (Kasuga et al., 1982; Gazzano et al., 1983; White et al., 1985; Ballotti et al., 1987; Pang et al., 1985). Further, with partially purified insulin receptor, insulin-stimulated phosphorylation of both tyrosine and serine on its receptor (Kasuga et al., 1982; Zick et al., 1983) as well as on exogenous substrates was shown (Gazzano et al., 1983; Ballotti et al., 1986). The serine kinase activity appears to be non-covalently associated with the receptor, and is removed during further purification, because highly purified receptor displayed only tyrosine kinase activity. The relationship between the two protein kinase activities associated with the receptor and their cellular role remain to be established. At least two possibilities exist, one in which both kinase activities serve separate cellular functions, and another one with sequential kinase activation (Gazzano et al., 1983). According to the first model the tyrosine kinase would be involved in insulin's growth-promoting action similar to tyrosine phosphorylations mediating cellular responses to growth factors and cellular and retroviral oncogene proteins (Hunter and Cooper, 1985). In contrast, the serine kinase activities would play a role in insulin's metabolic actions. All kinases involved in the control of intermediary metabolism are indeed serine- or threonine-specific. In the second model, the two types of kinases are activated sequentially. Insulin binding to receptor leads to activation of the constituent tyrosine kinase, which induces activation of the receptor-associated serine kinase(s) and this accounts for the generation of cellular responses to insulin.

SIGNAL TRANSDUCTION

A large body of evidence establishes the insulin receptor tyrosine kinase as fundamental property of the receptor, and indicates that this receptor enzymic function is essential for generation of the metabolic and growth promoting effects of insulin. The insulin receptor displays two functionnal domains, an extracellular insulin binding α-subunit, and an insulin responsive protein kinase contained within the intracellular domain of the β-subunit. At present it is not established how hormone recognition at the cell surface transmits a signal to the cytoplasmic receptor domain through a unique transmembrane stretch. The simplest mechanism would be that interaction of insulin with the receptor α-subunit triggers a conformational change which is propagated at the level of the contact region between α-β subunits resulting in activation of the receptor kinase. How the activated receptor kinase transduces the hormone signal is also not known presently. We favour the idea that the insulin-stimulated receptor kinase leads to phosphorylation of cellular protein substrates, which are likely to belong to two broad categories, those generating metabolic effects of insulin, and those resulting in growth-promoting effects.

REFERENCES

Ballotti R, Kowalski A, Le Marchand-Brustel Y, Van Obberghen E (1986) Biochem Biophys Res Commun 139: 179-185

Ballotti R, Kowalski A, White MF, Le Marchand-Brustel Y, Van Obberghen E (1987) Biochem J 241: 99-104

Ballotti R, Scimeca JC, Kowalski A, Van Obberghen E (Submitted)

Bernier M, Laird DM, Lane D (1987) Proc Natl Acad Sci 84: 1844-1848

Chou CK, Dull TJ, Russell DS, Gherzi R, Lebwohl D, Ullrich A, Rosen OM (1987) 262: 1842-1847

Debant A, Guerre-Millo M, Le Marchand-Brustel Y, Freychet P, Lavau M, Van Obberghen E (1987) Am J Physiol 252: E273-E277

Denton RM (1986) Adv Cyclic Nucleotide Protein Phosphorylation Res 20: 293-341

Ebina Y, Ellis L, Jarnagin K, Edery M, Graf L, Clauser E, Ou JH, Masiarz F, Kan YW, Goldfine ID, Roth RA, Rutter WJ (1985) Cell 40: 747-758

Gammeltoft S, Haselbacher G, Humbel RE, Fehlmann M, Van Obberghen E (1985) EMBO J 4: 3407-3412

Gammeltoft S, Kowalski A, Fehlmann M, Van Obberghen E (1984) FEBS Lett 172: 87-90

Gammeltoft S, Van Obberghen E (1986) Biochem J 235: 1-11

Gazzano H, Halban P, Prentki M, Ballotti R, Brandenburg D, Fehlmann M, Van Obberghen E (1985) Biochem J 226: 867-872

Gazzano H, Kowalski A, Fehlmann M, Van Obberghen E (1983) Biochem J 216: 575-582

Gherzi R, Russell DS, Taylor SI, Rosen OM, (1987) J Biol Chem 262: 16900-16905

Grigorescu F, White MF, Kahn CR (1983) J Biol Chem 258: 13708-13716

Grunberger G, Zick Y, Gorden P (1984) Science 223: 932-934

Haring HU, White MF, Kahn CR, Kasuga M, Lauris W, Fleischmann R, Murray M, Pawelek J (1984) 99: 900-908

Hunter T, Cooper JA (1985) Annual Rev Biochem 54: 897-930

Kadowaki T, Kasuga M, Akanum Y, Ezaki D, Takuku F (1984) J Biol Chem 259: 14208-14216

Kasuga M, Karlsson FA, Kahn CR (1982) Science 215: 185-187

Kasuga M, Zick Y, Blithe DL, Crettaz M, Kahn CR (1982) Nature 298: 667-669

Kasuga M, Zick Y, Blithe DL, Karlsson FA, Haring HU, Kahn CR (1982) J Biol Chem 257: 9891-9894

Kowalski A, Gazzano H, Fehlmann M, Van Obberghen E (1983) Biochem Biophys Res Commun 117: 885-893

Le Marchand-Brustel Y, Gremeaux T, Ballotti R, Van Obberghen E (1985) Nature 315: 676-679

Morgan DO, Roth RA (1987) 84: 41-45

Pang DT, Sharma BR, Schafer JA, White MF, Kahn CR (1985) J Biol Chem 260: 7131-7136

Petruzzelli LM, Ganguly S, Smith C.J., Cobb M.H., Rubin C.S., Rosen O.M. (1982) Proc Natl Acad Sci 79: 6792-6797

Petruzzelli L, Herrera R, Rosen OM (1984) Proc Natl Acad Sci 81: 3327-3331

Rees-Jones RW, Hedo JA, Zick Y, Roth J (1983) Biochem Biophys Res Commun 116: 417-422

Rees-Jones RW, Taylor SI (1985) J Biol Chem 260: 4461-4467

Roth RA, Cassel DJ (1983) Science 219: 299-301

Roth RA, Cassel DJ, Maddux DA, Goldfine ID (1983) Biochem Biophys Res Commun 115: 245-252

Sadoul JL, Peyron JF, Ballotti R, Debant A, Fehlmann M, Van Obberghen E (1985) Biochem J 227: 887-892

Shia MA, Pilch PF (1983) Biochemistry 22: 717-721

Tamura S, Fujita-Yamaguchi Y, Larner J (1983) J Biol Chem 258: 14749-14752

Tanti JF, Gremeaux T, Brandenburg D, Van Obberghen E, Le Marchand-Brustel Y (1986) Diabetes 35: 1243-1248
Ullrich A, Bell JR, Chen EY, Herrera R, Petruzzelli LM, Dull TJ, Gray A, Coussens L, Liao YC, Tsubokawa M, Mason A, Seeburg PH, Grunfeld C., Rosen OM, Ramachandran J (1985) Nature 313: 756-761
Van Obberghen E (1984) Biochem Pharmacol 33: 889-896
Van Obberghen E, Kowalski A (1982) FEBS Lett 143: 179-182
Van Obberghen E, Rossi B, Kowalski A, Gazzano H, Ponzio G (1983) Proc Natl Acad Sci 80: 945-949
White MF, Maron R, Kahn CR (1985) Nature 318: 183-186
White MF, Takayama S, Kahn CR (1985) 260: 9470-9478
Yu KT, Czech MP (1984) J Biol Chem 259: 5277-5286
Zick Y, Grunberger G, Podskalny JM, Moncada V, Taylor SI, Gorden P, Roth J (1983) Biochem Biophys Res Commun 116: 1129-1135

THE MACROPHAGE COLONY STIMULATING FACTOR, CSF-1, AND ITS RECEPTOR (c-fms)

Charles J. Sherr
Department of Tumor Cell Biology
St. Jude Children's Research Hospital
332 North Lauderdale
Memphis, Tennessee 38105
USA

The hematopoietic colony stimulating factors (CSFs) affect the proliferation, differentiation, and survival of progenitor cells in the bone marrow and prime mature blood cells to carry out their differentiated functions (Metcalf, 1985; Clark and Kamen, 1987; Sieff, 1987). The *in vitro* proliferation of normal myeloid progenitor cells in semisolid medium requires the continuous presence of CSFs, and the types of colonies obtained, whether granulocyte, macrophage, or erythroid, depend on the types of factors added to the cultures. Unlike pluripotent cytokines such as multi-CSF (interleukin-3, or IL-3) and granulocyte-macrophage CSF (GM-CSF) which affect the growth of colonies of multiple hematopoietic lineages, macrophage CSF (M-CSF or CSF-1), granulocyte CSF (G-CSF), and erythropoietin (Epo) stimulate the lineage-specific proliferation of macrophage, granulocyte, or erythroid precursors, respectively. Recently, each of these CSFs has been purified to homogeneity, and cDNA clones encoding the factors have been exploited to produce previously unrealizable quantities of CSFs for experimental and therapeutic use. CSFs are already proving their efficacy in clinical trials and have numerous potential applications in the treatment of blood cell disorders, malignancies, and in modulating the inflammatory response (Donohue et al., 1986; Welte et al., 1987; Eschbach et al., 1987; Groopman et al., 1987). Our interest in these factors stems from our observations that the receptor for CSF-1 is encoded by the c-*fms* proto-oncogene (Sherr et al., 1985), thereby underscoring the possibility that aberrations in CSF-1 receptor signaling might contribute to hematopoietic malignancy. The availability of molecularly cloned CSF-1 and CSF-1 receptor genes makes this system particularly amenable for studies of the regulation of hematopoietic cell proliferation.

The Macrophage Colony Stimulating Factor, CSF-1

The macrophage colony stimulating factor, CSF-1 (M-CSF), stimulates the proliferation and survival of committed monocyte-macrophage precur-

NATO ASI Series, Vol. H26
Cell to Cell Signals in Mammalian Development
Edited by S. W. de Laat et al.

sors and affects the differentiated functions of mature mononuclear phagocytes (Stanley et al., 1983). Human CSF-1 acts on both human and murine monocytes and macrophages, whereas murine CSF-1 does not stimulate human cells. Paradoxically, human recombinant CSF-1 is more active on murine than on human target cells (Ralph et al., 1986b), and it is therefore likely that human CSF-1 requires the presence of other factors for optimal stimulation. Unlike other hemopoietins, CSF-1 is constitutively present in serum, and its concentration in the circulation is homeostatically regulated through receptor mediated uptake and degradation by monocytes and tissue macrophages (Bartocci et al., 1987). Although CSF-1 is not a macrophage activating factor, it augments the production of other macrophage cytokines (eg. interleukin-1, tumor necrosis factor, interferon, and G-CSF), as well as plasminogen activator, prostaglandins, thromboplastin, and biocidal oxygen metabolites (Warren and Ralph, 1986; Ralph et al, 1986a; Lyberg et al., 1987). CSF-1 also protects murine macrophages from lytic infection by vesicular stomatatis virus (Lee and Warren, 1987), stimulates killing of *Candida albicans* (Karbassi et al., 1987), and promotes antibody-dependent tumor cell cytotoxicity in conjunction with other cytokines (P. Ralph and I. Nakoinz, personal communication).

CSF-1 is a homodimeric glycoprotein produced by fibroblasts and endothelial cells, which, when purified from urine or supernatants of cultured cells, is heterogeneous with molecular weights ranging from 45 to 80 kilodaltons (kd) (Stanley and Heard, 1977; Das and Stanley, 1982; Csejtey and Boosman, 1986; Wong et al, 1987; Rajavashisth et al., 1987). Removal of carbohydrate side chains does not affect its biological functions, but reduction of the disulfide linked dimers abrogates its activity (Das et al., 1981; Das and Stanley, 1982). Human CSF-1 is encoded by a single gene composed of 10 exons spanning 21 kilobases (kb) (Kawasaki et al., 1985; Ladner et al., 1987) located on the long arm of chromosome 5 (Pettenati et al., 1987). The gene produces a complex array of alternatively spliced mRNAs which encode different forms of the growth factor (Ladner et al., 1987).

The most ubiquitous CSF-1 mRNA is about 4 kb in length and encodes a 554 amino acid polypeptide (Wong et al., 1987). A shorter cDNA of 1.6 kb has also been cloned and encodes a polypeptide of only 256 amino acids (Kawasaki et al., 1985). Both the long and short CSF-1 precursors contain a 32 amino acid signal peptide (residues -32 to -1) followed by 149 amino acid residues which are presumed to include the biologically active growth factor moiety. The long clone encodes an additional unique segment of 298 amino acids followed by a second hydrophobic membrane spanning region and a short carboxylterminal tail. In contrast, the short clone lacks the 298 amino acid "spacer segment" due to alternative splicing in exon 6 which results in a coding region of 131 rather than 1025 base pairs in this segment of the mRNA (Ladner et al., 1987). Both mRNAs encode integral transmembrane glycoproteins which are cotranslationally glycosylated and rapidly dimerized through disulfide bonds (Rettenmier et al., 1987; Wong et al., 1987). The 554 amino acid precursor undergoes proteolysis prior to its secretion to generate a soluble homodimer which is rapidly secreted from cells (C.W. Rettenmier and M. Manos, personal communications). However, because the shorter 256 amino acid precursor lacks the preferred site(s) of proteolysis within the spacer segment, it is transported as a membrane-spanning glycoprotein to the plasma membrane and externalized at the cell surface (Rettenmier et al., 1987). Truncation of the open reading frame of the 1.6 kb cDNA 5' to the membrane-spanning region by insertion of a termination codon after amino acid 158 leads to the synthesis and efficient secretion of a biologically active homodimer, confirming that the minimal information necessary for activity is confined to the aminoterminal domain of both precursors (Heard et al., 1987a). Together, these findings suggest that the various forms of CSF-1 may play diverse roles, not only in endocrine or paracrine stimulation of receptor bearing cells, but possibly in cell to cell interactions.

Tissue-Specific Expression of the CSF-1 Receptor

The pleiotropic actions of CSF-1 in hematopoiesis are mediated through its interaction with a single class of high affinity receptors expressed on monocytes, macrophages, and their committed precursors (Guilbert and Stanley, 1980; Byrne et al., 1981; Stanley et al., 1983; Guilbert and Stanley, 1986). During the maturation of cells of this lineage, the number of CSF-1 receptors increases, so that mature mononuclear phagocytes express the highest numbers of receptors per cell (Guilbert and Stanley, 1986). Expression of CSF-1 receptors therefore represents a particularly useful differentiation-specific marker for blood cells of the monocyte-macrophage series (Sherr et al., 1985; Rettenmier et al., 1986). Transcripts of the CSF-1 receptor have been detected in spleen, brain, liver, and bone marrow of adult animals

(Rettenmier et al., 1985; Nienhuis et al., 1985), reflecting the distribution of tissue macrophages (Byrne et al., 1981).

CSF-1 receptors are also expressed on normal placental trophoblasts and on choriocarcinoma cell lines (Muller et al., 1983a,b; Rettenmier et al, 1986). Uterine glandular epithelial cells synthesize CSF-1 under the control of ovarian steroids (Bartocci et al., 1986; Pollard et. al, 1987), and the levels of the growth factor produced in the mouse uterus increase precipitously during pregnancy and are highest at parturition (Bartocci et al., 1986). This suggests that, apart from its role in hematopoiesis, CSF-1 may have an auxiliary function in placental development. The possibility that CSF-1 functions in additional biological contexts cannot be precluded.

Biochemical Characterization of the CSF-1 Receptor

The CSF-1 receptor is encoded by the c-*fms* proto-oncogene (Sherr et al., 1985) and is one of a family of growth factor receptors with an intrinsic tyrosine-specific protein kinase activity. In man, the CSF-1 receptor is an integral transmembrane glycoprotein of 150 kd composed of an extracellular aminoterminal ligand binding domain of 485 amino acids, a single hydrophobic membrane-spanning segment, and an intracellular carboxylterminal domain of 435 amino acids which specifies the receptor kinase activity (Coussens et al., 1986; Rettenmier et al., 1986). The murine and feline receptors have somewhat higher apparent molecular weights (ca. 165-170 kd) due primarily to differences in addition and processing of asparagine(N)-linked oligosaccharide chains (Rettenmier et. al, 1985; Sherr et al., 1985). The receptor becomes immobilized in the membrane of the endoplasmic reticulum (ER) during its synthesis and is cotranslationally glycosylated to yield an immature intracellular form of the glycoprotein of about 130 kd. During transport to the cell surface, modification of mannose-rich, N-linked oligosaccharide chains to complex sugars occurs within the Golgi apparatus, so that the immature and mature forms of the receptor can be readily distinguished by their mobility on denaturing polyacrylamide gels and by the composition of their carbohydrate side chains (Rettenmier et al., 1986). In the absence of extracellular CSF-1, the cell surface receptors turn over with a half life of 3-4 hours, whereas saturating quantities of CSF-1 induce internalization of receptor-ligand complexes into endosomes and their rapid degradation in lysosomes within minutes of ligand addition (Wheeler et al, 1986a; Guilbert and Stanley, 1986; Rettenmier et al., 1987; Downing et al., 1988). In the subsequent absence of ligand, receptor "downmodulation" is followed by a refractory period of several hours during which the receptor reaccumulates at the cell surface following de novo synthesis.

In macrophage membrane preparations incubated in vitro in the presence of [$\gamma^{32}P$]-ATP, manganese, and purified CSF-1, the CSF-1 receptor is phosphorylated on tyrosine (Sherr et al., 1985; Yeung et al., 1987). Exposure of SV40-immortalized, CSF-1 dependent macrophages to CSF-1 activates the receptor kinase in vivo and leads to autophosphorylation of the receptor on tyrosine (Downing et al., 1988). Tyrosine phosphorylation of the receptor is detectable within minutes of ligand stimulation, and only the mature cell surface form of the glycoprotein is phosphorylated in this manner. By Western blotting using antibodies to phosphotyrosine, CSF-1 binding to the receptor was shown to induce phosphorylation of a series of cellular substrates (Downing et al., 1988) whose physiologic role in the mitogenic response remains unclear. The fact that CSF-1 stimulation of macrophages induces the transcription of several other cellular genes, including the proto-oncogenes c-fos and c-myc (Bravo et al., 1987; Orlofsky and Stanley, 1987) demonstrates that intracellular events initiated through the action of the receptor kinase must ultimately affect the transcription of a family of CSF-1 responsive genes.

The CSF-1 receptor is also phosphorylated on serine residues, and addition of phorbol esters to macrophages downmodulates the receptor through the action of protein kinase C (Chen et al., 1983; Wheeler et al., 1986a). The mechanisms by which ligand and phorbol esters accelerate receptor turnover appear to be distinct. Cells chronically treated with phorbol esters reexpress cell surface CSF-1 receptors after de novo synthesis, and although protein kinase C is itself downmodulated under such conditions, CSF-1 receptors can be subsequently downmodulated by ligand (C. W. Rettenmier and C. J. Sherr, unpublished observations). The ability of CSF-1 receptors to be downmodulated through the action of protein kinase C suggests that other receptors which trigger the release of diacylglycerol from phosphatidylinositol diphosphate (PIP2) might "transmodulate" CSF-1 receptor expression. Similarly, if the CSF-1 receptor were itself coupled to a phospholipase C responsible for accelerated PIP2 turnover (Jackowski et al., 1986), activation of protein kinase C might accelerate the turnover of unoccupied receptor molecules.

The v-*fms* Oncogene Product

The v-*fms* oncogene of the Susan McDonough (SM) and Hardy-Zuckerman-5 (HZ-5) strains of feline sarcoma virus (FeSV) arose by transduction of cellular c-*fms* sequences by feline leukemia virus (FeLV) (Frankel et al., 1979; Donner et al., 1982; Besmer et al., 1986). The process of transduction involved the integration of FeLV 5' to the coding sequences of the feline c-*fms* gene (Wheeler et. al, 1986b) followed by the formation of spliced virus-initiated transcripts incorporating c-*fms* sequences as a cDNA copy. Subsequent recombination with FeLV during reverse transcription would account for the formation of FeSV strains containing the viral oncogene. The v-*fms* sequences of both viruses were recombined into the open reading frame of the retroviral *gag* gene, so that the full genome length mRNA encodes a *gag*-*fms* fusion protein of 180 kd ($gP180^{gag\text{-}fms}$) (Donner et al., 1982). Because transduction of the v-*fms* gene preserved c-*fms* sequences encoding the aminoterminal signal peptide of the CSF-1 receptor (Coussens et al., 1986; Wheeler et al., 1986b), the polyprotein is cleaved by "signal peptidase" near the *gag*-*fms* junction to generate an immature intracellular form of the glycoprotein ($gp120^{v\text{-}fms}$) (Barbacid et al., 1980; Ruscetti et al., 1980). The latter is modified during intracellular transport in a manner analogous to the CSF-1 receptor to yield a mature cell surface form of greater apparent molecular weight ($gp140^{v\text{-}fms}$) (Anderson et al., 1982; 1984).

Unlike the processing of the normal CSF-1 receptor, the transport of $gp120^{v\text{-}fms}$ through the Golgi complex occurs relatively inefficiently, so that at steady state, a majority of the v-*fms*-coded molecules remain sequestered in the ER (Anderson et al., 1984). The relative block in intracellular transport is not due to the presence of *gag* gene-coded sequences, since "gag-minus" retroviral vectors which initiate translation of the v-*fms* product at the conserved c-*fms* initiation codon similarly encode $gp120^{v\text{-}fms}$ molecules that are inefficiently processed in the Golgi complex (Wheeler et al., 1986b). The latter experiments also indicated that the *gag* gene-coded moiety is not required for v-*fms*-induced cell transformation. Because the v-*fms* gene product retains the complete ligand binding domain of the CSF-1 receptor (Hampe et al., 1984; Coussens et al., 1986), cells transformed by SM-FeSV express CSF-1 binding sites, and chemical cross-linking experiments performed with the radiolabeled ligand established that CSF-1 was specifically bound to $gp140^{v\text{-}fms}$ (Sacca et al., 1986).

Expression of gp140$^{v\text{-}fms}$ at the cell surface appears to be required for cell transformation, since mutations which further inhibit ER processing (Roussel et al., 1984) or drugs which prevent processing of N-linked oligosaccharides and transport of the glycoprotein to the plasma membrane (Nichols et al., 1985) inhibit v-fms-induced transformation. Alternatively, mutations which disrupt v-fms-coded tyrosine kinase activity abrogate the gene's transforming potential (Lyman et al., 1987). Taken together, these results suggest that the physiologic targets of the v-fms-coded kinase reside at the plasma membrane or within the endosome-lysosome compartment.

Unlike the normal CSF-1 receptor, gp140$^{v\text{-}fms}$ is constitutively phosphorylated on tyrosine in transformed cells (Tamura et al., 1986; Morrison et al., 1988; Downing et al., 1988) and is autophosphorylated in membrane preparations in the absence of ligand (Saccà et al., 1986). Moreover, the virus-coded receptor does not undergo rapid down modulation in response to human or murine CSF-1 or to phorbol esters (Wheeler et al., 1986a), in spite of the fact that human CSF-1 marginally enhances the autophosphorylation of gp140$^{v\text{-}fms}$ on tyrosine (Downing et al., 1988; M.F. Roussel, J.R. Downing, R.A. Ashmun, C.W. Rettenmier, and C.J. Sherr, manuscript submitted). The persistence of the v-fms coded glycoprotein at the cell surface and its apparent activity as a constitutive kinase in the absence of its ligand appear to contribute unregulated signals for cell growth. These properties of the oncogene product must be due to genetic alterations in v-fms which distinguish it from its normal c-fms progenitor and render its product capable of transforming cells.

Structural Differences Between the v-fms and c-fms gene products

Nucleotide sequencing and conceptual translation of the feline v-fms (Hampe et al., 1984) and human c-fms (Coussens et al., 1986) genes reveals scattered amino acid differences, most of which are most likely due to interspecies divergence. In addition, the carboxylterminal 40 amino acids of the normal CSF-1 receptor were replaced by 11 unrelated amino acids in the SM-FeSV v-fms product resulting in the loss of a single tyrosine residue (tyr^{969}). A frame shift in the analogous region of the HZ-5 FeSV v-fms gene also deleted this tyrosine residue (Besmer et al., 1986).

Although the carboxylterminal truncation was suspected to activate the latent transforming potential of the c-fms gene, possibly because tyr^{969} acts as a negative regulatory site of tyrosine phosphorylation, neither the full length c-fms gene nor C-terminally truncated or mutated analogs lacking tyr^{969} were active in transformation of mouse NIH/3T3 fibroblasts (Roussel et al., 1987). Therefore, the C-terminal truncation was in itself insufficient to induce transforming activity. By contrast, a chimeric v-fms/c-fms gene product containing a reconstructed c-fms C-terminus fused to the body of the v-fms gene was attenuated in its transforming activity when compared to v-fms itself. When tyr^{969} in the chimeric gene was mutated to phenylalanine, full transforming activity was restored. Similarly, in a cotransfection assay in which the wild-type c-fms and CSF-1 genes were cointroduced into NIH/3T3 cells, low efficiency transformation was obtained by an autocrine mechanism, whereas cotransfection of a mutant c-fms(phe^{969}) gene together with the CSF-1 gene induced transformation at a significantly higher efficiency (Roussel et al., 1987). Together, these results suggested that activation of c-fms to a fully transforming gene requires two independent genetic events: (i) an alteration in the body of the v-fms gene which is presumed to phenocopy the active receptor conformation induced by ligand, and (ii) deletion of tyr^{969} which may serve as a negative regulatory site of receptor phosphorylation.

We recently found that a chimeric gene specifying an aminoterminal portion of the human c-fms ligand binding domain fused to the remainder of the feline v-fms oncogene product encodes a functional CSF-1 receptor (Roussel et al., submitted). When expressed in NIH/3T3 cells, the chimeric gene product was nontransforming but was efficiently transported to the cell surface, was phosphorylated on tyrosine only in response to human CSF-1, underwent rapid ligand-induced downmodulation, and stimulated colony formation of the cells in semisolid medium in response to the exogenous growth factor. Since the chimeric gene contained the complete v-fms-coded tyrosine kinase domain, genetic alterations in the latter region of the molecule cannot be directly responsible for v-fms mediated transformation. Nor do such alteration in the cytoplasmic tyrosine kinase domain inherently abrogate the ability of the v-fms gene product to be rapidly transported to the cell surface or to undergo ligand-induced downmodulation. The simplest hypothesis is that the putative activating mutation(s) in v-fms are localized within the distal extracellular domain, thereby inducing a receptor conformation that is enzymatically active in the absence of CSF-1.

Transforming Potential of the v-*fms* Gene

The fact that the v-*fms* gene contains a functional ligand binding domain left open the possibility that v-*fms* transformed cells by an autocrine mechanism due to SM-FeSV mediated transduction of the oncogene into CSF-1 producing fibroblasts. Several lines of evidence argue against this possibility. First, SM-FeSV efficiently transforms murine fibroblasts and epithelial cell lines. Yet, murine CSF-1 binds to heterologous receptors on normal feline macrophages only at low affinity and is unable to support the growth or survival of feline cells. Similar low affinity binding was demonstrated between murine CSF-1 and the feline v-*fms* gene product, suggesting that transformation of mouse fibroblasts by SM-FeSV does not require the ligand. In agreement with this interpretation, neutralizing antibodies to murine CSF-1 did not affect the transformed phenotype of mouse NIH/3T3 cells transformed by SM-FeSV (Sacca et al., 1986). Moreover, the efficiency of transformation of mouse NIH/3T3 cells was not augmented when the cells were cotransfected with a human CSF-1 gene (Roussel et al., 1987). This inability to potentiate the transforming efficiency of v-*fms* is particularly significant since exogenous human CSF-1 induces a two-to-three-fold increase in the level of tyrosine phosphorylation of $gp140^{v\text{-}fms}$ (Downing et al., 1988; Roussel et al., manuscript submitted). Taken together with biochemical evidence that v-*fms* acts as a constitutive kinase, the results suggest that v-*fms* transforms fibroblasts by a nonautocrine mechanism. However, the possibility that the kinase activity of the v-*fms* gene product can be further up-regulated by feline CSF-1 has not been excluded, since, to date, purified feline CSF-1 has not been available for study. Such a mechanism for potentiation of v-*fms*-coded kinase activity might explain the propensity of both the SM and HZ-5 strains of FeSV produce fibrosarcomas in their natural host, the domestic cat.

The v-*fms* gene can also transform CSF-1 dependent murine macrophages and IL-3 dependent myeloid cell lines in culture, rendering them both factor independent for growth and tumorigenic in nude mice (Wheeler et al., 1986a; 1987). High levels of expression of the v-*fms* oncogene product were required for transformation, and in both cases, production of CSF-1 by the transformed cells was not detected. To determine whether v-*fms* might also transform hematopoietic cells in an *in vivo* setting, mouse bone marrow cells infected with SM-FeSV were used to repopulate the hematopoietic compartment of lethally-irradiated mice. Spleen cells from provirus-positive primary recipients expressed the v-fms-coded tyrosine kinase activity, and virus-marked clones had a proliferative advantage over provirus-negative cells during reconstitution. When these virus-infected splenocytes were used to repopulate lethally-irradiated secondary recipients, provirus-positive clonal B cell lymphomas and erythroleukemias emerged in some of the animals (Heard et al., 1987b). The fact that some secondary recipients inoculated with the same donor cells were repopulated by normal stem cells and did not develop disease suggested that spleen cells expressing v-fms did not have an obligate proliferative advantage and were not overtly malignant at the time of secondary transfer. In this system, then, introduction of the v-fms gene served as an initiating carcinogenic event which necessitated complementation by additional genetic changes occurring in vivo. Together with the results obtained with cultured hematopoietic cells, these results demonstrate that v-fms can contribute to transformation of hematopoietic target cells, including those that do not express CSF-1 receptors.

Possible Roles of CSF-1 and Its Receptor in Tumorigenesis

Our data suggest that alternative mechanisms involving fms gene expression could contribute to tumor formation. In cells which normally express CSF-1 receptors, activating mutations in the proto-oncogene analogous to those observed in v-fms might constitutively up-regulte the receptor kinase and provide persistent signals for cell growth in the absence of ligand. Alternatively, inappropriate expression of the CSF-1 gene in hematopoietic cells expressing its receptor, or transcription of the c-fms gene in CSF-1 producing cells, might contribute to factor independence through an autocrine mechanism. Thirdly, unscheduled expression of c-fms could render cells inappropriately responsive to CSF-1, leading to their clonal expansion and increasing their probability of undergoing further genetic changes leading to malignancy.

A precedent for the latter mechanism has recently been documented in murine myeloid leukemias induced by the Friend murine leukemia virus (F-MuLV). Integration of F-MuLV 5' to CSF-1 receptor coding sequences induces high levels of c-fms expression in immature myeloid cells and leads to their clonal expansion and eventual neoplastic conversion (Gisselbrecht et al., 1987). At least some of the leukemic cells have been found to synthesize CSF-1 (Heard et al., 1984; 1986), whereas others have lost the unrearranged CSF-1 allele (Gisselbrecht et al., 1987), suggesting that additional genetic events leading to overt leukemia might, in some cases, involve the CSF-1 or c-fms genes themselves.

In man, both the c-fms and CSF-1 genes are cytogenetically linked on the long arm of human chromosome 5 at bands q33.3 and q33.1, respectively (Pettenati et al., 1987). This region of human chromosome 5 is frequently deleted in therapy-related acute myelogenous leukemias (AML) as well as in certain myelodysplastic and refractory anemia syndromes. We have used monoclonal antibodies directed to the human CSF-1 receptor to document receptor expression in cases of both adult and pediatric AML (Ashmun et al., 1987). Although we anticipated that the expression of CSF-1 receptors would be restricted to malignancies of the mononuclear phagocyte series, about 40% of human AMLs were found to express ostensibly normal CSF-1 receptors even in the absence of other morphological or immunophenotypic parameters characteristic of monocytic differentiation. Expression of c-fms in a subset of AMLs lacking monocytic characteristics might potentially be due to rearrangements 5' to CSF-1 receptor coding sequences, or alternatively, to the synthesis or post-translational modification of trans-acting factors that regulate expression of the c-fms promoter. To date, the promoter of the c-fms gene has not been identified, and a detailed molecular characterization of this region will be necessary to distinguish between these mechanisms. Such studies should indicate whether aberrant c-fms expression is a causative factor in human myeloid malignancies and will facilitate a further understanding of how the CSF-1 receptor gene is regulated during normal myelopoiesis.

Acknowledgments

The author gratefully thanks the many colleagues and collaborators who have shared in this work, and in particular, acknowledges Drs. Martine F. Roussel and Carl W. Rettenmier who have made persistent contributions to these studies throughout the last five years. This work was supported by U.S. Public Health Service Grants CA-38187 and CA-20810 from the National Cancer Institute and by the American Lebanese Syrian Associated Charities of St. Jude Children's Research Hospital.

Literature references

Anderson SJ, Furth M, Wolff L, Ruscetti SK, Sherr CJ (1982) Monoclonal antibodies to the transformation-specific glycoprotein encoded by the feline retroviral oncogene v-fms. J Virol 44:696-702

Anderson SJ, Gonda MA, Rettenmier CW, Sherr CJ (1984) Subcellular localization of glycoproteins encoded by the viral oncogene v-fms. J

Virol 51:730-741

Ashmun RA, Look AT, Furman WL, Rettenmier CW, Seremetis S, Sherr CJ (1987) Expression of the human c-fms proto-oncogene product (CSF-1 receptor) on leukemic blasts. Blood 70 (Suppl 1) 274a

Barbacid M, Lauver AV, Devare SG (1980) Biochemical and immunological characterization of polyproteins coded for by the McDonough, Gardner-Arnstein, and Snyder-Theilen strains of feline sarcoma virus. J Virol 33:196-207

Bartocci A, Pollard JW, Stanley ER (1986) Regulation of colony-stimulating factor 1 during pregnancy. J Exp Med 164:956-961

Bartocci A, Mastrogiannis DS Migliorati G, Stockert RJ, Wolkoff AW, Stanley ER (1987) Macrophages specifically regulate the concentration of their own growth factor in the circulation. Proc Natl Acad Sci USA 84:6179-6183

Besmer P, Lader E, George PC, Bergold PJ, Qiu F-H, Zuckerman EE, Hardy WD (1986) A new acute transforming feline retrovirus with fms homology specifies a C-terminally truncated version of the c-fms protein that is different from SM-feline sarcoma virus v-fms protein. J Virol 60:194-203

Bravo R, Neuberg M, Burckhardt J, Almendral J, Wallich R, Müller R (1987) Involvement of common and cell type-specific pathways in c-fos gene control: Stable induction by cAMP in macrophages. Cell 48:251-260

Byrne PV, Guilbert LJ, Stanley ER (1981) Distribution of cells bearing receptors for a colony-stimulating factor (CSF-1) in murine tissues. J Cell Biol 91:848-853

Chen BD, Lin HS, Hsu S (1983) Tumor-promoting phorbol esters inhibit the binding of colony stimulating factor (CSF-1) to murine peritoneal exudate macrophages. J Cell Physiol 116:207-212

Clark SC, Kamen R (1987) The human hematopoietic colony-stimulating factors. Science 236:1229-1237

Coussens L, Van Beveren C, Smith D, Chen E, Mitchell RL, Isacke CM, Verma IM, Ullrich A (1986) Structural alteration of viral homologue of receptor proto-oncogene fms at carboxyl terminus. Nature 320: 277-280

Csejtey J, Boosman A (1986) Purification of human macrophage colony stimulating factor (CSF-1) from medium conditioned by pancreatic carcinoma cells. Biochem Biophys Res Commun 138:238-245

Das SK, Stanley ER (1982) Structure-function studies of a colony stimulating factor (CSF-1). J Biol Chem 257:13679-13684

Das SK, Stanley ER, Guilbert LJ, Forman LW (1981) Human colony-stimulating factor (CSF-1) radioimmunoassay: resolution of three subclasses of human colony-stimulating factors. Blood 58:630-641

Donner L, Fedele LA, Garon CF, Anderson SJ, Sherr CJ (1982) McDonough feline sarcoma virus: characterization of the molecularly cloned provirus and its feline oncogene (v-fms). J Virol 41:489-500

Donohue RE, Wang EA, Stone DK, Kamen R, Wong GC, Sehgal PK, Nathan DG, Clark SC (1986) Stimulation of haematopoiesis in primates by continuous infusion of recombinant human GM-CSF. Nature 321:872-875

Downing JR, Rettenmier CW, Sherr CJ Ligand-induced tyrosine kinase activity of the CSF-1 receptor in a murine macrophage cell line. Mol Cell Biol (in press)

Eschbach JW, Egrie JC, Downing MR, Browne JK, Adamson JW (1987) Correction of the anemia of end-stage renal disease with recombinant human erythropoietin. New Engl J Med 316:73-78

Frankel AE, Gilbert JH, Porzig KJ, Scolnick EM, Aaronson SA (1979) Nature and distribution of feline sarcoma virus nucleotide sequences. J Virol 30:821-827

Gisselbrecht S, Fichelson S, Sola B, Bordereaux D, Hampe A, André C, Galibert F, Tambourin P (1987) Frequent c-fms activation by proviral insertion in mouse myeloblastic leukaemias. Nature 329:259-261
Groopman JE, Mitsuyasu RT, DeLeo MJ, Oette DH, Golde DW (1987) Effect of recombinant human granulocyte-macrophage colony stimulating factor on myelopoiesis in the acquired immunodeficiency syndrome. New Engl J Med 317:593-598
Guilbert LJ, Stanley ER (1980) Specific interaction of murine colony-stimulating factor with mononuclear phagocytic cells. J Cell Biol 85:153-159
Guilbert LJ, Stanley ER (1986) The interaction of ^{125}I-colony-stimulating factor-1 with bone marrow-derived macrophages. J Biol Chem 261:4024-4032
Hampe A, Gobet M, Sherr CJ, Galibert F (1984) The nucleotide sequence of the feline retroviral oncogene v-fms shows unexpected homology with oncogenes encoding tyrosine-specific protein kinases. Proc Natl Acad Sci USA 81:85-89
Heard JM, Fichelson S, Sola B, Martial MA, Varet B, Levy JP (1984) Multistep virus-induced leukemogenesis in vitro: Description of a model specifying three steps within the myeloblastic malignant process. Mol Cell Biol 4:216-220
Heard JM, Roussel MF, Rettenmier CW, Sherr CJ (1987a) Synthesis, post-translational processing, and autocrine transforming activity of a carboxylterminal truncated form of colony stimulating factor-1. Oncogene Research 1:423-440
Heard JM, Roussel MF, Rettenmier CW, Sherr CJ (1987b) Multilineage hematopoietic disorders induced by transplantation of bone marrow cells expressing the v-fms oncogene. Cell 51:663-673
Heard JM, Sola B, Martial MA, Fichelson S, Gisselbrecht S (1986) Long-term culture of bone marrow-derived preleukemic cells from F-MuLV-infected mice. Blood 68:193-199
Jackowski S, Rettenmier CW, Sherr CJ, Rock CO (1986) A guanine nucleotide-dependent phosphatidylinositol-4,5-diphosphate phospholipase C in cells transformed by the v-fms and v-fes oncogenes. J Biol Chem 261:4978-4985
Karbassi A, Becker JM, Foster JS, Moore RN (1987) Enhanced killing of Candida albicans by murine macrophages treated with macrophage colony stimulating factor: Evidence for augmented expression of mannose receptors. J Immunol 139:417-421
Kawasaki ES, Ladner MB, Wang AM, Van Arsdell J, Warren MK, Coyne MY, Schweickart VL, Lee MT, Wilson KJ, Boosman A, Stanley ER, Ralph P, Mark DF (1985) Molecular cloning of a complementary DNA encoding human macrophage-specific colony stimulating factor (CSF-1). Science 230:291-296
Ladner MB, Martin GA, Noble JA, Nikoloff DM, Tal R, Kawasaki ES, White TJ (1987) Human CSF-1: Gene structure and alternative splicing of mRNA precursors. EMBO J 6:2693-2698
Lee MT, Warren MK (1987) CSF-1-induced resistance to viral infection in murine macrophages. J Immunol 138:3019-3022
Lyberg T, Stanley ER, Prydz H (1987) Colony stimulating factor-1 induces thromboplastin activity in murine macrophages and human monocytes. J Cell Physiol 132:367-370
Lyman SD, Rohrschneider LR (1987) Analysis of functional domains of the v-fms-encoded protein of Susan McDonough strain feline sarcoma virus by linker insertion mutagenesis. Mol Cell Biol 7:3287-3296
Metcalf D (1985) The granulocyte-macrophage colony stimulating factors. Science 229:16-22

Morrison DK, Browning PJ, White MF, Roberts TM (1988) Tyrosine phosphorylations _in vivo_ associated with v-_fms_ transformation. Mol Cell Biol 8:176-185

Müller R, Slamon DJ, Adamson ED, Tremblay JM, Müller D, Cline MJ, Verma IM (1983a) Transcription of c-_onc_ genes c-_ras_^{ki} and c-_fms_ during mouse development. Mol Cell Biol 3:1062-1069

Müller R, Tremblay JM, Adamson ED, Verma IM (1983b) Tissue and cell type specific expression of two human c-_onc_ genes. Nature 304:454-456

Nichols EJ, Manger R, Hakomori S, Herscovics A, Rohrschneider LR (1985) Transformation by the v-_fms_ oncogene product: Role of glycosylational processing and cell surface expression. Mol Cell Biol 5: 3467-3475

Nienhuis AW, Bunn HF, Turner PH, Gopal TV, Nash WG, O'Brien SJ, Sherr CJ (1985) Expression of the human c-_fms_ proto-oncogene in hematopoietic cells and its deletion in the 5q$^-$ syndrome. Cell 42: 421-428

Orlofsky A, Stanley ER (1987) CSF-1-induced gene expression in macrophages: Dissociation from the mitogenic response. EMBO J 6:2947-2952

Pettenati MJ, Le Beau MM, Lemons RS, Shima EA, Kawasaki ES, Larson RA, Sherr CJ, Diaz MO, Rowley JD (1987) Assignment of _CSF-1_ to 5q33.1: Evidence for clustering of genes regulating hematopoiesis and for their involvement in the deletion of the long arm of chromosome 5 in myeloid disorders. Proc Natl Acad Sci USA 84:2970-2974

Pollard JW, Bartocci A, Arceci R, Orlofsky A, Ladner MB, Stanley ER (1987) Apparent role of the macrophage growth factor, CSF-1, in placental development. Nature 330:484-486

Rajavashisth TB, Eng R, Shadduck RK, Waheed A, Ben-Avram CM, Shively JE, Lusis AJ (1987) Cloning and tissue-specific expression of mouse macrophage colony-stimulating factor mRNA. Proc Natl Acad Sci USA 84:1157-1161

Ralph P, Warren MK, Ladner MD, Kawasaki ES, Boosman A, White TJ (1986) Molecular and biological properties of human macrophage growth factor, CSF-1. Cold Spring Harbor Symp Quant Biol 51:679-683

Ralph P, Warren MK, Lee MT, Csejtey J, Weaver JF, Broxmeyer HE, Williams DE, Stanley ER, Kawasaki ES (1986a) Inducible production of human macrophage growth factor, CSF-1. Blood 68:633-639

Rettenmier CW, Chen JH, Roussel MF, Sherr CJ (1985) The product of the c-_fms_ proto-oncogene: a glycoprotein with associated tyrosine kinase activity. Science 228:320-322

Rettenmier CW, Roussel MF, Ashmun RA, Ralph P, Price K, Sherr CJ (1987) Synthesis of membrane-bound colony-stimulating factor-1 (CSF-1) and downmodulation of CSF-1 receptors in NIH 3T3 cells transformed by cotransfection of the human CSF-1 and c-_fms_ (CSF-1 receptor) genes. Mol Cell Biol 7:2378-2387

Rettenmier CW, Sacca R, Furman WL, Roussel MF, Holt JT, Nienhuis AW, Stanley ER, Sherr CJ (1986) Expression of the human c-_fms_ proto-oncogene product (colony-stimulating factor-1 receptor) on peripheral blood mononuclear cells and choriocarcinoma cell lines. J Clin Invest 77:1740-1746

Roussel MF, Dull TJ, Rettenmier CW, Ralph P, Ullrich A, Sherr CJ (1987) Transforming potential of the c-_fms_ proto-oncogene (CSF-1 receptor). Nature 325:549-552

Roussel MF, Rettenmier CW, Look AT, Sherr CJ (1984) Cell surface expression of v-_fms_-coded glycoproteins is required for transformation. Mol Cell Biol 4:1999-2009

Ruscetti SK, Turek LP, Sherr CJ (1980) Three independent isolates of feline sarcoma virus code for three distinct _gag_-X polyproteins. J Virol 35:259-264

Sacca R, Stanley ER, Sherr CJ, Rettenmier CW (1986) Specific binding of the mononuclear phagocyte colony-stimulating factor CSF-1 to the product of the v-fms oncogene. Proc Natl Acad Sci USA 83:3333-3335

Sherr CJ, Rettenmier CW, Sacca R, Roussel MF, Look AT, Stanley ER (1985) The c-fms proto-oncogene product is related to the receptor for the mononuclear phagocyte growth factor, CSF-1. Cell 41:665-676

Sieff CA (1987) Hematopoietic growth factors. J Clin Invest 79: 1549-1557

Stanley ER, Heard PM (1977) Factors regulating macrophage production and growth. Purification and some properties of the colony stimulating factor from medium conditioned by mouse L cells. J Biol Chem 252: 4305-4312

Stanley ER, Guilbert LJ, Tushinski RJ, and Bartelmez SH (1983) CSF-1 – A mononuclear phagocyte lineage-specific hemopoietic growth factor. J Cell Biochem 21:151-159

Tamura T, Simon E, Niemann H, Snoek GT, Bauer H (1986) Gp140$^{v\text{-}fms}$ molecules expressed at the surface of cells transformed by the McDonough strain of feline sarcoma virus are phosphorylated in tyrosine and serine. Mol Cell Biol 6:4745-4748

Warren MK, Ralph P (1986) Macrophage growth factor CSF-1 stimulates human monocyte production of interferon, tumor necrosis factor, and colony stimulating activity. J Immunol 137:2281-2285

Welte K, Bonilla MA, Gillio AP, Boone TC, Potter GK, Gabrilove JL, Moore MAS, O'Reilly RJ, Sousa LM (1987) Recombinant human granulocyte colony-stimulating factor: Effects on hematopoiesis in normal and cyclophosphamide-treated primates. J Exp Med 165:941-948

Wheeler EF, Askew D, May S, Ihle JN, Sherr CJ (1987) The v-fms oncogene induces factor-independent growth and transformation of the interleukin-3-dependent myeloid cell line FDC-P1. Mol Cell Biol 7:1673-1680

Wheeler EF, Rettenmier CW, Look AT, Sherr CJ (1986a) The v-fms oncogene induces factor independence and tumorigenicity in CSF-1 dependent macrophage cell line. Nature 324:377-380

Wheeler EF, Roussel MF, Hampe A, Walker MH, Fried VA, Look AT, Rettenmier CW, Sherr CJ (1986b) The amino-terminal domain of the v-fms oncogene product includes a functional signal peptide that directs synthesis of a transforming glycoprotein in the absence of feline leukemia virus gag sequences. J Virol 59:224-233

Wong GG, Temple PA, Leary AC, Witek-Giannotti JS, Yang Y-C, Ciarletta AB, Chung M, Murtha P, Kriz R, Kaufman RJ, Ferenz CR, Sibley BS, Turner KJ, Hewick RM, Clark SC, Yanai N, Yokota H, Yamada M, Saito M, Motoyoshi K, Takaku F (1987) Human CSF-1: Molecular cloning and expression of 4 kb cDNA encoding the human urinary protein. Science 235:1504-1508

Yeung YG, Jubinsky PT, Sengupta A, Yeung DCY, Stanley ER (1987) Purification of the colony-stimulating factor 1 receptor and demonstration of its tyrosine kinase activity. Proc Natl Acad Sci USA 84: 1268-1271 [illegible] Stanley EM, Smart [illegible], Bachrach [illegible] W (1980) [illegible] the [illegible] [illegible] cycle including [illegible] and [illegible] of the [illegible] [illegible]. [illegible] Data Nucl Sci [illegible]

Stein [illegible] Rettenmier [illegible], [illegible] K, Roussel [illegible] (1982) The c-fms proto-oncogene product is related to the receptor for the mononuclear phagocyte growth factor, CSF-1. Cell [illegible]

[illegible] (1977) [illegible] growth factors. [illegible]

Stanley ER, [illegible] (1977) [illegible] phage [illegible] on the colony [illegible] [illegible] [illegible] conditioned [illegible] [illegible] [illegible]

Stanley ER, Guilbert LJ, Tushinski RJ, Bartelmez SH (1983) CSF-1 — A mononuclear phagocyte lineage-specific hemopoietic growth factor. J Cell Biochem [illegible]

[illegible] Sefton BM, Hunter T, Beemon K, Eckhart W (1980) Evidence that the phosphorylation of tyrosine is essential for cellular transformation by Rous sarcoma virus. Cell [illegible]

Stephen [illegible] (1980) [illegible]

Signal Transduction by Growth Factor Receptors

W.H. Moolenaar
The Netherlands Cancer Institute
Plesmanlaan 121
1066 CX Amsterdam
The Netherlands

Introduction

The proliferation of cells in vivo and in culture is tightly regulated by polypeptide growth factors. Like all polypeptide hormones, growth factors initiate their action by binding to specific, high affinity receptor molecules on the cell surface. Interest in growth factors and their receptors has been dramatically intensified by the discovery that at least some of these molecules show a striking structural homology with certain viral oncogene products. For example, the sis oncogene encodes a platelet-derived growth factor (PDGF) molecule, while the erb-B oncogene product is a truncated form of the receptor for epidermal growth factor (EGF) (Waterfield et al, 1983; Schlessinger 1986). Inappropriate expression of the cellular counterparts of viral oncogenes is thought to be responsible for the initiation and maintenance of malignant growth. Therefore, the study of the mode of action of growth factors will undoubtedly yield important insight into the mechanisms underlying carcinogenesis.

Signal pathways in growth factor action

Understanding the mode of action of growth factors requires the identification of intracellular signals that are essential for the stimulation of DNA synthesis and cell division. Following growth factor binding, the activated receptor mediated a cascade of rapid biochemical and physiological changes in the cell, which ultimately (after 10-20 h) lead to enhanced DNA synthesis. As a rule, the growth factor has to be present

NATO ASI Series, Vol. H26
Cell to Cell Signals in Mammalian Development
Edited by S. W. de Laat et al.

throughout the entire "pre-replicative" G0/G1 phase (usually 8-10 h) for commitment to DNA synthesis to occur. It appears that there are at least three potential signal pathways for mediating growth factor action: (a) activation of a protein tyrosine kinase activity that is intrinsic to the receptors for EGF and PDGF (b) the phospholipase C-mediated breakdown of inositol phospholipids generating various signal molecules, and (c) increases in cytoplasmic free Ca^{2+} and pH (pHi).

There is increasing evidence that in addition to the "classical" mitogens such as EGF and PDGF, certain Ca^{2+}-mobilizing neurotransmitters can also function as growth factors. For example, the peptides bombesin, substance P and vasopressin are mitogenic for quiescent fibroblasts (Zachary et al. 1987), while our own results indicate that bradykinin and histamine, acting through H-1 receptors, stimulate the growth of certain human carcinoma cells (Tilly et al. 1987, and unpublished results).

Inositol lipid hydrolysis

There is now a general agreement that some, but not all, growth factors receptors evoke the phospholipase C-mediated breakdown of inositol phospholipids resulting in the formation of diacylglycerol which binds to and directly activates protein kinase C (Nishizuka, 1984) and inositol-1,4,5-trisphosphate (IP_3(1,4,5)) which triggers the release of Ca^{2+} from internal stores (Berridge & Irvine, 1984), as well as various other inositol phosphates (Fig. 1).

We have analyzed and compared the actions of EGF and bradykinin or histamine on both inositol phosphate formation and calcium signalling in responsive human carcinoma cells (Tilly et al., 1987). It appears that EGF is a rather weak inducer of phospholipase C activity. In A431 cells, which overexpress the EGF receptor, EGF evokes a rather small and variable release of intracellularly stored Ca^{2+}, while accumulation of IP_3(1,4,5) is barely detectable. In contrast, EGF-induced increases in other inositol phosphates such as IP_1 and IP_2 are readily detectable.

Bradykinin and histamine, on the other hand, are among the most potent activators of phospholipase C, resulting in large Ca^{2+} signals and prominent increases in inositol phosphate levels.

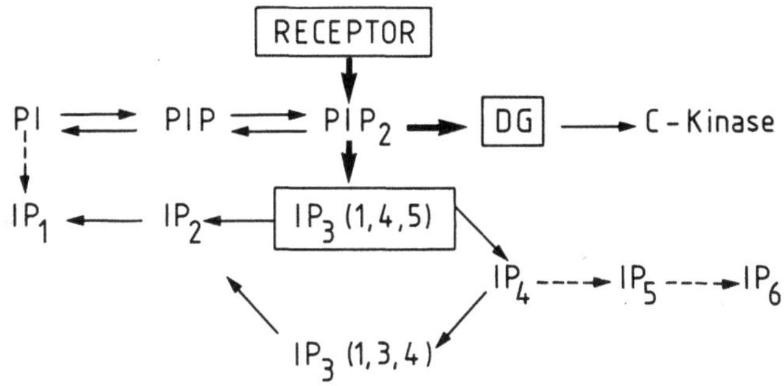

Fig.1 Metabolic pathways responsible for the formation of inositol phosphates from the breakdown of phosphatidylinositol-4,5-bisphosphate (PIP_2). See text for details.

Bradykinin is mitogenic for skin fibroblasts (see review by Zachary et. al., 1987) while our own results (unpublished) indicate that bradykinin can also function as a mitogen for human A431 carcinoma cells. Bradykinin stimulation of A431 cells rapidly activates phospholipase C, resulting in the formation of inositol phosphates and a transient rise in $[Ca^{2+}]i$ (Tilly et al, 1987; Hepler et al, 1987). Almost immediately after hormone addition the levels of IP_3 (1,4,5) starts to increase, reaching a peak value of -10 times the basal level within 10-15 sec which coincides with the $[Ca^{2+}]i$ peak. The bradykinin-induced formation of IP_3 (1,4,5) is a very transient phenomenon, and the results support the view that the (1,4,5) isomer is rapidly metabolized either to IP_2 through phosphatase action of to IP_3 (1,3,4) via phosphorylation/dephosphorylation with IP_4 as intermediate. The accumulation of inositol phosphates in response to bradykinin levels off in less than 2 min. This apparent desensitization of the response is not observed with EGF as a stimulus (Tilly et al, manuscripts submitted). In general, the bradykinin-dependent pattern of stimulation is significantly different from that observed in EGF-stimulated A431 cells, suggesting that there are separate mechanisms of inositol lipid hydrolysis involved.

A hitherto unrecognized growth factor is histamine, which stimulates the growth of human HeLa cells by activating the Ca^{2+}-mobilizing H1 receptor (Tilly et al, manuscript in preparation). Histamine is formed by decarboxylation of the amino acid histidine and is an active mitogenic amine involved in a number of biological processes including neurotransmission, inflammation and allergic reactions. Recently a number of studies have suggested a possible role of histamine in tumor development and cell proliferation (for review see Bartholeyns and Fozard, 1985). In a recent series of experiments, we have examined the action of the H1 receptor on inositol phosphate metabolism and growth stimulation in human HeLa cells. Histamine induces a dose-dependent accumulation of inositol phosphates that parallels its mitogenic activity. Both the formation of inositol phosphates and the stimulation of cell proliferation by histamine are blocked by the H1 receptor antagonist pyrilamine (Figure 2). Histamine-induced inositol phosphate formation in HeLa cells last for at least several hours. This would imply that the H1 receptor also mediates a prolonged production of diacylglycerol and concomitant activation of protein kinase C. Since a number of other mitogens, including thrombin and bombesin, similarly evoke phosphoinositide turnover, it seems that any agonist capable of stimulating phosphoinositide hydrolysis may ultimately function as a growth factor.

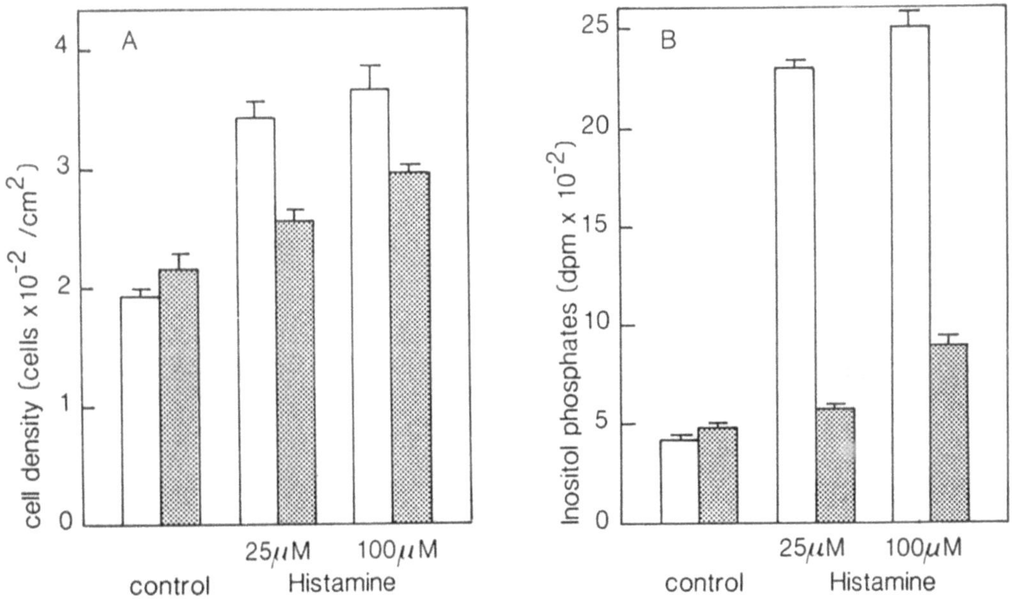

Fig.2 Histamine induced cell proliferation (A) and inositol phosphate formation (B) in HeLa cells in the presence (closed bars) or absence (open bars) of 250 nM pyrilamine. Inositol phosphates were determined as in Tilly et al. (1987). In A histamine and pyrilamine were present for 48 hr prior to cell counting.

Tyrosine kinase activation

To explore the importance of the receptor tyrosine kinase in the action of EGF, we have used transfected NIH-3T3 cells expressing either the normal human EGF-receptor or a receptor mutated at Lys 721, a key residue in the presumed ATP-binding region (collaboration with dr J. Schlessinger). The wild type receptor responds to EGF by causing inositol phosphate formation (Fig. 3), rises in Ca^{2+} and pHi and DNA synthesis. In marked contrast, the kinase-deficient mutant fails to evoke any of these responses (Moolenaar et al., 1988; Fig. 3).

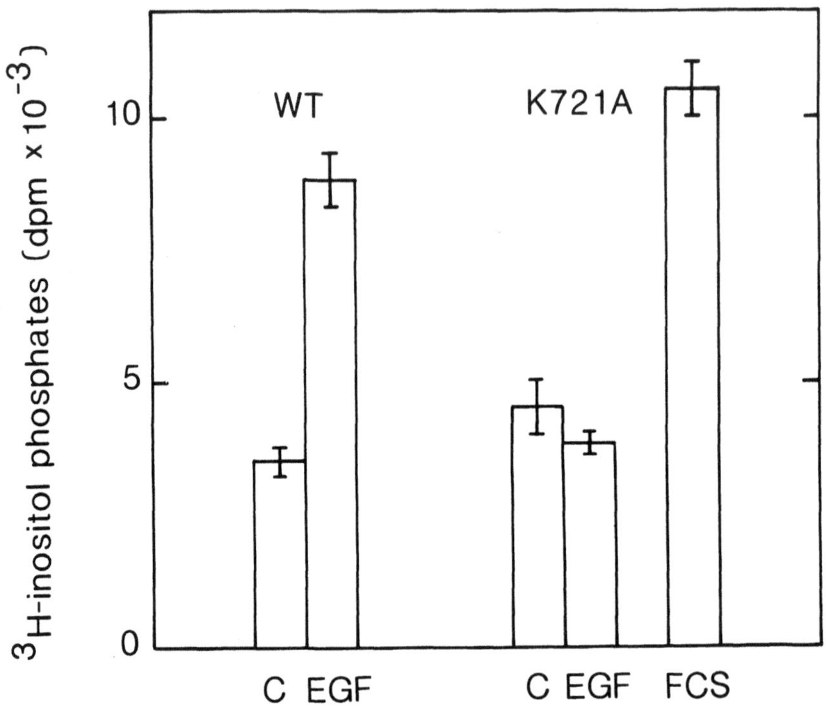

Fig. 3 Inositol phosphate formation induced by EGF interacting with the wild type (WT) or kinase-defective (K721A) human EGF receptor expressed in 3T3 cells. FCS, fetal calf serum (7.5% v/v). For further details see Moolenaar et al. (1988).

It thus appears that activation of the receptor tyrosine kinase is a crucial signal that initiates the multiple postreceptor effects of EGF that lead to DNA synthesis. In other words, EGF-induced phosphorylation of specific cellular substrates on tyrosine residues is a critical step in the pleiotropic response to EGF. The inability of the kinase-deficient receptor to stimulate inositol lipid hydrolysis (Fig. 3) is particularly intriguing, since it suggests that one or more of the key regulatory proteins in the phosphoinositide signalling cascade (e.g. G-proteins, phospholipase-C) may serve as a substrate for the EGF-R kinase. Such a mechanism contrasts to the accepted mode of action of Ca^{2+}-mobilizing hormones, such as histamine and bradykinin, whose receptors are linked to G-protein intermediates, without any evidence for tyrosine phosphorylation being required for the activation process.

Activation of Na^+/H^+ exchange

Most, if not all, mitogens activate the otherwise quiescent Na^+/H^+ exchanger in the plasma membrane of their target cells (Moolenaar, 1986). This Na^+/H^+ exchanger, which is amiloride-sensitive, is normally involved in the close regulation of pHi, while under appropriate conditions growth factors activate the Na^+/H^+ exchanger to raise steady-state pHi by 0.2 - 0.3 unit. This rise in pHi is thought to be permissive for DNA synthesis (Chambard & Pouyssequr, 1986; Moolenaar, 1986).

How does receptor occupancy lead to activation of Na^+/H^+ exchange? The underlying kinetic mechanism is now fairly well understood. By comparing the pHi-dependence of Na^+/H exchange in quiescent and stimulated cells, it turns out that growth factors and activators of kinase C, such as phorbol esters, act by increasing the sensitivity of the Na^+/H^+ exchanger for cytoplasmic H^+. This alkaline shift in pHi sensitivity of 0.2 - 0.3 unit may reflect an increase in the apparent affinity of the regulatory or "modifier" site for H^+. One could imagine that some ionizable group at the regulatory site acquires a greater pKa because its immediate environment becomes more negatively charged, for example by phosphorylation either via kinase C or through the intrinsic receptor tyrosine kinase.

The involvement of the Na^+/H^+ exchanger in such a carefully regulated process as cell growth raises the intriguing possibility that the exchanger in autonomously growing tumour cells may be subject to uncontrolled activation. Indeed, recent work by Bierman et al. (1987) on pHi regulation in embryonal carcinoma (EC) cells shows that the Na^+/H^+ exchanger in mouse P19 EC cells is constitutively activated and fails to respond to any extracellular stimulus. These pluripotent cells have no requirement for exogenous growth factors and exhibit a highly transformed phenotype. Upon differentiation, however, the cells loose their transformed phenotype and they become dependent on the presence of external growth factors. This differentiation process, induced by either retinoic acid or DMSO, is accompanied by "deactivation" of the Na^+/H^+ exchanger and a resultant drop in steady-state pHi of up to 0.5 unit (Bierman et al., manuscript in preparation). Kinetic analysis indicates that the relatively high resting pHi of undifferentiated EC cells is attributable to an alkaline shift in the pHi sensitivity of the Na^+/H^+ exchange rate as compared to that in the differentiated cells. It thus seems as if signal pathways normally utilized by growth factor receptors to activate the Na^+/H^+ exchanger, are constitutively operative in autonomously growing, undifferentiated EC cells. The biochemical nature of this pathway remains to be elucidated.

Concluding remarks

In recent years much has been learned about the various molecular events that are induced by growth factor-receptor interaction. Although uncertainty still exists about the biological significance of many of the receptor-linked "early events", it is obvious that site-directed mutagenesis of cloned receptor and G-protein cDNA's followed by transfection into appropriate target cells will be of great help in analyzing the role of the various signal pathways in growth control.

Another challenge for future research is to examine how the signals generated by growth factor receptors are modified or overridden by the action of transforming oncogene products, some of which are known to be tyrosine protein kinases or G-like proteins.

Acknowledgements

Research related to this paper was carried out by A.J. Bierman, L.H.K. Defize and B.C. Tilly at the Hubrecht Laboratory in Utrecht, the Netherlands. Mrs Gea de Jong-Meijerink is acknowledged for preparing the manuscript.

References

Berride MJ, Irvine RF (1984) Inositol trisphosphate, a novel second messenger in cellular signal transduction. Nature 312:315-321

Bartholeyns J, Fozard JR (1985) Role of histamine in tumor development. Trends Pharmacol Sc 6:123-125

Bierman AJ, Tertoolen LGJ, de Laat SW, Moolenaar, WH (1987) The Na^{+}/H^{+} exchanger is constitatively activated in P19 embryonal carcinoma cells. J Biol Chem 262:9621-9628

Hepler JR, Nakahata N, DiGuiseppi J, Herman B, Earp JS, Harden TK (1987) Inositol phosphate production by EGF in A431 cells. J Biol Chem 262:2951-2956

Moolenaar WH (1986) Effects of growth factors on intracellular pH regulation. Ann Rev Physiol 48:363-376

Moolenaar WH, Bierman AJ, Tilly BC, Verlaan I, Honegger AM, Ullrich A, Schlessinger J (1988) A point mutation at the ATP-binding site of the EGF-receptor abolishes signal transduction. EMBO J 7: 707-710

Nishizuka Y (1984) The role of protein kinase C in cell surface signal transduction and tumour promotion. Nature 304:645-648

Schlessinger J (1986) Allosteric regulation of the EGF receptor kinase. J Cell Biol 103:2067-2072

Tilly BC, Van Paridon PA, Verlaan I, Wirtz KWA, de Laat SW, Moolenaar WH (1987) Inositol phosphate formation in bradykinin-stimulated A431 cells. Biochem J 244:129-135

Waterfield MD, Scrace GT, Whittle N, Johnson A, Wasteson A, Westermark B, Heldin CH, Huang JS, Deuel TF (1983) PDGF is structurally related to the putative transforming protein of simian sarcoma virus. Nature 304:35-39

Zachary I, Woll PJ, Rozengurt E (1987) A role for neuropeptides in the control of cell proliferation. Dev Biol. 124:295-308

Growth and Differentiation Factors of Embryonic Stem Cells.

John K. Heath, Austin G. Smith, Alan J. Wills and Dylan R. Edwards
Dept. of Biochemistry, University of Oxford
South Parks Rd. Oxford OX1 3QU U.K.

Introduction

Morphogenetic processes in vertebrate development require coordinated control of both cell proliferation and differentiation. It is becoming apparent that soluble polypetide regulatory factors, and their associated cellular response systems, may play a key role in these processes. This paper reviews the current status of two polypeptide regulatory factors implicated in the control of proliferation and differentiation in early mammalian development.

Attempts to directly analyse the expression and action of growth and differentiation factors in mammalian development are hampered by two severe technical problems. Firstly, the embryo is very small, comprising a few thousand cells at the stage crucial developmental events occur. This means it is very difficult, given current technology, to biochemically purify regulatory peptides, which may be present at very low concentrations, directly from embryos themselves. Secondly the embryo is composed of multiple cell types in close association located in an inaccessible milieu in utero. Experimental dissection of possible intercellular signalling processes in normal development is consequently difficult.

It is for these reasons that alternative tactics for the identification and analysis of regulatory factors involved in early developmental processes have been developed.

The first approach relies on the existence of cultured cell lines derived from pluripotential stem cells of the early mouse embryo. These can be created either by induction of teratocarcinoma tumours and subsequent recovery of malignant embryonal carcinoma (EC) cells into culture, or more recently, by isolation in vitro of embryonic stem (ES) cell lines directly from embryos without an intervening tumourigenic phase.

It is important to appreciate that extrapolation of results obtained from the analysis of EC and ES cell lines to the normal embryo depends on the extent to which these cells can be shown to respond to normal developmental cues and signals. This can be tested by examining the developmental potential of the cells when re-incorporated into the normal embryo. From these experiments it is clear that EC cells are, in general, defective in their ability to

NATO ASI Series, Vol. H26
Cell to Cell Signals in Mammalian Development
Edited by S. W. de Laat et al.

differentiate correctly in vivo whereas ES cell lines can contribute to all tissues of the adult animal including the germ cell lineage. Thus EC cells, because of their embryonic origins, may be a valuable source of potential embryonic regulatory factors but are a less reliable guide to the possible responses of their normal embryonic counterparts. ES cells, on the other hand, appear at present a useful system for the identification of novel embryonic bioregulatory factors, since they are demonstrably capable of responding correctly to signals that exist in normal embryos.

A second tactic that has been recently developed in the search for embryonic bioregulatory factors involves taking advantage of vertebrate embryo species with more accesible embryonic stages, especially amphibians, where the biological effects of exogenously applied growth factors can be tested directly. One justification for this approach lies in the thought that evolutionarily conserved developmental control mechanisms may involve evolutionarily conserved molecular participants. It is this conjecture that underlies the analysis of 'homeobox' containing genes in vertebrates discussed elsewhere in this volume. The second justification is that the approach works: as dicussed below, mammalian growth and differentiation factors do indeed function to control developmental events in amphibian embryogenesis.

Embryonal Carcinoma derived growth factors

Certain EC cell lines, such as PC13 EC, are blocked in their capacity to undergo differentiation in vitro or in vivo and can consequently be propagated as a homogeneous EC cell population in culture. Differentiation may be induced in these cells by the action of exogenous inducers such as retinoic acid (RA). In the case of PC13, RA-induced differentiation generates an apparently homogenous cell type PC13 END. The PC13 EC cell system therefore represents a simple two component differentiation system with an RA mediated switch between the EC and END cell state. The PC13 EC cell system has been extensively studied with respect to changes in the control of cell proliferation accompanying differentiation in vitro (reviewed Heath and Rees 1985). One of the striking features of this analysis has been the discovery that the expression of specific growth factors, and their associated response systems, is subjected to developmental regulation. Thus EC differentiation into END cells is accompanied by the expression of insulin-like growth factors (IGFs) which have both paracrine and autocrine growth effects on END cells and their EC progenitors (Heath and Shi 1986). This identification of IGF expression by an early embryo derived cell type led to a closer examination of IGF expression in embryos themselves, and it was observed that IGF-like molecules were expressed by extra-embryonic mesoderm and amnion in early postimplantation development. Subsequent in situ hybridisation studies using oligonucleotide probes have both confirmed and extended these observations, revealing a specific tissue-limited expression of IGFs in the early postimplantation rodent embryo (Beck et. al.1987).

The ability to propagate PC13 EC cells as a homogeneous population in vitro, and the development of growth factor free defined culture conditions (reviewed Heath 1987) permitted the identification of a potent mitogenic activity secreted by the EC cells, whose expression ceased as differentiation ensued (Isacke and Deller 1984). The major mitogenic activity in PC13 EC cell conditioned medium was subsequently isolated and shown to be a single chain 17,500 molecule termed ECDGF (embryonal carcinoma-derived growth factor). At the time these early purification studies of ECDGF were being performed, definitive purification of acidic and basic fibroblast growth factor (FGF) from bovine tissue sources was achieved (Esch et. al 1985). A characteristic feature of the FGF family of growth factors is their strong affinity for heparin-derived oligo-saccharides. Indeed, affinity chromatography on heparin conjugated supports proved to be a key step in their eventual isolation. It was of some interest therefore that ECDGF was also found to bind strongly to heparin and heparin affinity chromatography could be used effectively in the purification of ECDGF (Heath 1987, Van Veggel et. al. 1987). This similarity bewtween FGFs and ECDGF in heparin affinity and apparent molecular weight naturally led to a more detailed appraisal of their structural and functional similarities. The results of these investigations are outlined in table 1.

The hallmark of acid and basic FGF is their mitogenic action on endothelial cells. ECDGF, although sharing at least some target cells with the FGFs, has not been found to be mitogenic for bovine, human or rodent endothelial cells of arterial, venous or capillary origin. This suggests that either ECDGF operates through a distinct cell surface receptor system from the FGFs, or at least two receptor systems exist with differing affinities for ECDGF and the FGFs, only one of which is expressed on endothelial cells. ECDGF is a secreted factor, indeed its presence in conditioned medium provided the means for its identification. Both acid and basic fibroblast growth factor seem to be closely cell-associated in cultured cell lines and are not readily detected in conditioned medium. This may be due to the apparent absence of a secretory signal sequence in the primary acid or basic FGF translation product (Abraham et. al. 1987).

Detailed analysis of the chromatographic behaviour of ECDGF and acidic and basic FGF also reveals subtle differences between the three factors; thus ECDGF elutes at earlier salt concentrations on heparin-sepharose affinity columns than basic FGF, but elutes at higher salt concentrations than either acid or basic FGF on mono-S cation-exchange matrices. ECDGF is also significantly more retarded when analysed by reverse phase chromatography than either acid or basic FGF. It is accordingly possible to separate ECDGF from either acidic or basic FGF by a variety of different methods. Furthermore, the amino acid composition of ECDGF (unpublished observations) is significantly different from either acidic or basic FGF. Finally an extensive series of antibodies generated against peptides derived from either the acidic or basic FGF amino acid sequence have been tested for cross-reaction with ECDGF by western blotting. With the exception of antibodies directed against residues 16-24 of basic FGF, no cross reaction has been observed.

All these findings therefore indicate that ECDGF is related to, but distinct from, the two 'prototype' heparin binding growth factors.

Definitive evidence regarding the exact relationship between ECDGF and other heparin binding growth factors depends upon definition of the primary amino acid sequence of ECDGF. This has unfortunately proved difficult to obtain, since ECDGF is chemically modifed at the N-terminus, rendering it inaccessible to Edman degradation, and is not currently available in sufficent quantities for the ready preparation of peptides.

It is worth remarking, however that at least two new members of the FGF 'family' have been recently identified. The Hst oncogene was isolated by standard NIH/3T3 tranfection procedures from a human stomach tumour (Taira et. al.1987). An apparently identical gene Ks has also been recently isolated from NIH/3T3 cells transfected with DNA taken from a Karposi's sarcoma tumour (Delli Bovi et. al.1987). The Hst/Ks gene encodes a functional mitogen with considerable sequence homology to the basic FGF prototype gene. A second oncogene, initially identified in mammary tumours by retrovirus insertion is Int-2. DNA sequence analysis of the Int-2 gene also reveals significant sequence homology with the basic FGF prototype gene (Dickson and Peters 1987), although this has yet to be shown to encode a functional mitogen. A significant feature of both Ks/Hst and Int-2 is the presence of a secretory signal sequence in the predicted primary translation product. Furthermore, the product of the Ks/Hst gene can be shown to be secreted into the medium of cells when the gene is expressed under the control of a heterologous promotor (Delli-Bovi et. al.1987). There are good precidents for supposing, therefore, that the FGF family of growth factors is larger than hitherto recognised, and at least some of these growth factors can be secreted into the culture medium, rather than remaining cell associated. The relationship between ECDGF and these novel FGF-like growth factors remains to be fully determined.

Mesoderm Induction

Aside from their mitogenic properties, a novel and striking property of heparin binding growth factors is their ability to induce meosdermal differentiation in isolated animal pole explants of Xenopus laevis embryos (Slack et. al. 1987). The differentiation of mesodermal tissues (principally muscle) in Xenopus appears to depend upon an inductive interaction between animal pole cells and the underlying vegetal endoderm, since isolated animal pole explants fail to generate muscle unless recombined with vegetal tissues. This effect of vegetal tissues can be, at least partially, reproduced with exogenously added (basic or acidic) FGF or ECDGF (Slack et. al. 1987). The case for FGF-like factors being the primary mesoderm-inducing agents is further strengthened by the finding that a Xenopus homologue of the mammalian basic FGF gene exists, and RNA transcripts hybridising with this gene have been found to be expressed in embryos at stages where mesoderm induction is taking place (Kimmelman and Kirschner 1987). However it is not at present clear whether Xenopus FGF itself or the equivalent of another member of the heparin binding growth factor family such as ECDGF is the primary agent of Xenopus mesoderm induction in vivo. Nevertheless this unexpected biological property of ECDGF, along with its derivation from embryo-derived EC cells may indicate that it plays a role in mesoderm differentiation processes in the mouse.

This involvment of FGF-like factors in mesoderm induction has several important consequences for further studies of growth factor involvement in mammalian development. Firstly, in general terms it reveals that agents initially identified as growth factors may also have profound effects on cell differentiation in particular circumstances. Secondly, as discussed above, it shows that growth factors from mammalian cells may function in other species, in this case Xenopus laevis. This not only opens up the possibility of analysing the possible biological function of growth factors

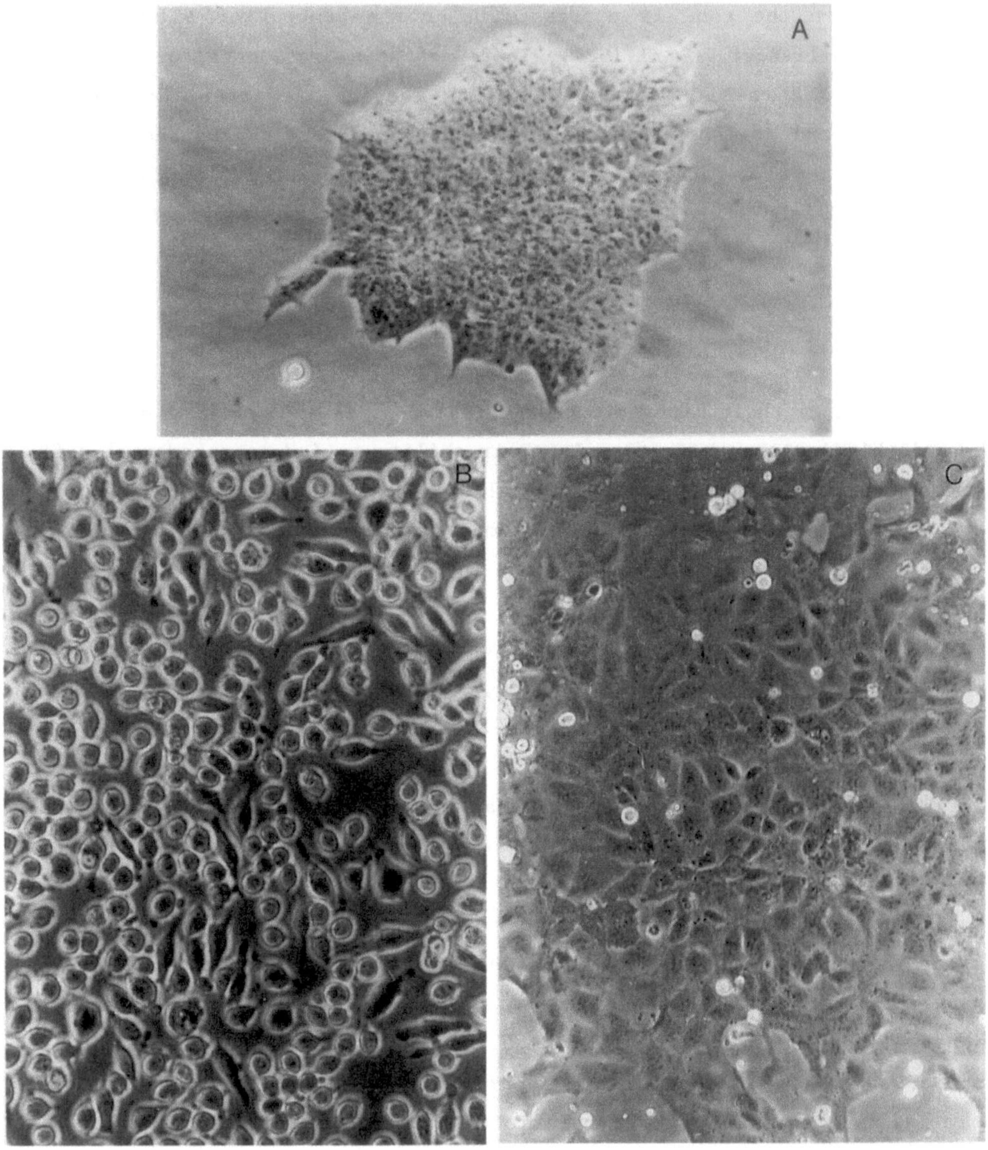

Fig. 1: Bipotential differentiation of CPl ES cells in vitro.
CPl ES cells were grown in the presence of DIA (A), induced to differentiate in the presence of 10^{-7}M RA (B) or induced to differentiate by removal of DIA (C).

in mammalian embryos by experiments in other species, but also conversely suggests that 'morphogens' and other extracellular bioregulatory factors found in Xenopus, such as the XTC cell derived mesoderm-inducing factor (Smith 1987) may have a biological role in mammalian development. Finally, the accessibility of Xenopus embryos at early developmental stages compared to their mammalian counterparts suggests that they could have potential as an experimental system for the identification of novel types of extracellular growth and differentiation regulatory factors.

Embryonic stem cell differentiation

As discussed above, murine embryo-derived ES cells offer certain advantages over EC cells as an in vitro test system for the analysis of the response of early embryonic cells to growth and differentiation factors. However, ES cells readily undergo spontaneous differentiation in vitro unless cultured in the presence of a live heterologous feeder cell layer. This suggests that ES cells will enter a differentiation programme unless 'prevented' from doing so by a signal which can be provided by the feeder cells. This contrasts with the situation observed in many teratocarcinoma cell lines such as PC13, where entry into the differentiation programme is blocked unless an inducer, such as RA, is present. It would obviously be of interest to understand the biochemical basis of the feeder cell effect on ES cell differentiation, since it would appear to affect mechanisms controlling a cellular decision to enter a differentiation programme, and these mechanisms would, at least in part, appear to be the target for the changes underlying the loss of differentiation potential and acquisition of the malignant phenotype associated with EC cells. Indeed, rather as transformed fibroblasts lose the requirement for exogenous growth factor signals to maintain multiplication, it could be argued that EC cells lose the requirement for exogenous differentiation inhibitory factors.

Isacke and Deller (1983) have demonstrated that the feeder cell effect is probably multifactorial, involving both cell contact dependant and independant mechanisms. However, it is of some significance that the principal biological effect of feeder cells on ES cell differentiation can be reproduced by conditioned medium from various cell lines (Smith and Hooper 1983, Koopman and Cotton 1984). The medium from one cell line, Buffalo Rat Liver (BRL) has a particularly potent differentiation-inhibiting effect on multipotenital EC and pluripotential ES cell cells in vitro (Smith and Hooper 1987). The

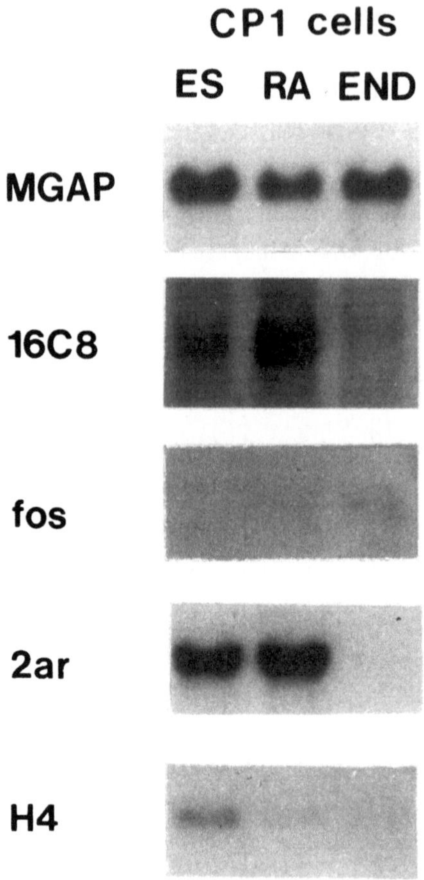

Fig. 2: Changes in gene expression resulting from bipotential differentiation of CP1 ES cells in vitro.
CP1 ES cells were grown in the presence of DIA (ES), induced to differentiate in the presence of 10^{-7}M RA (RA) or induced to differentiate by removal of DIA (END). Total cytoplasmic RNA was prepared from each cell type and subjected to Northern blot analysis. The probes were; MGAP: murine glyceraldehyde 3-phosphate dehydrogenase, 16C8: murine tissue inhibitor of metalloproteinase cDNA, fos: c-fos cDNA, 2ar: murine osteopontin cDNA, H4: histone H4 cDNA.

effect of BRL conditioned medium is fully reversible, in that the cells can be maintained indefinitely in an undifferentiated state (in the absence of feeder cells) in the presence of BRL conditioned medium, but will rapidly differentiate at any time they are returned to normal unconditioned medium. A particularly important point is that the effect of prolonged culture in BRL conditioned medium in the absence of feeder cells does not compromise the ability of ES cells to colonise the embryo in a chimaera or give rise to functional gametes. Hooper et. al. (1987) have, for example, succesfully used culture in BRL conditioned media to facilitate the isolation of hypoxanthine phosphoribosyl transferase deficient ES cells which were used to generate ES cell derived germ line chimeras transmitting HPRT-deficiency to progeny. The effect of BRL conditioned medium on ES cell differentiation could not be reproduced by a variety of growth factors or growth modulators (such as IGF-2 or TGF-beta) which were known to be secreted by BRL cells (Smith and Hooper 1987). The biological action of BRL cell conditioned medium depends upon an apparently novel secreted diferentiation regulatory factor (termed differentiation inhibitory activity, DIA) which acts to specifically supress ES cell differentiation in vitro. Further analysis of the structure of DIA and its mechanism of action action will undoubtedly have important consequences for our knowledge of the mechanisms controlling ES cell differentiation. Furthermore BRL conditioned medium has already found use as a practical tool for the experimental manipulation of ES cell proliferation and differentiation in vitro.

Mixing Differentiation Regulatory Signals

The regulation of ES cell and EC cell differentiation, as described above, would appear to depend upon exogenous signals; DIA supresses differentiation whereas agents like RA induce differentiation. It is possible that the actual pattern of differentiation may depend upon the balance between differentiation-inducing and differentiation-suppressing extracellular signals. This can be investigated by exposing ES cells to combinations of signals and observing their effects upon differentiation. The results of this type of experiment are unexpected (Fig.1). In the presence of BRL medium as a source of DIA activity, CP1/86 ES cell differentiation is suppressed, and the cells grow as tight colonies (Fig 1a). Removal of DIA from cells grown in monolayer results in induction of differentiation and concomitant change in cell morphology and pattern of gene expression (Fig 1b, Fig2). Addition of RA to cells grown in the presence of DIA results in the induction of differentiation, but the cell type formed is morphologically distinct from that formed by removal of DIA (Fig. 1c) resembling parietal endoderm in appearance. This is confirmed by examination of specific gene expression (fig 2). The conclusion from these experiments is that RA not only overrides the differentiation inihibitory action of DIA but also influences the differentiation programme that is chosen.

It is likely, however that at least some this response to extracellular differentiation regulatory signals also depends upon both the target cell itself, spatial relationships between cells or other forms of local cellular interactions. Firstly, ES cell lines such as CCE exhibit alternate patterns of differentiation in response to RA induction (not shown) than CP1/86 and secondly, differentiation of ES cells in aggregates or high cell densities yields a broader spectrum of differentiated cell types than occurs in low density monolayer culture.

Conclusions

Intercellular signalling by extracellular bioregulatory factors can control both differentiation and cell multiplication in early mammalian development. The exploitation of EC and ES cell lines has provided both the sources and the means to identify these agents. The ability of at least some of these signalling molecules to cross taxonomic boundaries argues that, like classical endocrine hormones, extracellular growth and differentiation factors may have a fundamental role in widely differing developmental strategies.

Acknowledgements

Research in the author's laboratory was supported by the Cancer Research Campaign.

REFERENCES

Beck, F, Samani, N.J, Penschow, J, Thorley, B, Tregear, G, and Coghlan, J (1987) Histochemical localisation of IGF-I and IGF-II mRNA in the developing rat embryo. Development. 101 175-184

Delli-Bovi, P, Curatola, A.M, Kern F, Greco, A, Ittmann, M, and Basilico, C. (1987) An oncogene isolated by transfection of Karposi's sarcoma DNA encodes a growth factor that is a member of the FGF family. Cell 50 729-737

Dickson, C, and Peters, G. (1987) Potential oncogene product related to

growth factors. Nature 326 833
Esch, F, Baird, A, Ling, N, Ueno, N, Hill, F, Denroy, L, Klepper, R, Gospodaroeicz, D,Bohlen P, and Guillman, R, (1985) Primary structure of bovine pituitary basic fibroblast growth factor (FGF) and comparison with the amino-terminal sequence of bovine brain acidiac FGF Proc. Natn. Acad. Sci. (U.S.A.) 82 6507-6511
Giminez-Gallego, G, Rodkey, J, Bennett, C, Rios-Candelore, M, DiSalvo, J, and Thomas, K. (1985) Brain-derived acidic fibroblast growth factor: complete amino acid sequence and homologies. Science 230 1385-1388
Heath, J.K. & Rees, A.R. (1985) Growth factors in mammalian embryogenesis. In: "Growth factors in biology and medicine" (D.Evered & M. Stoker eds.) Ciba Symp. 116, 21
Heath, J.K. & Shi, W-K (1986) Developmentally regulated expression of insulin-like growth factors by differentiated murine teratocarcinomas and extra-embryonic mesoderm. J.E.E.M. 95, 312
Heath, J.K. (1987) Purification of embryonal carcinoma derived growth factor and analysis of teratocarcinoma cell multiplication. In "Teratocarcinomas and Embryonic Stem Cells: a practical approach" (E. Robertson ed.) IRL press pp. 183-206
Hoper, M, Hardy, K, Handyside, A, Hunter, S, and Monk, M. (1987) HPRT-deficent (Lesch-Nyhan) mouse embryos derived from germline colonisation by cultured cells. Nature 326 292-295
Isacke, C.M. and Deller, M.J. (1983) Teratocarcinoma cells exhibit growth cooperativity in vitro. J. Cell Physiol. 117 407-414
Kimelman, D, and Kirschner, M. (1987) Synergistic induction of mesoderm by FGF and TGF-Beta and the identification of an mRNA coding for FGF in the early Xenopus embryo. Cell 51 869-877
Koopman, P, and Cotton, R. (1984) A factor produced by feeder cells which inhibits embryonal carcinoma differentiation. Exp. Cell res. 154, 233-242
Slack, J, Darlington, B, Heath, J.K. & Godsave, S. (1987) Mesoderm induction in xenopus by Heparin building growth factors. Nature 326, 197-200
Smith, A.G, and Hooper M.L. (1987) Buffalo rat liver cells produce a diffusible activity which inhibits the differentiation of murine embryonal cvarcinoma and embryonic stem cells. Dev. Biol. 121,1-9
Smith, J (1987) A mesoderm-inducing factor is produced by a Xenopus cell line. Development 99, 3-14
Smith, T.A. and Hooper, M.L. (1983) Medium conditioned by feeder cells inhibts the differentiation of embryonal carcinoma cell cultures. Exp. Cell res. 145, 458-462
Taira, T, Yoshida, Y, Miyagawa, K, Sakamoto, H, Tereda, M, and Sugimira, T. (1987) cDNA sequence of human transforming gene hst and identification of coding sequences required for transforming activity. Proc. Natn. Acad. Sci. (U.S.A.) 84 2980-2984
van Veggel, J, Oostward, T, deLaat, S, and van Zoelen, E. (1987) PC13 embryonal carcinoma cells produce a heparin-binding growth factor. Exp. Cell res. 169 280-286

Regulation of Growth Factors and their Receptors in Early Murine Embryogenesis

C.L. Mummery, J. van den Eijnden-van Raaij, A. Feijen, H-C Tsung and W. Kruijer
Hubrecht Laboratory
Netherlands Institute for Developmental Biology
Uppsalalaan 8
3584 CT UTRECHT
The Netherlands

Introduction

Elucidation of the molecular mechanisms underlying embryonic growth is essential to understanding the regulation of development as well as cellular proliferation and its impairment in malignancy. Growth factors are likely to be intimately involved in the regulation of these processes but their identification and characterization in embryogenesis are hampered by the complexity of events at each developmental step and the limited availability of embryonic material for analysis. Embryonal carcinoma (EC) cells are the undifferentiated stem cells of spontaneous or experimentally-induced teratocarcinomas and resemble the pluripotent cells in the inner cell mass (ICM) of preimplantation embryos in their biochemical and antigenic properties and capacities for differentiation. Not all EC cell lines are equally capable of participating in normal development, however, and often form abnormal chimaeras when injected into normal blastocysts. Embryonic stem (ES) cells established in vitro directly from isolated ICM's (Martin 1981; Evans and Kaufman 1981) readily form normal chimaeras and as such are much closer to their embryonic origins. Further, EC cell lines require chemical induction of differentiation by substances such as retinoic acid whereas ES cells will differentiate extensively when grown in the absence of the feeder cells normally required to maintain them in an undifferentiated state. The feeder-dependence of ES cell lines is often experimentally inconvenient but recently a differentiation inhibiting activity (DIA, see Heath, this volume) has been indentified in the conditioned medium of Buffalo rat liver cells which also maintains ES cells in an undifferentiated state.

NATO ASI Series, Vol. H26
Cell to Cell Signals in Mammalian Development
Edited by S. W. de Laat et al.

Although growth factor expression, secretion and receptors in EC cells have been the subject of several studies (reviewed Jacobowits 1986) the results have been inconclusive as they have frequently depended on the particular cell line studied. We are now approaching this problem more systematically and are comparing a series of EC cell lines of different origin and with different growth properties, for example in terms of feeder-dependence, with differentiating ES cells. Some results of these studies are described below.

Growth factor expression in EC and ES cells

Comparison of EC cells of different embryonic and tumour origin with ES cells for expression and growth factor mRNA leads to results which emphasize the differences between these various cell types rather than illustrating the properties they have in common. Although all EC and ES cells investigated so far are consistent in the changes of TGFβ and IGF-II expression observed during differentation, PDGF-A, PDGF-B and IGF-I expression appear to depend on the particular cell line investigated. In the following section, examples of Northern blot analyses of mRNA from several EC cell lines and a newly isolated ES cell line hybridized with cDNA probes for some of these factors will be given. Most detailed analysis of changes during differentiation has been obtained from P19 EC cells (McBurney et al. 1982), since they can be induced in a controlled manner to form mesodermal derivatives, including beating muscle, when aggregated with DMSO, neuronal derivatives when aggregated with retinoic acid (RA) and endodermal derivatives when treated in monolayer with RA. In addition, they can be synchronized by mitotic selection; they then become sensitive to the differentation inducing effects of RA in the G_1 phase of the cell cycle if exposed for a pulse of only two hours after replating (Mummery et al. 1987). This provides a means of clarifying the role of growth factors in the changes in growth phenotype during differentation and in particular the loss in tumour-forming ability.

TGFβ expression during the differentiation of P19 EC cells

The genes for both human and mouse have recently been cloned and mRNA transcripts found in both neoplastic and non-neoplastic cells, generally as a single band of 2.5 kb (Derynck et. al 1985, 1986). Northern blot analysis of mRNA from P19 EC cells showed in addition to the usual 2.5 kb transcript an extra band of 1.8 kb (fig. 1.). Following induction of differentation of monolayer cultures by RA, this smaller transcript is rapidly and significantly reduced; 24 hours after RA addition less then 50% remains while after 4 days it is no longer visible (fig. 1.). This is related to the rate at which P19 cells in monolayer form endoderm-like cells; rehybridization of the blot with a probe for ENDO-A, a cytokeratin expressed in endoderm and trophoblast but not in the inner cell mass nor, later, in mesoderm (Duprey et al. 1985), shows a significant increase in this mRNA 24 hrs after RA addition which reaches a maximum after 2-3 days (fig. 1.). When P19 EC cells are induced to form derivatives of the other primitive germ layers, specific loss of the 1.8 kb TGFβ is also observed (fig. 2.); significant reduction does not occur during aggregation, a phase of differentation essential in determining the cell types formed by a particular inducer, but only occurs after aggregates have reattached to an appropriate substrate. Rehybridization of the blot in figure 2 with the ENDO-A probe shows that although aggregation of P19 cells, with or without RA and DMSO, induces endoderm formation, a significant EC component remains since the aggregates also express 1.8 kb TGFβ transcript. After replating, the endodermal component from aggregates treated with RA and DMSO is minimal (lanes 6 and 8) although aggregates formed in the absence of chemical inducers do show significant levels of endoderm differentiation (lanes 5 and 7). Rehybridization of the β-actin probe to this blot shows that lanes contain similar amounts of mRNA except lane 3 (aggr. RA); this would explain the apparently lower signal for ENDO-A and TGFβ in these cells.

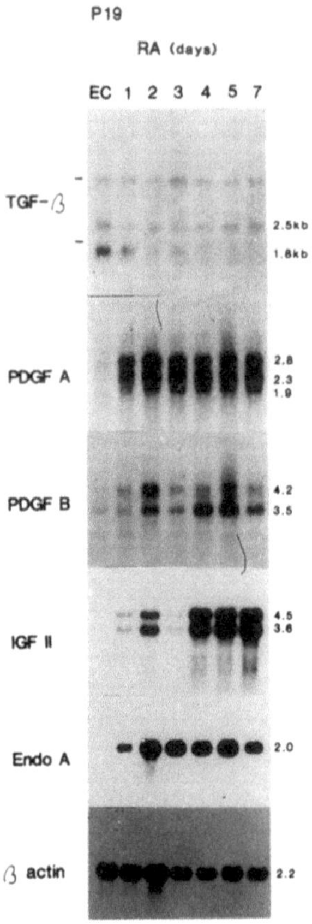

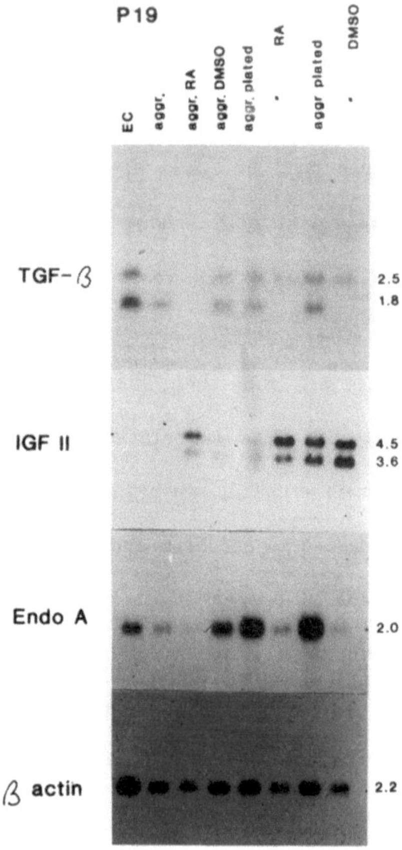

Fig. 1. Northern blot analysis of mRNA from P19 EC cells treated in monolayer with RA (10^{-6}M). RNA was isolated and poly (A^{+}) RNA selected as described previously (van den Eijnden–van Raaij et al.). Probes were obtained by nick translation of DNA fragments with ^{32} P–dCTP (>3000 mMol) and DNA polymerase I to a specific activity of 0.5–1 x 10^{9} cpm per μg. The following plasmids were used to isolate DNA fragments for nick translation:– PDGF–A: a 1.3 kb ECoR 1 fragment from the A–chain clone D1 (Betsholtz et al. 1985); PDGF–B, (c–sis): a 1.0 kb Sst I–Xba I fragment from pC60 (Gelman et al. 1981); TGFβ: a 1.0 kb EcoR1 fragment of pbC1 (Derynck et al., 1985); IGF–II : a 1.4 kb Pst I/ EcoR1 fragment of pUC 12; ENDO A: a 0.25 kb fragment in pBR 322.

Fig. 2. Northern blot analysis of mRNA from P19 EC cells induced to form neuronal or mesodermal derivatives by aggregation with RA (10^{-6}M) or DMSO (1%) respectively. RNA was isolated after 4 days of aggregation in bacteriological dishes in the absence (aggr.) or presence (aggr. RA; aggr. DMSO) of inducer, 3 days after replating to tissue culture dishes when neurite outgrowth is maximal in the presence of RA (aggr. plated RA) and 8 days after replating when areas of beating muscle cells are evident (aggr. plated DMSO). At each time point, parallel control cultures in the absence of inducer (aggr. plated) were included. Hybridization was as in the legend to fig. 1.

Synchronized P19 EC cells

During RA induced differentation, not only is the growth rate of EC cells reduced but they rapidly lose their tumorigenic potential (Rayner and Pulsford 1984). In EC cells synchronized by mitotic selection, we have shown previously that an increase in cell cycle duration from 12.5 to 24 hours (Mummery et al. 1984) and the loss of colony forming ability in soft agar (Mummery et al. 1987), an in vitro analogue for tumorigenicity in vivo, occur after two cell cycles in the continuous presence of RA. Although exponentially growing EC cells require at least 24 hours RA exposure to become committed to this program of differentiation, synchronized P19 EC cells are exquisitely sensitive to RA-induced commitment in the G1 phase of the cell cycle (Mummery et al. 1987). Only 2 hours exposure to RA is then sufficient to induce irreversible loss of the capacity to form colonies in soft agar (fig.3) and to lose expression of EC specific antigenic markers; similar exposure of synchronized cells in S phase has no effect.

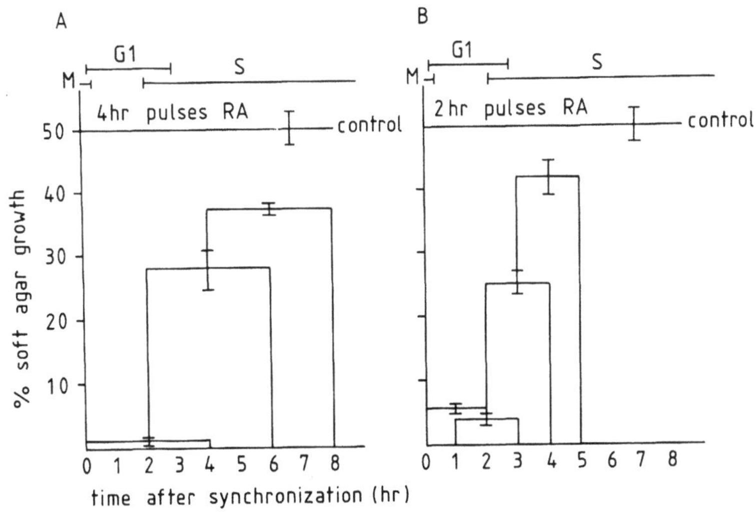

Fig. 3. Effect of pulsed RA treatment on the commitment of synchronized P19 EC cells to differentiation: soft agar growth. Cells were synchronized at t=0 by mitotic selection (Mummery et al. 1984, 1987), and RA (10^{-6}M) added for 2 or 4 hr pulses at the time indicated. After removal of RA, cultures were incubated further in DF + 7.5% FCS before plating in soft agar at t= 75 hr in DF + 7.5% FCS. Data show are the percentage of cells plated which had formed macroscopic colonies after 10 days. Each point shows the mean of 4 separate dishes (± SEM). Control cultures were synchronized at t=0 but were further untreated until t = 75 hr.

Figure 4 shows that 2 hours RA exposure in G1 phase also induces irreversible loss of the 1.8 kb TGFβ transcript, while cells exposed in S phase still express both the 1.8 kb and 2.5 kb mRNA. Although the cells derived by G1 phase exposure to RA resemble those derived by continuous exposure morphologically, they do not express identical differentiation markers, for example plasminogen activator production (Mummery et al. 1987). However, rehybridization of the blot in figure 4 shows that they do express high levels of ENDO-A mRNA. This compliments earlier observations (Mummery et al. 1987) that the cytoskeleton of these cells is frequently labelled by the antibody TROMA-1 (Kemler et al. 1981). Loss of the 1.8 kb transcript is therefore an early indication of differentiation in P19 EC cells and its expression highly correlated with anchorage independence for growth.

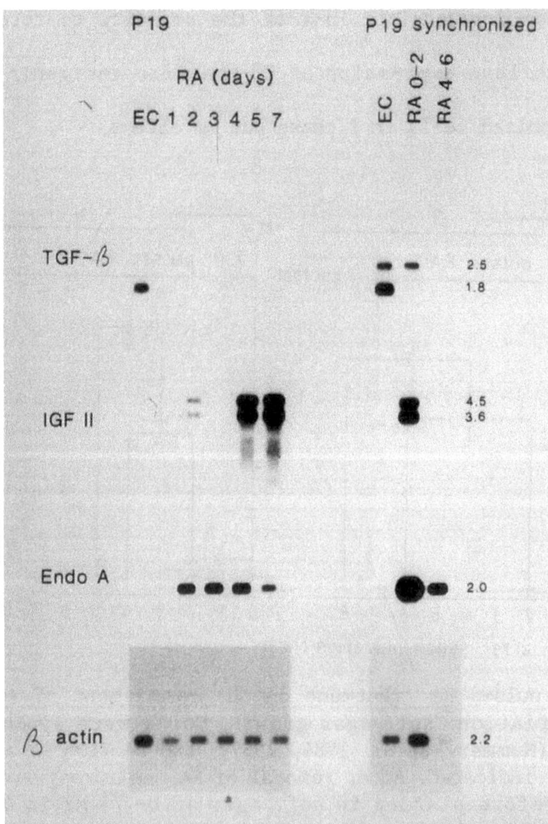

Fig. 4. Northern blot analysis of mRNA from synchronized P19 EC cells and treated with RA in G1 or S-phase as in the legend to figure 3. Hybridization with growth factor probes was as indicated in the legend to fig. 1.

TGFβ expression in feeder dependent EC and ES cells

Although the pluripotent EC cell line P19 is a useful model for differentiation in vitro, it is unable to form normal chimaeric mice (Rossant et al. 1984). Further, P19 is derived from a tumour formed by a 7.5 day embryo, so the nature of the stem cell and its exact embryonic equivalent is unclear. The extent to which specific properties may be extrapolated to embryonic development may therefore be limited. Certain EC cells, such as PSA-1 and P10, are available which do participate in normal development. Similarly, embryonic stem (ES) cells, isolated directly from the inner cell mass of preimplantation embryos also efficiently form chimaeric mice. They are, however, dependent for maintenance of their undifferentiated state on coculture with growth inactivated feeder cells which has limited their experimental accessibility. The identification of DIA (Smith & Hooper 1987) in medium conditioned by Buffalo rat liver cells, which efficiently inhibits differentiation of these cells, has now solved this problem. By culturing PSA-1, P10 and ES cells in this medium, we have been able to isolate mRNA in the absence of feeder cell contamination. The expression of the TGFβ in these cell lines is shown in figure 5. The feeder-dependent EC cells express both the 1.8 and 2.5 kb transcripts at similar levels as the feeder-independent lines. More interestingly, the ES cells also express both transcripts. Even after several passages in BRL conditioned medium prior to isolation of mRNA, these cells retained their diploid karyotype, expression of antigens recognized by αECMA-7 antibodies, and capacity for extensive differentiation in vitro and in vivo (fig. 6.). They therefore still closely resemble the pluripotent stem cells of the developing mouse embryo, which suggests that the unusual 1.8 kb TGFβ transcript is also expressed during early embryonic development.

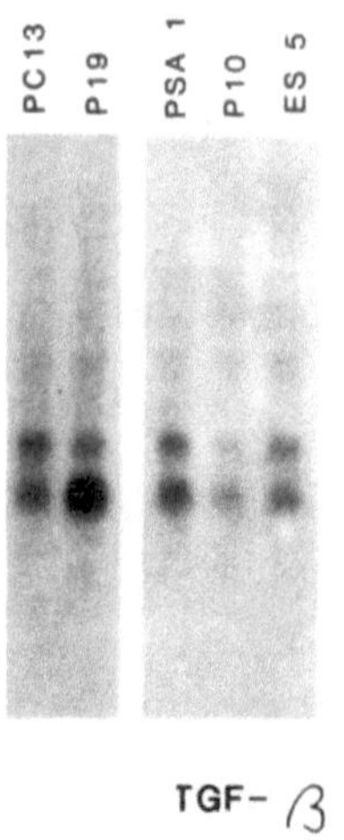

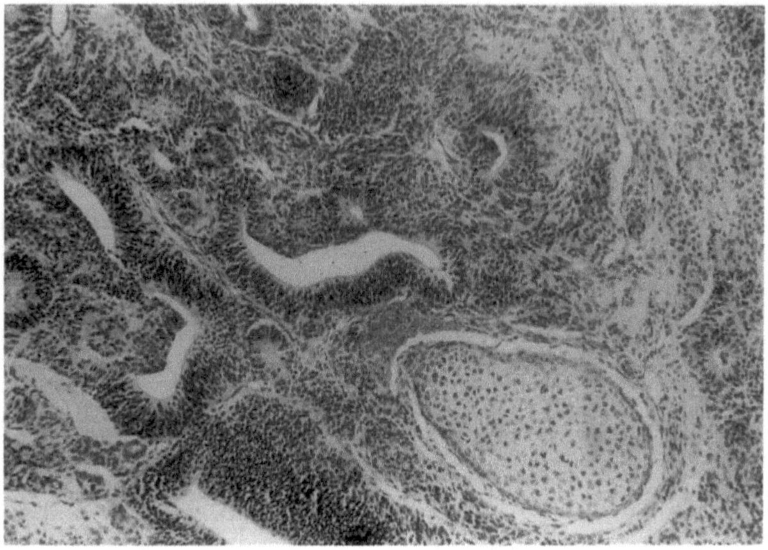

Fig. 5. TGFβ expression in feeder dependent EC and ES cells grown in BRL cell conditioned medium. Northern blot analysis was carried out as in the legend to fig. 1. Feeder independent P19 and PC 13 EC are shown for comparison.

Fig. 6. Teratocarinoma formed by injection of $1x10^6$ ES cells into 129 mice after culture for 10 passages in BRL conditioned medium. Cartilage, differentiated epithelium and muscle cells are clearly evident.

TGFβ expression during differentiation of F9 and C1003 EC cells

To establish unequivocally that the 1.8 kb TGFβ transcript is a general characteristic of undifferentiated EC cells no longer expressed after differentiation, we investigated TGFβ expression in two other cell lines with well established patterns of differentiation. Figure 7 shows the results in C1003 treated with RA to give endodermal derivatives or deprived of serum to form nerve cells. The 1.8 kb mRNA is absent after differentiation.

Rehybridization of the blot with NF-68, which codes for neurofilament proteins, (Lewis & Cowan 1985) and ENDO-A, as above, confirm the identity of the cells used for mRNA isolation. Similarly, when F9 EC cells are treated in monolayer with RA to form parietal endoderm-like derivatives, the 1.8 kb transcript is significantly reduced (fig. 8). In all cases equal amounts of mRNA were present on the blots, as demonstrated by rehybridization with β-actin.

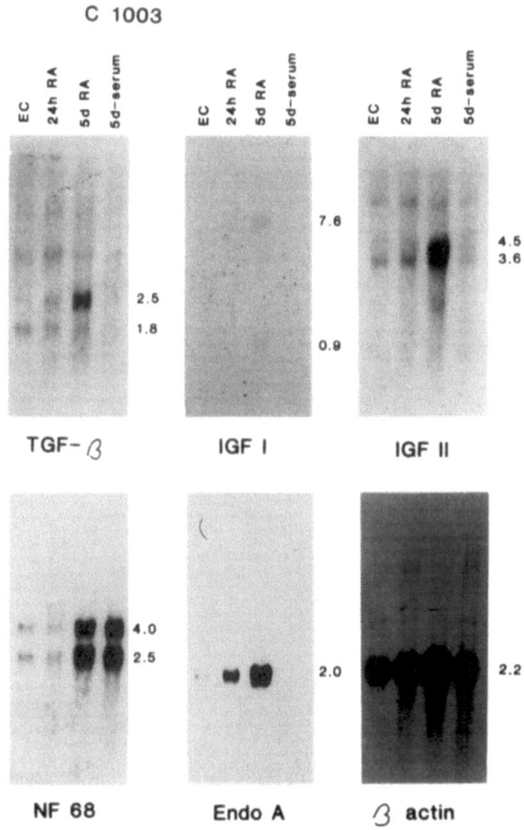

Fig. 7. Growth factor expression during the differentiation of C1003 EC cells. Endoderm formation was induced by the addition of RA (10^{-6}M) to monolayer cultures and neuronal cells by culture in medium without serum, as described by Darmon et al. (1982) RNA isolation and hybridization was as described in the legend to fig.1.

Expression of PDGF and IGF-II during differentiation of EC cells

Under conditions of growth and differentiation similar to those used for TGFβ expression, we also studied the expression of several other growth factor genes, namely PDGF A and B and IGF-I and II. In P19 EC cells treated in monolayer with RA, both PDGF A and B were induced to almost maximal levels within 24 hours (fig. 1.). In addition, PDGF A and B were expressed in the cells derived from G1 phase exposure of synchronized P19 to RA but not in the cells derived from S phase exposure. Exactly opposite results were found in F9 where RA significantly reduced expression of PDGF-A, as also observed in the human teratocarcinoma cell line Tera-2 (Wiema et al., submitted). Expression of IGF-II on the other hand, was rapidly induced in P19 (fig. 1.), C1003 and F9 (fig. 7. and 8.) EC cells treated with RA. Thus, as TGFβ, all lines respond consistently to RA-addition although in neural derivatives of C1003, resulting after serum deprivation, IGF-II is not expressed. IGF-I was barely detectable in EC cells either before or after differentiation although exceptionally in C1003 treated with RA for 5 days of a 7.8 kb transcript was evident (fig. 7.).

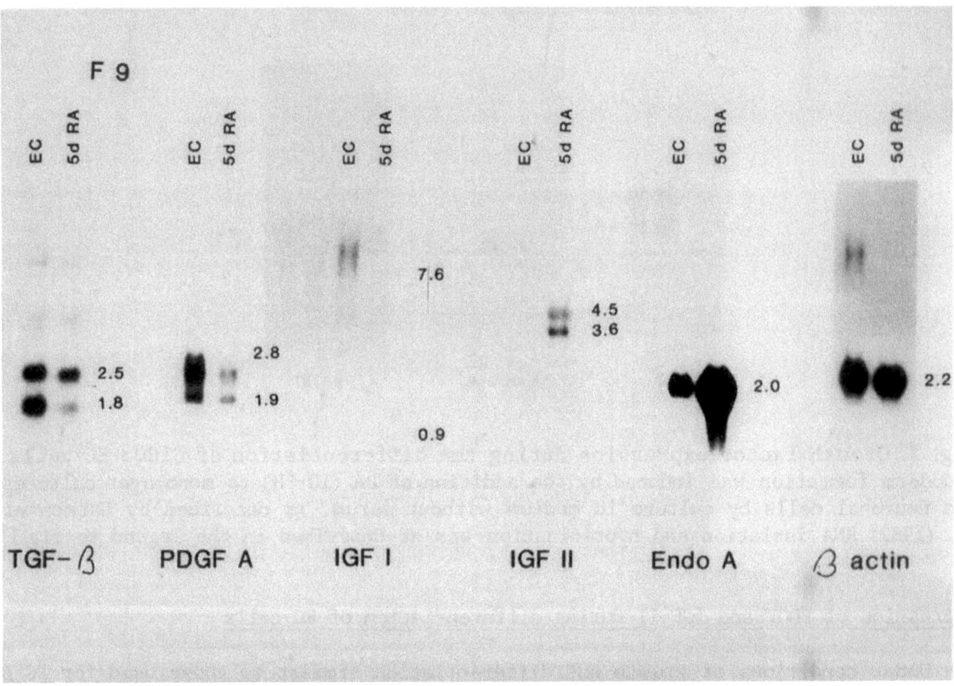

Fig. 8. Growth factor expression during the differentiation of F9 EC cells. Endoderm formation was induced by the addition of RA (10^{-6}M) to monolayer cultures for 5 days.

<u>Growth factor receptors during differentiation</u>

Not only is the production of growth factors differentially regulated during the differentiation of EC and ES cell but also the expression of receptors for these factors. We have studied three factors in particular; PDGF, EGF and TGFβ. Undifferentiated cells do not express surface receptors for any of these factors (table 1) but, depending on the inducer used, receptors are frequently expressed after differentiation. Both EGF-R mRNA (fig. 9) and EGF binding (Mummery et al. 1987) are detectable in P19 EC cells 4 to 5 days after RA addition and in three cloned differentiated derivatives of P19 which resemble cells of different germ

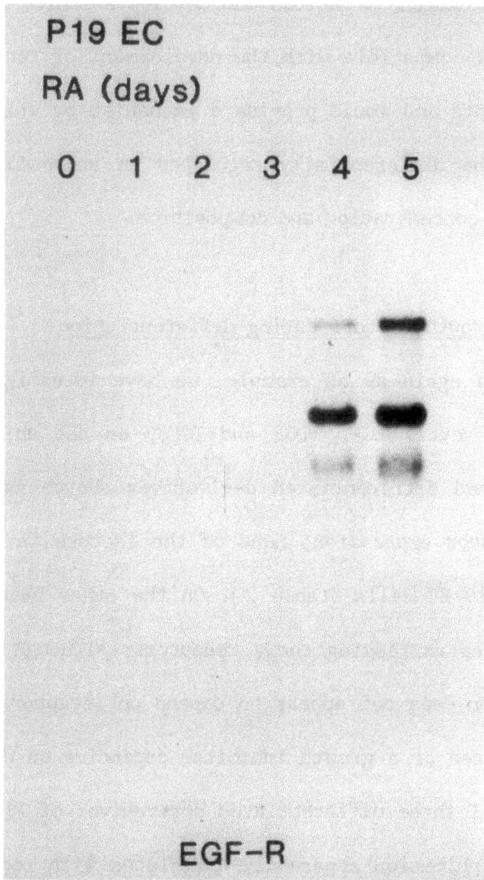

Fig. 9. EGF receptor expression in P19 EC cells treated in monolayer by retinoic acid.

layers (Mummery et al. 1985, 1986) (table 1). On the other hand, only the mesodermal derivatives of P19 express PDGF receptors (table 1); receptors for TGFβ, are also expressed by all three differentiated cell types although receptor affinity differs significantly between each cell type. In general cells producing biologically active growth factors do not express their receptors; since EC cells grow in the complete absence of exogenous (serum) growth factors and do not express receptors for at least three major factors, it would seem likely that their growth is regulated by an autocrine mechanism. Following differentiation, EC cells become dependent for growth on exogenous factors and develop a finite lifespan (Rayner and Pulsford 1984); this is compatible with the development of receptors for at least some major growth factors and would provide a mechanism by which growth of differentiated cells could be differentially regulated in an environment constant in terms of growth factor concentration and composition.

Biological effects of growth factors during differentiation

Taking P19 EC cells again as an example, we have investigated the biological effects of the three factors EGF, PDGF and TGFβ, on the undifferentiated cells compared with the cloned differentiated derivatives above. As might be expected from the lack of receptor expression, none of the factors is either mitogenic or growth inibitory for P19 EC cells (table 2.). On the other hand, EGF and PDGF are mitogenic for cell types expressing their receptors, although in the case of EGF, the extent of simulation does not appear to depend on receptor number or affinity. TGFβ can act as a mitogen or a growth inhibitor depending on the target cell type. It is inhibitory for all three differentiated derivatives of P19, but in this case the extent of growth inhibition apparently correlates with receptor affinity; the cell line with the least receptors, EPI–7 but highest affinity is most strongly inhibited in growth (table 2.).

Table 1 Growth factor receptors in P19 EC cells and three cloned differentiated derivatives END-2, EPI-7 (Mummery et al. 1985) and MES-1 (Mummery et al. 1986).

Cell	P19 EC	END-2	EPI-7	MES-1
Receptor nr. per cell (affinity)				
EGF	900 (0.5 nM)	60,000 (0.76 nM)	39,500 (0.9 nM)	12,000 (30 pM) 180,000 (12.3 nM)
PDGF	—	1,000 (0.5 nM)	—	115,000 (0.3 nM)
TGFβ	—	6,000 (75 pM)	1,200 (40 pM)	12,000 (100 pM)

Table 2 Effects of EGF, PDGF and TGFβ on the growth of P19 EC cells and three clones, differentiated derivatives

	P19 EC	END-2	EPI-7	MES-1
*EGF (25 ng/ml) (stimulation)	1.0	5.1	3.1	75.7
*PDGF (10 ng/ml) (stimulation)	1.0	1.1	0.9	19.7
**TGFβ (1 ng/ml) (inhibition)	0	50%	75%	40%

* Cumulative ^{3}H-Tdr incorporation following stimulation of quiescent cells - units relative to control.

** % inhibition of ^{3}H-TdR incorporation after addition to exponentially growing cells.

Discussion

Although an inventory of growth factor production, receptor expression and biological response of cells present during early development is far from complete, a model of mutual cell interaction based on growth factors would seem to be entirely plausible. In such a model, certain cells such as those of the ICM would produce factors for their own use via an autocrine mechanism, as well as for other cells expressing appropriate receptors. In this way, a complex pattern of growth regulation could be achieved. Further since certain factors, such as TGFβ, are also known to induce expression of various other genes, such as those for extracellular matrix factors, in appropriate target cells, they may act indirectly to affect processes such as cell migration in embryonic development. Factors such as PDGF may also effect cell migration directly, since some cells respond chemotactically to its presence. As more specific antibodies in particular become available, it will become more straightforward to characterize developing embryos for growth factor production and receptor expression directly and by using appropriate antagonists, to interfere with their action and ascertain their exact function.

References

Betsholtz C, Johnsson A, Heldin CH, Westermark B, Lind P, Urdlea MS, Eddy R, Shows TB, Philpott K, Melloral , Knott TJ, Scott J (1985) cDNA sequence and chromosomal localization of human PDGF A chain and its expression in tumour cell lines. Nature 320:695–699

Darmon M, Stalicup WB, Pittman QJ (1982) Induction of neural differentiation by serum deprivation in cultures of the embryonal carcinoma cell line 1003. Exp Cell Res 138:73–78

Derynck R, Jarret JA, Chen EY, Eaton DH, Beu JR, Assoian RK, Roberts A, Sporn M, Goeddez DV (1985) Human transforming growth factor-beta cDNA sequence and expression in tumour cell lines. Nature 316:701–705

Derynck R, Jarrett JA, Chen EY, Goeddel DV (1986) The murine transforming growth factor-beta precursor. J Biol Chem 261:4377–4379

Duprey P, Morello D, Vasseur M, Babinet C, Londamine H, Brûlet P, Jacob F (1985). Expression of the cytokeratin ENDO-A gene during early mouse embryogenesis. PNAS 82:8535–8539

Evans MJ, Kaufman MH (1981) Establishment in culture of pluripotential cells from mouse embryos. Nature 292:154–156

Gelmann EP, Wong-Staal F, Kramer RA, Gallo RC (1981) Molecular cloning and comparative analyses of the genomes of the simian sarcoma virus and its associated helper virus. PNAS 78:3373–3377

Jakobovits A (1986) The expression of growth factors and growth factor receptors during mouse embryogenesis. Kahn P, Graf T (eds) Springer Verlag Heidelberg 9–17

Kemler R, Brûlet P, Schnebelen MT, Gaillard J, Jacob F (1981) Reactivity of monoclonal antibodies against intermediate filament proteins during embryonic development. J E E M 64:45

Lewis S, Cowan NJ (1985) Genetics, evolution and expression of the 68,000-mol-wt neurofilament protein: isolation of a cloned cDNA probe. Cell Biol 100:843–850

Martin GR (1981) Isolation of a pluripotent cell line from early mouse embryos cultured in medium conditioned by teratocarcinoma stem cells. PNAS 78:7634–7638.

McBurney MW, Jones-Villeneuve EMV, Edwards MKS, Anderson PJ (1982) Control of muscle and neuronal differentiation on a cultured embryonal carcinoma cell line. Nature 299:165–167

Mummery CL, Van den Brink CE, Van der Saag PT, De Laat SW (1984) The cell cycle, cell death and cell morphology during retinoic acid induced differentiation of embryonal carcinoma cells. Dev Biol 104:297–307

Mummery CL, Feijen A, Van der Saag PT, Van den Brink CE, De Laat SW (1985) Clonal variants of differentiated P19 embryonal carcinoma cells exhibit epidermal growth factor receptor kinase activity. Dev Biol 109:402–410

Mummery CL, Feijen A, Van den Brink CE, Moolenaar WH, De Laat SW (1986) Establishment of a differentiated mesodermal line from P19EC cells expressing functional PDGF and EGF receptors. Exp Cell Res 165:229–242

Mummery CL, Van den Brink CE, De Laat SW (1987) Commitment to differentiation induced by retinoic acid in P19 embryonal carcinoma cells in cell cycle dependent. Dev Biol 121:10–19

Rayner MJ, Pulsford JAJ (1984) Acquisition of a limited lifespan by differentiating cells derived from PC13 embryonal carcinoma cells. J Cell Sci 72:227–240

Rossant J, Papaioannou VE (1984) The relationship between embryonic, embryonal carcinoma and embryo-derived stem cells. Cell Diff 15:155–161

Smith AG, Hooper ML (1987) Buffalo rat liver cells produce a diffusible activity which inhibits the differentiation of murine embryonal carcinoma and embryonic stem cells. Dev Biol 121:1–9

Van den Eijden-van Raaij AJM, Koornneef I, Van Oostwaard ThMJ, Feijen A, Kruijer W, De Laat SW, Van Zoelen EJJ. Simultaneous purification of a PDGF-like growth factor and a type β transforming growth factor secreted by mouse neuroblastoma cells: a general strategy utilizing exclusively volatile solvents and resulting in a high protein recovery. JCB submitted

Mesodermal Differentiation of Embryonal Carcinoma Cells in Coculture with Visceral Endoderm Cell Lines

A.H. Piersma*, A. Willemse, C.E. van den Brink, S.W. de Laat and C.L. Mummery
Hubrecht Laboratorium
Netherlands Institute for Developmental Biology
Uppsalalaan 8
3584 CT UTRECHT
The Netherlands
Phone: (31).30.510 211

SUMMARY

In mammalian embryogenesis mesoderm is formed from embryonic ectoderm in an area adjacent to visceral endoderm. We have established an in vitro model for this process using P19 embryonal carcinoma (EC) cells and various endodermal cell lines, which were characterized with specific antibodies. In coculture with END-2 cells, a visceral endoderm cell line derived from P19, the EC cells formed aggregates and differentiated extensively, with mesoderm-derived beating muscle and adipocytes being the cell types most evident. Differentiation was less pronounced when P19 cells were cocultured with EPI-7 and PSA-5E cells which have some but not all of the visceral endoderm characteristics expressed by END-2. Parietal endoderm (PYS-2) and mesoderm (MES-1) cell lines neither permitted aggregation nor differentiation, whereas Dif 5 cells, reported to have both visceral and parietal endoderm characteristics, allowed aggregation but not differentiation. Mesodermal differentiation of P19 EC cells therefore specifically occured in coculture with visceral endoderm-like cells. Aggregation is a necessary but not sufficient step in this process. Differentiation was not observed on fixed or lysed END-2 monolayers to which END-2 conditioned media were added. However, when P19 cells were aggregated over a monolayer of END-2 cells covered with an agar layer to prevent contacts between the two

*Present address and correspondence:
Department of Teratology
National Institute for Public Health and Environmental Protection
P.O. Box 1
3720 BA BILTHOVEN
The Netherlands
Phone: (31).30.742526

NATO ASI Series, Vol. H26
Cell to Cell Signals in Mammalian Development
Edited by S.W. de Laat et al.

cell types, they formed beating muscle significantly earlier after replating on tissue culture substrate than when aggregated similarly without an underlayer or over MES-1 monolayers. This finding suggested that a visceral endoderm-derived soluble mediator stimulated mesoderm formation. This coculture system, in which no chemical differentiation inducer is needed, provides a valuable in vitro model of mesoderm induction in mammalian embryogenesis.

INTRODUCTION

Cellular interactions between different embryonic tissues mediate the formation of new tissue types. The formation of mesoderm, occurring around day 7 of murine development, may be considered as such an induction. This germ layer is formed by the embryonic ectoderm in a specific area adjacent to the visceral endoderm (3). The mechanism of mesoderm formation and the nature of inducing signals remain unknown. These aspects of developmental processes are difficult to study in the embryo.

Embryonal carcinoma cells (EC), the stem cells of teratocarcinomas, provide model systems for embryonic differentiation processes. The differentiation of EC cells in culture is dependent on cell conformation as well as on the presence of inducing agents (4,6,10,19). Although differentiation of embryonic cells can thus be mimicked in culture, the cellular interactions leading to differentiation in the embryo have not been reproduced in these models.

In the present study we have extended the characterization of a number of endoderm-like cell lines using specific antibodies. Furthermore the ability of these cell lines to promote mesodermal differentiation of P19 EC cells was investigated in a coculture system. Mesoderm formation was found to correlate specifically with visceral endoderm characteristics of the promoting cells. This coculture system provides a unique in vitro counterpart of the development of mesodermal tissue between endoderm and ectoderm in the mammalian embryo.

MATERIALS AND METHODS

Cell culture

All cells were grown on tissue culture plastics in a 1:1 mixture of Dulbecco's minimal essential medium (DMEM) and Ham's F12 medium (DF) containing 7.5% foetal calf serum (FCS) and buffered with $NaHCO_3$ (44 mM) in a humidified atmosphere containing 7.5% CO_2. Confluent cultures of END–2, EPI–7 (11), MES–1 (12), PSA5E (1), PYS–2 (9) and Dif 5 (14) were treated with mitomycin–c (10 μg/ml) for at least three hours, washed with phosphate buffered saline (PBS), and inoculated with 5.10^5 P19 cells per 25 cm^2 flask in 5 ml DF+7.5% FCS. Medium was changed three times a week. Fixed feeder layers were prepared by treatment with methanol for ten minutes at -20^oC. Feeder cell matrix was prepared by incubation in 0.5 mM EGTA for 30 minutes at 37^oC. Cocultures where feeder and P19 were physically separated were prepared by adding a layer of 1% agar in DF+7.5% FCS (7 ml per 100 mm Φ dish) over confluent feeders. Subsequently P19 suspensions were inoculated on top of the agar layer at 5.10^5 cells per dish in 7 ml DF+7.5% FCS, and allowed to form aggregates for four days. Aggregates were then collected and the contents of each dish were plated in a 25 cm^2 flask coated with 0.1g/l gelatin, and maintained for further observation. Controls included 1% DMSO added to the medium throughout the experiment, and the same procedure without a feeder layer in the dish. Conditioned media were collected from confluent END–2 cultures refreshed three days earlier with 5 ml DF+7.5% FCS per 25 cm^2 flask, and used the same day.

Immunofluorescence

Cells grown to confluence on gelatin–coated glass cover slips were fixed with methanol: glacial acetic acid (20:1) for 10 minutes at -20^oC. Fixed cells were incubated with antibodies then the appropriate FITC–labelled conjugate for 60' at 20^oC. After each incubation cells were washed with PBS+ 1% newborn calf serum to remove unbound reagent. Antibodies and conjugates were diluted in PBS+ 1% newborn calf serum. Antibodies used were as follows: rabbit-anti-mouse laminin (Bethesda Research Laboratories, diluted 1:100), rabbit-anti-human fibronectin (S.K. Brahma, University of Utrecht, 1:100); monoclonal TROMA-1 (anti-ENDO A cytokeratin, diluted 1:10) (8); monoclonal anti-brushin (1:10) and anti-FT1 (1:100) (13); monoclonal anti-SSEA-1 (1:300), (18); monoclonal MF20 (anti-myosin, 1:500), (2).

RESULTS

Characterization of cell lines

Five endodermal cell lines and one mesodermal cell line were characterized with specific antibodies (Table 1). Laminin was expressed by all cell lines, whereas all except Pys-2 produced fibronectin. TROMA-1 stained cytokeratin in Pys-2 and END-2 and in about 10% of PSA-5E and EPI-7 cells in monolayer. Dif 5 and MES-1 were negative for TROMA-1. Brushin, a marker for primitive and visceral endoderm, was expressed by PYS-2 and END-2 only. FT-1, specific for visceral endoderm, stained about 10% of PSA-5E and EPI-7 cells in a monolayer, and all END-2 cells (figure 1), whereas MES-1, PYS-2 and Dif 5 were negative. Thus visceral endoderm characteristics were found in END-2, and to some extent in EPI-7 and PSA-5E, but not in Dif 5, PYS-2 and MES-1.

Table 1 Characterization of cell lines using specific antibodies.

cell line	anti-laminin	anti-fibronectin	TROMA-1	anti-brushin	anti FT-1
END-2	+	+	+	+	+
EPI-7	+	+	+/-	-	+/-
PSA-5E	+	+	+/-	-	+/-
Dif 5	+	+	-	-	-
PYS-2	+	-	+	+	-
MES-1	+	+	-	-	-

Methanol / acetic acid-fixed confluent monolayers of cell lines grown on gelatin-coated glass cover slips were stained with antibodies and FITC-conjugated anti-immunoglobulins as described in materials and methods.
+ : all cells positive - : all cells negative +/- : around 10% of cells positive.

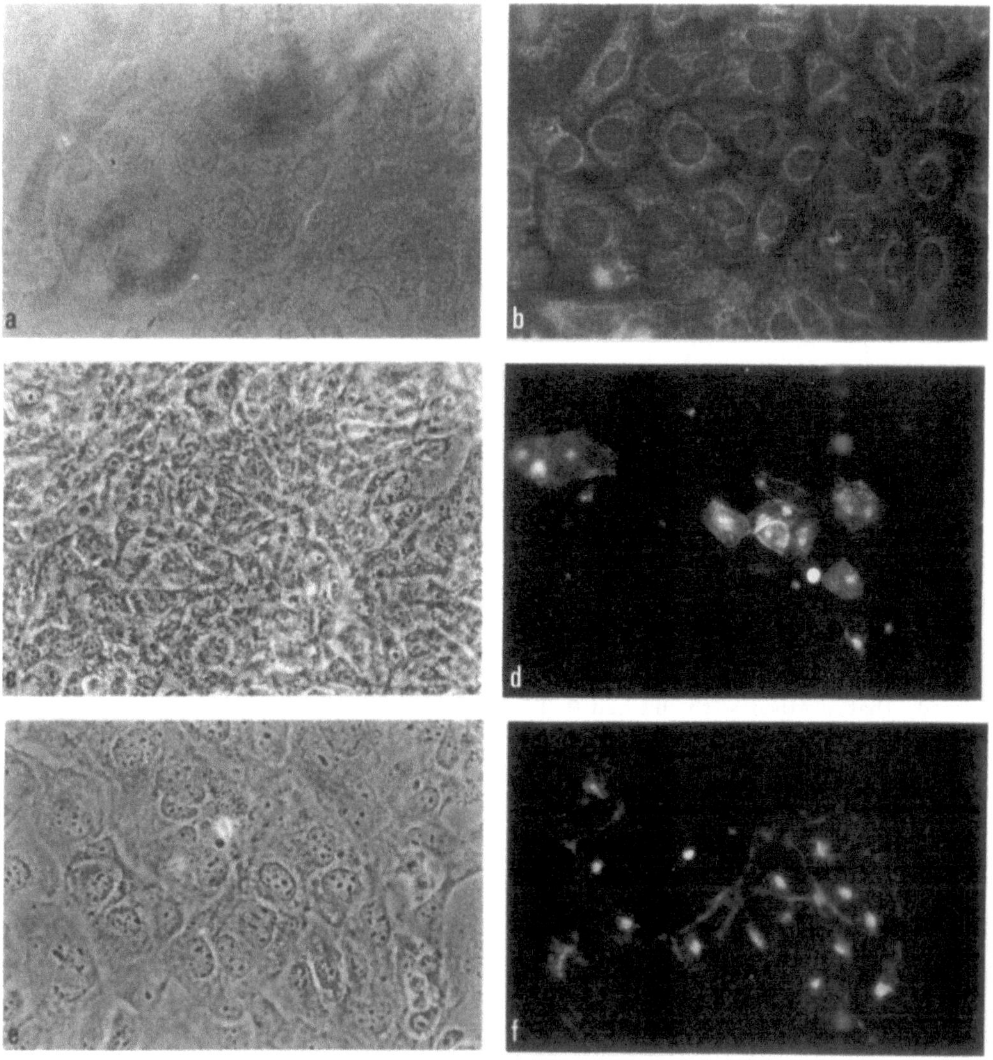

Fig. 1. Cunfluent cultures of END-2 (a, b), EPI-7 (c, d), and PSA-5E (e, f); phase contrast (a, c, e); FT-1 fluorescence (b, d, f).

Mesoderm formation in cocultures

Monolayer cultures of the P19-derived END-2 cell line were treated with mitomycine-C to prevent further proliferation, and subsequently single cell suspensions of P19EC cells were added. During the first week of coculture P19 cells formed undifferentiated aggregates attached to the END-2 monolayer. Figure 2a-d shows typical cultures on day 5 of coculture with SSEA-1 positive P19 aggregates and TROMA-1 positive END-2 cells. During the second week P19 aggregates differentiated into a variety of cell types, including epithelioïd, fibroblastoid and endodermlike cells, growing as monolayers or multiple cell layers over the END-2 cells. Some neurone-like cells were formed which disappeared within a week.

Approximately two weeks after the initiation of coculture progressively greater areas of contractile muscle cells were observed. These foci were stainable with anti-myosin antibodies (Figure 2 e-f). After three to four weeks adipocytes were formed, that stained with Oil red O (Figure 2 g-h).

For comparison cultures were studied with MES-1 and EPI-7, both P19-derived cell lines, as well as with PSA-5E, Dif 5, and PYS-2. Results are summarized in Table 2. Cocultures of P19 cells with EPI-7 and PSA-5E developed in a manner similar to the END-2 cocultures with the exception that beating muscle was rarely observed. In coculture with PYS-2 and MES-1 however, P19 cells did not form aggregates, but formed a second monolayer over the supporting layer of differentiated cells. These P19 cells remained undifferentiated.

Mitomycine-C treated feeder layers gave results comparable to untreated monolayers except in the case of Dif 5 cells. Mitomycine-C treated Dif 5 monolayers lysed and detached within a week, while the P19 cells inoculated formed an EC monolayer. However on non-mitomycine-C treated Dif 5 monolayers P19 cells formed large aggregates that adhered to the Dif 5 layer but formed no outgrowths even after four weeks. The surface of these aggregates stained with SSEA-1 (Figure 4) indicating they consisted of EC cells. These results suggest that mesoderm

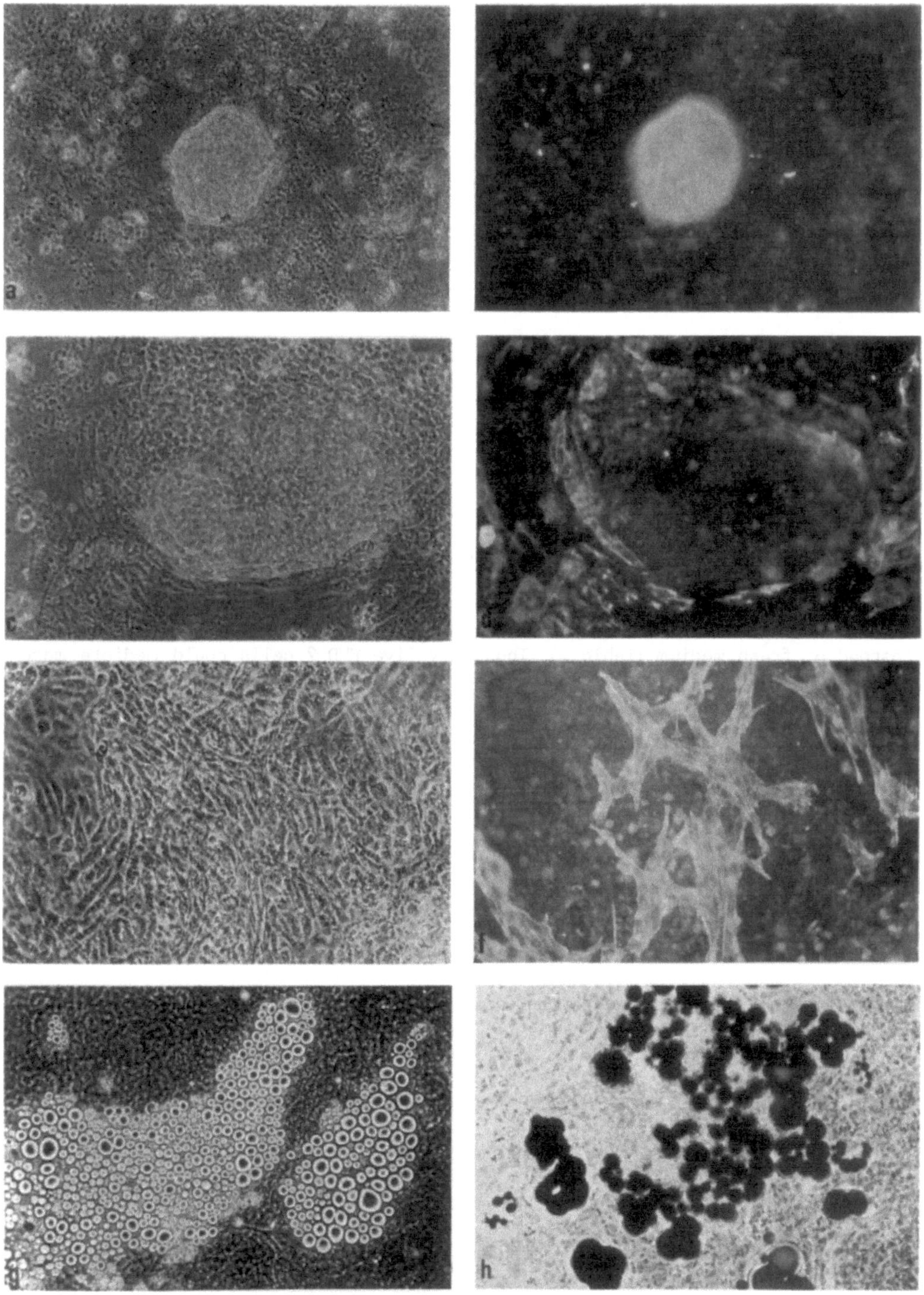

Fig. 2. Cocultures of P19 cells on mitomycin-C-treated monolayers of END-2 cells. SSEA-1 stained aggregate at day 5 (a,b); TROMA-1-stained feeder layer at day 5 (c,d); MF20 stained muscle cells at day 14 (c,f); oil Red O-stained adipocytes at day 30 (g); adipocytes in culture at day 30 (h). a,c,e,g,h: phase contrast; b,d,f: fluorescence. x250.

formation, as indicated by muscle and adipocyte differentiation, specifically occurs on visceral endoderm-like cells. Furthermore, aggregation is a necessary but not sufficient prerequisite for mesoderm formation.

The nature of mesoderm formation

To determine the requirements for muscle and adipose differentiation more precisely, cocultures were examined in which the END-2 feeder layer was either fixed with methanol or lysed with EGTA to leave only an extracellular matrix. Under these conditions, P19 cells did not aggregate, but formed a monolayer of EC cells. The same was found when such cultures were fed with END-2 conditioned medium instead of fresh medium (Table 2). Thus only live END-2 cells could mediate mesoderm formation in this system.

To find out whether diffusible factors played a role in this system, P19 cells were aggregated for 4 days in conditioned medium from END-2 cultures. After plating on tissue culture plastics such aggregates formed beating muscle foci more rapidly than when aggregated in fresh medium in only two out of five experiments (data not shown). This lack of consistent effect may be due to instability of the hypothesized factor. P19 cells were therefore also allowed to aggregate for four days on feeder layers covered with an agar layer to prevent intercellular contacts but not diffusion of soluble factors. The resulting aggregates were plated on tissue culture substrates where they adhered and formed outgrowths. Aggregates that had formed over END-2 monolayers or in the presence of DMSO gave rise to beating muscle foci significantly earlier than aggregates that had formed over MES-1 monolayers or in the absence of a feeder layer (Table 3). This is suggestive of a differentiation-promoting factor produced by the END-2 feeder layer.

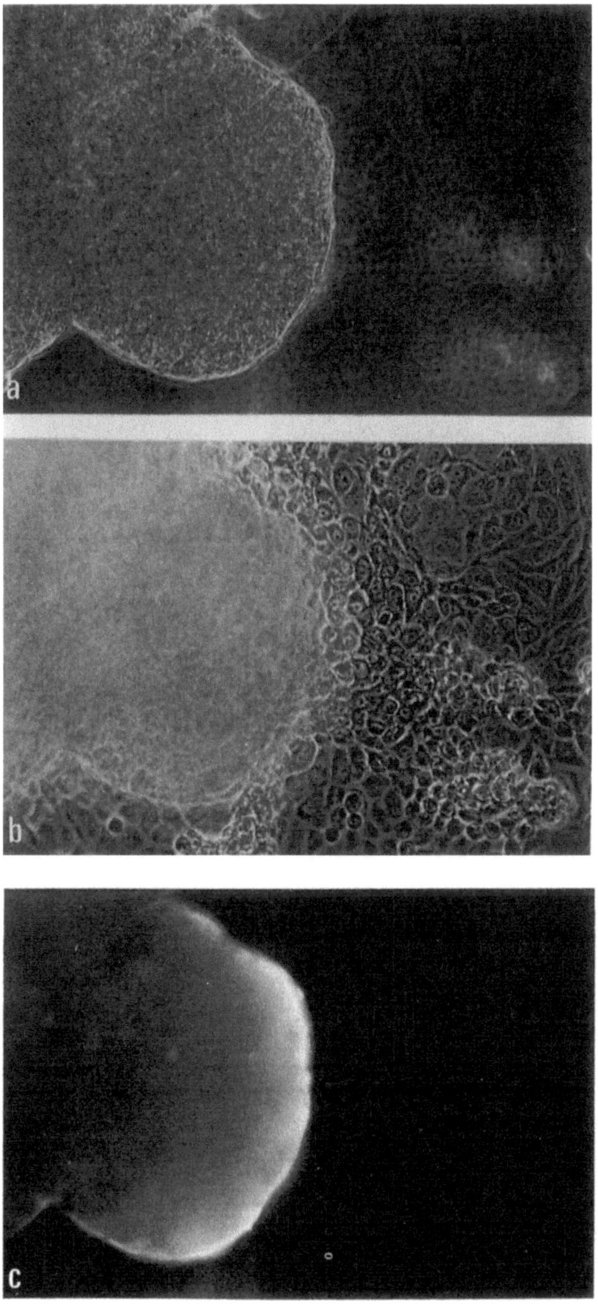

Fig. 3. Coculture of P19 cells on a monolayer of Dif 5 cells at day 30. Phase contrast with aggregate focussed (a); phase contrast with feeder focussed (b); SSEA-1 fluorescence (c). x250.

Table 2 Effect of various feeder layers on differentiation of P19 EC cells.

feeder cell treatment		P19 differentiation in coculture	
		aggregation	differentiation
END-2	M+/-	+	+ (486)
EPI-7	M+/-	+	+ (44)
PSA-5E	M+/-	+	+ (8)
Dif 5	M -	+	-
Dif 5	M +	-	-
PYS-2	M+/-	-	-
MES-1	M+/-	-	-
END-2	EGTA(+/-CM)	-	-
END-2	MEOH(+/-CM)	-	-

Monolayers of feeder cells were treated with mitomycin-C (M+) or not (M-), lysed with EGTA, or fixed with with methanol (MeOH) and inoculated with 5.10^5 P19 EC cells per 25 cm^2 in 5 ml of either DF +7.5% FCS or fresh END-2 conditioned medium (CM). Cocultures were maintained for four weeks and aggregation and differentiation were scored. Differentiation: +: extensive morphological differentiation, with number of beating muscle foci per 10cm^2 at day 14 in brackets; -: no differentiation.

DISCUSSION

In the present study EC cells and their differentiated derivatives were used to establish an in vitro model for mesoderm formation in murine embryogenesis. To that end several endoderm-like cell lines were characterized. These cell lines were used to establish cocultures with P19 EC cells. After inoculation as single cell suspensions on monolayers of visceral endoderm-like cell lines (END-2, EPI-7, PSA-5E) P19 cells formed aggregates that differentiated into a variety of cell types, including mesoderm-derived muscle and adipose cells. Differentiation did not occur in coculture with parietal endoderm (PYS-2) or mesoderm (MES-1) cell lines. Dif 5, which expresses characteristics of both visceral and parietal endoderm (14) did promote aggregation but not differentiation of P19 cells. Two explanations for this finding can be envisaged. First, Dif 5 cells, although allowing aggregation and extensive growth of the aggregates, may fail to produce a differentiation inducer necessary for mesoderm formation. Second, Dif 5 cells may produce an inhibitor of differentiation by which the otherwise spontaneously occurring differentiation of P19 aggregates is blocked. The differentiation-promoting ability of the various cell lines correlates with their visceral endoderm characteristics, as is most obvious from the FT-1 staining. Thus mesodermal differentiation in this system occurs specifically on visceral endoderm like cells, thereby providing an excellent model for mesoderm formation in embryogenesis, where embryonic ectoderm (the embryonic equivalent of EC cells) forms mesoderm in the area adjacent to visceral endoderm (cf. END-2 cells).

Table 3 Differentiation of P19-aggregates established in cocultures without cellular contacts between the two cell types.

Inducer in coculture	Days after plating when beating muscle was first observed			Number of beating muscle foci at day 14.**
	exp 1	exp 2	exp 3	
END-2	9	10	10	43
DMSO 1%	9	10	9	15
none	15	14	14	3
MES-1	N.D.*	N.D.	14	5

Monolayers of END-2 or MES-1 or empty dishes were covered with an agar layer and subsequently inoculated with $5x10^5$ P19 cells in DF+7.5% FCS with or without 1% DMSO. After four days, aggregates were taken and plated in 25 cm^2 flasks coated with gelatin. Cultures were maintained and checked for beating muscle daily.

* N.D.: not determined

** Mean of 3 representative cultures

Importantly chemical induces such as DMSO (10) are not required for inducing differentiation in this coculture system. The substratum formed by END-2 cells has intriguing properties, allowing concomittantly aggregation of P19 cells and their adherence to the feeder layer surface. After fixation or lysis of the feeder layer this capacity was apparently lost and not recovered upon addition of END-2 conditioned medium. Similarly, Smith et.al. (17) did not find muscle differentiation-promoting activity in conditioned medium of endoderm. However in cocultures in which an agar layer separated the P19 cells from the differentiated cell lines (table 3) END-2 did promote differentiation relative to controls, suggesting a soluble mediator diffusing through the agar barrier. Over such agar layers, P19 would form aggregates irrespective of the underlying cell line. Also in controls some of these aggregates gave rise to beating muscle, although 4 to 6 days later than when formed over END-2, or in the presence of DMSO. Smith et.al. (17) also noted muscle cell development in P19 aggregates that were not treated with DMSO. Their aggregates had an outer layer of visceral endoderm which may have promoted muscle formation. Similarly the development of beating muscle in our control aggregates formed on agar after 14 to 15 days may be explained as a consequence of spontaneous formation of visceral endoderm on these aggregates. Taken together the available data imply that in differentiation of P19 aggregates visceral endoderm formation precedes mesoderm formation and promotes it via a diffusible mediator. Accordingly, in cocultures the END-2 cells promoted mesoderm differentiation directly, thereby shortening the time scale of beating muscle formation by 4-6 days (table 3).

Promotion of growth but not differentiation of EC cells by endoderm cells has been described before (5,7). Several reports describe mesoderm-inducing factors in Xenopus. Smith (16) reported a mesoderm-inducing factor produced by a Xenopus cell line that promoted muscle and notochord differentiation in animal pole regions of the Xenopus blastula. Similarly in the mouse, cocultures between EC cells and embryo-derived visceral endoderm may reveal whether the soluble mediator in our system has a physiological counterpart. Slack et.al. (15) showed that heparin-binding growth factors induce mesoderm in Xenopus embryos. In our system neither basic fibroblast growth factor nor heparin had an effect (not shown). Further study of the factor involved here awaits a medium containing mesoderm-inducing activity.

ACKNOWLEDGEMENTS

We thank Dr. S.K. Brahma, R. Kemler, T. Muramatsu, D. Solter and D. Fischman for generously providing us with antibodies, and Dr. M. Buckingham for the pMHC32 probe. The valuable suggestions of Dr. M.W. McBurney are gratefully acknowledged. Ms S.N. Rusch, Ms D. Steggink and Ms A. Brest van Kempen are thanked for expert secretarial assistance.

REFERENCES

Adamson ED, Evans MJ, Magrane GG (1977) Biochemical markers of the progress of differentiation in cloned teratocarcinoma cell lines. Eur J Biochem.79:607–615

Bader D, Masaki T, Fischman DA (1982) Immunochemical analysis of myosin heavy chain during avian myogenesis in vivo and in vitro. J Cell Biol 95:763–770

Beddington R (1983) The origin of foetal tissues during gastrulation in the rodent. In: Development in mammals vol.5 Ed. M.H. Johnson. Elsevier Science Publishers

Darmon M, Bottenstein J, Sato G, (1981) Neural differentiation following culture of embryonal carcinoma cells in a serum-free defined medium. Devel Biol 85:463–473

Germain EL, Littlefield JW (1986) Endoderm-secreted factor stimulates growth of embryonal carcinoma stem cells. In Vitro Cell and Devel Biol 22:107–112

Hogan BLM, Barlow DP, Tilly R (1983) F9 teratocarcinoma cells as a model for the differentiation of parietal and visceral endoderm in the mouse embryo. Cancer Surveys 2:115–140

Isacke CM, Deller MJ (1983) Teratocarcinoma cells exhibit growth cooperativity in vitro. J Cell Phys 117:407–414

Kemler R, Brûlet P, Schnebelen MT, Gaillard J, Jacob F (1981) Reactivity of monoclonal antibodies against intermediate filament proteins during embryonic development. J Embryol exp Morph 64:45–60

Lehman JM, Speers WC, Swartsendruber DE, Pierce GB (1974) Neoplastic differentiation: characteristics of cell lines derived from a murine teratocarcinoma. J Cell Physiol 84:13–28

McBurney MW, Jones-Villeneuve EMV, Edwards MKS, Anderson PJ (1982) Control of muscle and neuronal differentiation in a cultured embryonas carcinoma cell line. Nature 299:165–167

Mummery CL, Feyen A, Van der Saag PT, Van den Brink CE, De Laat SW (1985) Clonal variants of differentiated P19 embryocarcinoma cells exhibit epidermal growth factor receptor kinase activity. Devel Biol 109:402–410

Mummery CL, Feyen A, Moolenaar WH, Van den Brink CE, Laat SW (1986) Establishment of a differentiated mesodermal line from P19 EC cells expressing functional PDGF and EGF receptors. Exp Cell Res 165:229–242

Muramatsu T (1984) Cell surface glycoproteins as markers in monitoring in vitro differentiation of embryonal carcinoma cells. Cell Differentiation 15:101–108

Nagarajan L, Jetten AM, Anderson WB (1983) A new differentiated cell line (Dif.5) derived by retinoic acid treatment of F9 teratocarcinoma cells capable of extracellular matrix production and growth in the absence of serum. Exp Cell

Res 147:315–327
Slack JMW, Darlington BG, Heath JK, Godsave SF (1987) Mesoderm induction in early Xenopus embryos by heparin-binding growth factors. Nature 326:197–200
Smith JC (1987) A mesoderm-inducing factor is produced by a Xenopus cell line. Development 99:3–14
Smith SC, Reuhl KR, Craig J, McBurney MW (1987) The role of aggregation in embryonal carcinoma cell differentiation. J Cell Physiol 131:74–84
Solter D, Knowles BB (1978) Monoclonal antibody defining a stage-specific mouse embryonic antigen (SSEA-1). P N A S 75:5565–5569
Strickland S, Mahdavi V (1978) The induction of differentiation in teratocarcinoma stem cells by retinoic acid. Cell 15:393–403

Platelet-derived Growth Factor and its Role in Mitogenesis and Transformation

B. Westermark*, C.-H. Heldin+

*Dept.of Pathology
University Hospital
S-751 85 Uppsala
Sweden

Introduction

Human platelet-derived growth factor (PDGF) is an M_r 30,000 protein made up as dimers of A and B chains (Johnsson et al. 1982) which are encoded by separate genes (Betsholtz et al. 1986a and references therein). The two chains are synthesized as prepropeptides that are processed by dimerization and proteolytic cleavage. Within the amino acid sequences of the mature chains, there is a 60% amino acid sequence homology indicating that the two genes have evolved through gene duplication and evolutionary divergence. The possible evolutionary advantage of the development of two closely related PDGF genes is not understood. However, it has recently become evident that all three possible dimeric forms of PDGF exist and that they may display differences in their functional activities.

PDGF was orginally identified as a constituent of platelet α-granules. The storage site in conjunction with its restricted target cell specificity (connective tissue- and glia-derived cells) led to the widely accepted view that its physiological role is to stimulate cell proliferation in the acute phase of wound healing. In recent years, PDGF has been found be synthesized by a large number of normal cell types, such as endothelial cells (Dicorleto and Bowen-Pope, 1983; Barrett et al. 1984), macrophages (Shimokado et al. 1985; Martinet et al. 1986), placental

+Ludwig Institute for Cancer Research
Biomedical Center
Box 595
S-751 23 Uppsala
Sweden

NATO ASI Series, Vol. H26
Cell to Cell Signals in Mammalian Development
Edited by S. W. de Laat et al.

cytotrophoblasts (Goustin et al. 1985), vascular smooth muscle cells (Seifert et al. 1984) and foreskin fibroblasts (Paulsson et al. 1987). It is therefore assumed that PDGF may be more widely implicated in the regulation of proliferation and perhaps also in other cellular processes than implied from its presence in platelets. Most of the current interest in PDGF is generated by its structural homology with the oncogene product of simian sarcoma virus (SSV), since this infers a role of PDGF-like growth factors in neoplastic transformation and tumorigenesis.

Structural aspects on platelet-derived growth factor

The amino acid sequence of the PDGF A chain has been deduced from cDNA clones isolated from a human glioma cell library (Betsholtz et al. 1986a; Rorsman et al. 1988). Two types of clones were isolated which differed in the 3' protein-coding sequences. One type, represented by clone D1, predicts a 211-amino-acid PDGF A chain precursor with a highly basic carboxyterminus. The other type encodes a product that is 15 residues shorter and has an alternative sequence of the three most carboxyterminal amino acid residues (Rorsman et al. 1988). Characterization of genomic A chain sequences showed that the two types of transcripts probably arise by alternate splicing; exon 6 is only present in clone D1 and contains the coding sequence for the 18 carboxyterminal residues of the long A chain, whereas the carboxyterminus of the short A chain variant is encoded by exon 7. The overall organization of the B chain gene is similar to that of the A chain gene (Rao et al. 1986; Rorsman et al. 1988), as expected from their common origin. The human B chain mRNA has been identified as a single 3.5 kb species (Rao et al. 1986) whereas the A chain mRNA occurs as three species, ranging from 1.9 to 2.8 kb (Betsholtz et al. 1986a). This is most likely due to an alternative utilization of polyadenylation signals, several of which are found in the 3' end of the A chain gene (Rorsman et al. 1988).

Using different sources for isolation, all three dimeric forms of PDGF have been indentified. Recent studies have shown that human platelet PDGF is assembled as a heterodimer of A and B chains (PDGF-AB) (Hammacher et al. 1988a). Conclusive evidence was obtained by the use of immobilized metal affinity chromatography, which resolves all three dimeric forms of PDGF. Structural analysis of PDGF-like growth factors released by human osteosarcoma (Heldin et al. 1986) and melanoma (Westermark et al. 1986) cell lines has shown that they are homodimers of A chains (PDGF-AA). Porcine platelet-PDGF is a homodimer of B chains (PDGF-BB) (Stroobant and Waterfield, 1983).

Interestingly, even in cells expressing both A and B chain mRNA at high levels, such as osteosarcoma and glioma cells (Betsholtz et al. 1986a), virtually only PDGF-AA is found in the conditioned medium (Heldin et al. 1986; Hammacher et al. 1988b). In the glioma-conditioned medium, small amounts of PDGF-AB were found and only traces of PDGF-BB. Thus, in cells expressing both transcripts, all three dimeric forms of PDGF can be synthesized although at markedly different efficiencies. The preponderance of PDGF-AA is most likely due to a relatively low efficiency in B chain translation; the B chain transcript has a long GC rich 5′ segment which adversely affects translation (Rao et al. 1988).

Structural relationship between the PDGF B chain gene and the oncogene v-sis of simian sarcoma virus

The transforming potential of the B chain of PDGF became apparent through the finding of a close structural homology with the v-sis oncogene product $p28^{sis}$ of SSV (Devare et al. 1983; Waterfield et al. 1983; Doolittle et al. 1983). Molecular cloning of the cellular homolog of v-sis, i.e. the proto-oncogene c-sis, clearly established its identity with the structural gene encoding the PDGF B chain precursor (Josephs et al. 1984; Johnsson et al. 1984). The virus has acquired most of the protein coding sequences of the cellular gene, however, leaving out the entire first exon encoding the 5′ untranslated segment as well as the signal peptide. The loss of signal sequences is apparently compensated for by the upstream viral env segment that encodes a hydrophobic leader peptide (King et al. 1985). Most likely, the loss of 5' untranslated sequences in v-sis is of importance for the transforming activity of SSV since it can be expected to increase the efficiency of the translation of the v-sis message.

Evidence that SSV-transformation is mediated by a homodimer of PDGF B chains

A body of evidence has been presented in favor of the notion that SSV-transformation of cultured cells is mediated by an autocrine growth factor that is structurally similar to PDGF-BB (reviewed in Westermark et al. 1987). The primary translation product of v-sis is dimerized and proteolytically trimmed to yield a 24 kDa product (Robbins et al. 1983). The autocrine hypothesis is strengthened by the target cell specificity of SSV; the virus transforms PDGF-responsive cells only (Leal et al. 1985). Moreover, SSV-transformation can be blocked and reverted by PDGF-antibodies (Johnsson et al. 1985) or by suramin, a drug that displaces PDGF from its receptor (Betsholtz et al. 1986b). The morphological and other phenotypic properties of SSV-transformed cells in culture are entirely compatible with the view that SSV generates a powerful autocrine growth signal but does not evoke any other cellular response and does not immortalize the infected cells (Johnsson et al. 1986). To explain the development of fully malignant tumors, i.e. fibrosarcomas and gliomas, in experimental animals administered with SSV (Deinhardt, 1980), one has to invoke additional genetic lesions afflicting the infected cell population.

Production of PDGF by human tumor cells in culture

Several human tumor cell lines have been found to express PDGF A and/or B chain mRNA (reviewed in Heldin et al. 1987). Many of these, such a glioma and sarcoma cell lines, are derived from cells that normally respond to PDGF. The production of PDGF receptors in these cells has been confirmed. Many of the cell lines express PDGF receptor transcripts and specifically bind ^{125}I-PDGF (Nistér et al. 1988a). It is thus possible that the endogenous growth factor has an autocrine function and plays a role in the pathogenesis of the neoplastic process. However, attempts to inhibit growth of glioma or sarcoma cells using PDGF antibodies have so far been futile. This finding does not invalidate the autocrine hypothesis in the patho-genesis of sarcoma and glioma since autocrine stimulation may be an early event in tumor development. Subsequent changes in the cell genome may abolish the growth factor requirement. Precedence for such a situation has been found in hemopoietic cells: expression of a GM-CSF cDNA vector in factor dependent cells renders the cells factor-independent and inhibitable by GM-CSF antibodies. Upon culture, the transfected cells give rise to clones which still produce the growth factor but their growth is not inhibited by antibodies (Laker et al. 1987).

Production of PDGF is not an exclusive property of tumor cell lines that are derived from PDGF-responsive tissues. Both melanoma (Westermark et al. 1986) and mammary carcinoma (Perez et al. 1987) produce PDGF at a relatively high frequency. If the corresponding cells produce PDGF in vivo, the growth factor may contribute to the stimulation of stroma growth.

PDGF production in normal cells

Smooth muscle cells (Seifert et al. 1984) and placental cytotrophoblasts (Goustin et al. 1985) produce PDGF and express PDGF receptors. It is thus possible that PDGF production in these cells serves an autocrine function, although formal proof for this hypothesis has not been produced. Recently it was observed that stimulation of fibroblasts with mitogens, including PDGF, leads to the expression of PDGF A chain mRNA and production of functional A chain dimes (Paulsson et al. 1987). This finding suggests the possibility that a positive feed-back mechanism operates in growth stimulation.

Endothelial cells (Dicorleto and Bowen-Pope, 1983) and activated macrophages

(Shimokado et al. 1985; Martinet et al. 1986) are examples of cells that produce PDGF but do not carry the cognate receptor. It therefore seems possible that PDGF released by these cells has a function as a paracrine growth factor in inflammation and tissue repair.

A role of paracrine PDGF in the development of the central nervous system has recently been suggested. Cultured type 1 astrocytes derived from opticus nerve were found to release a PDGF-like growth factor that stimulates the growth of the bipotential O-2A progenitor cells (Richardson et al. 1988).

Functional activities of different dimeric forms of PDGF

Recent studies using PDGF-AB purified from human platelets and PDGF-AA obtained from the conditioned medium of human glioma cells indicated differences in functional activity. Thus, PDGF-AA had lower mitogenic activity, lower ability to stimulate actin reorganization and membrane ruffling and had no chemotactic activity on fibroblasts (Nistér et al. 1988b).

In order to compare the transforming activities of the two PDGF genes we have recently constructed retrovirus vectors containing full length cDNAs corresponding to the A and B chain coding sequences. Infection of Rat 1 cells with these retro-virus constructs demonstrated a significant difference in transforming activity of the two genes. In spite of similar levels of mRNA expressed and amount of dimers produced by infected cells, only cells infected with the B construct attained a transformed phenotype, comparable to that seen in SSV-transformed cultures (Bywater et al. 1988).

On the basis of the finding described above, we conclude that a full mitogenic response is only elicited by PDGF dimers containing the B chain (PDGF-AB, PDGF-BB). This notion has been confirmed using recombinant PDGF-dimers obtained from yeast or CHO cell-based expression systems (Östman et al. in preparation).

Evidence for two separate PDGF receptor types on human fibroblasts

The differences in functional activities between the various dimeric forms of PDGF may indicate that there exists more than one type of PDGF receptor. This view has recently gained support from cross-competition experiments (Heldin et al. 1988). The previously identified PDGF receptor, displaying protein tyrosine kinase activity, was found to bind only PDGF-AB and PDGF-BB; it was therfore denoted PDGF receptor type B. A novel PDGF receptor species was identified which binds all three dimeric forms of PDGF; it was denoted PDGF receptor type A. The different abilities of the dimeric forms of PDGF in inducing cell growth and transformation indicate that the major mitogenic effect of PDGF is mediated by the B type receptor.

Conclusion and perspective

Not long ago, PDGF was thought to exist as one defined molecule and function via the binding to a single class of receptors. We now know that PDGF occurs as several isoforms. Alternate splicing of the A chain transcript may give rise to two different A chains. Moreover, the A and B chain can be assembled as homodimers or as a heterodimer. The various dimers have different functional activities which are exerted via the binding to at least two separate receptor types.

Characterization, molecular cloning and tissue distibution of the A type of PDGF receptor will help us to better understand the biological role of PDGF-AA, which is enigmatic at the moment.

REFERENCES

Barrett TB, Gajdusek CM, Schwartz SM, McDougall JK, Benditt EP (1985) Expression of the sis gene by endothelial cells in culture and in vivo. Proc Natl Acad Sci USA 81:6772-6774

Betsholtz C,. Johnsson A, Heldin C-H, Westermark B, Lind P, Urdea MS, Eddy R, Shows TB, Philpott K, Mellor AL, Knott TJ, Scott J (1986a) cDNA sequence and chromosomal localization of human platelet-derived growth factor A-chain and its expression in tumor cell lines. Nature 320:695-699

Bywater M, Rorsman F, Bongcam-Rudloff E, Mark G, Heldin C-H, Westermark B, Betsholtz C (1988) Expression of recombinant PDGF A and B chain homodimers in Rat 1 cells and human fibroblasts reveals differences in protein processing and autocrine effects. Submitted

Betsholtz C, Johnsson A, Heldin C-H, Westermark B (1986) Efficient reversion of SSV-transformation and inhibition of growth factor-induced mitogenesis by suramin. Proc Natl Acad Sci USA 83:6440-6444

Deinhardt F (1980)The biology of primate retroviruses. In Klein G (ed) Viral Oncology. New York Raven Press pp 359-398

Devare SG, Reddy EP, Law JD, Robbins KC, Aaronson SA (1983) Nucleotide sequence of the simian sarcoma virus genome: demonstration that its acquired cellular sequences encode the transformation gene product p28 ^{sis}. Proc Natl Acad Sci USA 80:731-735

Dicorleto PE, Bowen-Pope DF (1983) Cultured endothelial cells produce a platelet-derived growth factor-like protein. Proc Natl Acad Sci USA 80:1919-192

Doolittle RF, Hunkapiller MW, Hood LE, Devare SG, Robbins KC, Aaronson SA, Antoniades HN (1983) Simian sarcoma virus oncogene, v-sis, is derived from the gene (or genes) encoding a platelet-derived growth factor. Science 221:275-277

Goustin AS, Betsholtz C, Pfeifer-Ohlsson S, Persson H, Rydnert J, Bywater M, Holmgren G, Heldin C-H, Westermark B, Ohlsson R (1985) Co-expression of sis and myc proto-oncogenese in human placenta suggest autocrine control of trophoblast growth. Cell 41:301-312

Hammacher A, Hellman U, Johnsson A, Östman A, Gunnarsson K, Westermark B, Wasteson A, Heldin C-H (1988a) PDGF purified from human platelets is a heterodimer of one A and one B chain. Submitted

Hammacher A, Nistér M, Westermark B, Heldin C-H (1988b) A human glioma cell line secretes three structurally and functionally different dimeric forms of PDGF. Submitted

Heldin C-H, Johnsson A, Wennergren S, Wernstedt C, Betsholtz C, Westermark B (1986) A human osteosarcoma cell line secretes a growth factor structurally related to a homodimer of PDGF A chains. Nature 319:-514

Heldin C-H, Betsholtz C, Claesson-Welsh L, Westermark B (1987) Subversion of growth regulatoy pathways in malignant transformation. Biochim Biophys Acta 907:219-244

Heldin C-H, Bäckström G, Östman A, Hammacher A, Rönnstrand L, Rubin K, Nistér M, Westermark B (1988) Different dimeric forms of PDGF differ in binding to separate receptors on human fibroblasts. Submitted

Johnsson A, Heldin C-H, Wasteson A, Westermark B, Deuel TF, Huang JS, Seeburg PH, Gray A, Ullrich A, Scrace G, Stroobant P, Waterfield MD (1984) The c-sis gene encodes a precursor of B chain of platelets-derived growth factor. EMBO J 3:921-928

Johnsson A, Betsholtz C, Heldin C-H, Westermark B ((1985) Antobodies against platelet-derived growth factor inhibit acute transformation by simian sarcoma virus.Nature 317:438-440

Johnsson A, Betsholtz C, Heldin C-H, Westermark B (1986) The phenotypic characteristics of simian sarcoma virus-transformed human fibroblasts suggest that the v-sis gene product acts solely as a PDGF receptor agonist in cell transformation. EMBO J 5:1535-1541

Josephs SF, Ratner L, Clarke MF, Westin EH, Reitz MS, Wong-Staal F (1984) Transforming potential of human c-sis nucleotide sequences encoding platelet-derived growth factor. Sience 225:636-639

King CR, Giese NA, Robbins KC, Aaronson SA (1985) In vitro mutagenesis of the v-sis transforming gene defines functional domains of its growth factor-related product. Proc Natl Acad Sci USA 82:5295-5299

Laker C, Stocking C, Bergholz U, Hess N, De Lamarter JF, Ostertag W (1987) Autocrine stimulation after transfer of the granulocyte/macrophage colony-stimulating factor gene and autonomous growth are distinct but independent steps in the oncogenic pathway. Proc Natl Sci USA 84:8458-8462

Leal F, Williams LT, Robbins KC, Aaronson KC (1985) Evidence that the v-sis gene product transforms by interaction with the receptor for platelet-derived growth factor. Science 230:327-330

Martinet Y, Bitterman PB, Mornex J-F, Grotendorst G, Martin GR, Crystal RG (1986) Activated human monocytes express the c-sis proto-oncogene and release a mediator showing PDGF-like activity. Nature 319:158-160

Nistér M, Libermann T, Betsholtz C, Pettersson M, Claesson-Welsh L, Heldin C-H, Schlessinger J, Westermark B (1988a) Expression of messenger RNAs for platelet-derived growth factor and transforming growth factor-α and their receptors in human malignant glioma cell lines. Cancer Res in press

Nistér M, Hammacher A, Mellström K, Siegbahn A, Rönnstrand L, Westermark B, Heldin C-H (1988b) A glioma-derived PDGF A chain homodimer has different functional activities than a PDGF AB heterodimer purified from human platelets. Cell in press

Paulsson Y, Hammacher A, Heldin C-H, Westermark B (1987) Possible autocrine feedback in the prereplicative phase of human fibroblasts. Nature 328:715-717

Peres R, Betsholtz C, Westermark B, Heldin C-H (1987) Frequent expression for growth factors for mesenchymal cells in human mammary carcinoma. Cancer Res 47:3425-3429

Rao CD, Igarashi H, Chiu I-M, Robbins KC, Aaronson SA (1986) Structure and sequence of the human c-sis/platelets-derived growth factor 2 (SIS/PDGF 2) transcriptional unit. Proc Natl Acad Sci USA 83:2392-2396

Rao CD, Pech M, Robbins KC, Aaronson SA (1988) The 5′ untranslated sequence of c-sis/platelet-derived growth factor 2 transcript is a potent translational inhibitor. Mol Cell Biol 8:284-292

Richardson WD, Pringle N, Mosley MJ, Westermark B, Dubois-Dalcq M (1988) A role for platelet-derived growth factor in normal gliogenesis in the central nervous system. Cell in press

Robbins KC, Antoniades HN, Devare SG, Hunkapiller MW, Aaronson SA (1983) Structural and immunological similarities between simian sarcoma virus gene product(s) and human platelet-derived growth factor. Nature 305:605-608

Rorsman F, Bywater M, Knott TJ, Scott J, Betsholtz C (1988) Structural characterization of the human platelet-derived growth factor A-chain cDNA and gene: alternative exon usage predicts two different precursor proteins. Mol Cell Biol in press

Seifert RA, Schwartz SM, Bowen-Pope DF (1984) Developmentally regulated production of platelet-derived growth factor-like molecules. Nature 311:669-671

Shimokado K, Raines EW, Madtes DK, Barrett TB, Benditt EP, Ross RA (1985) A significant part of macrophage-derived growth factor consists of at least two forms of PDGF. Cell 43:277-286

Stroobant P, Waterfield MD (1983) Purification and properties of porcine platelet-derived growth factor EMBO J 3:2963-2967

Waterfield MD, Scrace GT, Whittle N, Stroobant P, Johnsson A, Wasteson A, Westermark B, Heldin C-H, Huang JS, Deuel TF (1983) Platelet-derived growth factor is structurally related to the putative transforming protein $p28^{sis}$ of simian sarcoma virus. Nature 304:35-39

Westermark B, Johnsson A, Paulsson Y, Betsholtz C, Heldin C-H, Herlyn M, Rodeck U, Koprowski H (1986) Human melanoma cells of primary and metastatic origin express the genes encoding the constituent chains of PDGF and produce a PDGF-like growth factor. Proc Natl Acad USA 83:7197-7200

Westermark B, Betsholtz C, Johnsson A, Heldin C-H (1987) Acute transformation by simian sarcoma virus is mediated by an externalized PDGF-like growth factor. In Viral Carcinogenesis. Kjeldgard NO, Forchhammer J (eds) Munksgaard, Copenhagen, pp 445-457.

The Role of Polypeptide Growth Factors in Phenotypic Transformation of Normal Rat Kidney Cells

E.J.J. van Zoelen
Hubrecht Laboratory
Netherlands Institute for Developmental Biology
Uppsalalaan 9
3584 CT UTRECHT
The Netherlands

Summary

Using growth factors-defined and serum-free culture media, we have shown that phenotypic transformation of normal rat kidney cells does not require specific polypeptide growth factors but more generally reflects the ability of these cells to respond to multiple growth factors. Epidermal growth factor and platelet-derived growth factor contribute to this process by a direct mitogenic action on these cells, modulating agents like transforming growth factor-β and retinoic acid have an indirect growth stimulating effect. Some aspects of the mechanism whereby these modulating agents can exert their action will be discussed.

Introduction

Transformed cells differ in many respects from their normal counterparts. Apart form there tumorigenic properties in vivo, transformed cells show an increased nutrient requirement and an enhanced rate of hexose uptake. On the other hand they have a reduced serum dependence and can be cultured relatively simply in defined serum-free media. Furthermore transformed cells exhibit an altered morphology, characterized by disorganized actin cables, have lost their density dependence of proliferation and can be cultured under anchorage-independent conditions in semi-solid media (Ozanne et al., 1982; Kaplan et al., 1982). In recent years it has become clear that transformed cells produce and secrete polypeptide hormones in particular the so-called transforming growth factors (TGFs), which are able to induce reversibly all of the above characteristics of transformed cells in non-

NATO ASI Series, Vol. H26
Cell to Cell Signals in Mammalian Development
Edited by S. W. de Laat et al.

transformed indicator cells in vitro (phenotypic transformation; Ozanne et al., 1982; Heldin and Westermark, 1984; Goustin et al., 1986). In particular the ability of these polypeptide growth factors to induce non-transformed cells to grow in semi-solid media such as soft agar has received much attention, since anchorage-independent proliferation is considered to be the best in vitro correlate of tumorigenicity in vivo (Shin et al., 1975; Cinofe and Fidler, 1980). Because of the reversibility of this process, phenotypic transformation can also be used as a model for studying both cellular alterations resulting from malignant transformation and loss of malignancy following cellular differentiation. In vivo, phenotypic transformation may be an important factor in tumor outgrowth, because polypeptide growth factors secreted by transformed cells may induce uncontrolled proliferation of the surrounding normal cells in a paracrine fashion. Detailed characterization of the polypeptide growth factor requirements for phenotypic transformation is therefore essential for understanding the cellular mechanisms underlying this process.

Normal rat kidney (NRK) cells have been widely used as a non-transformed indicator cell line for studying the role of polypeptide growth factors in this process. This cell line has gained its popularity from the original observations of De Larco and Todaro (1978) that murine sarcoma virus-transformed fibroblasts secrete so-called sarcoma growth factors, which are able to induce a transformed phenotype in these cells. Later studies revealed that this sarcoma growth factor activity was caused by the combined activity of two secreted growth factors, transforming growth factor (TGF)-α and TGFβ (Anzano et al., 1982). TGFα is structurally homologous to epidermal growth factor (EGF), and exerts it action through the EGF-receptor. Its presence is mainly restricted to transformed and embryonic tissue (for a recent review see Derynck (1986)). Because of the similarity in activity between EGF and TGFα, EGF is used in the majority of studies on the role of polypeptide growth factors in phenotypic transformation. TGFβ is a so-called modulating growth factor which depending on the target cell line used can either act as a growth inhibitor or as a growth stimulator by potentiating the transforming activity of other growth factors (for a recent review see Sporn et al., (1987)). The observation that a specific combination of EGF and TGFβ is required for phenotypic transformation of NRK cells (Anzano et al., 1982), contrasts with the observations made on a variety of other non-transformed indicator cell lines. In the case of AKR-2B cells, TGFβ by itself is sufficient to induce anchorage-independent proliferation (Moses et al., 1985). Based on studies in a variety of other mesenchymal cell lines, Kaplan and Ozanne (1983) have shown that phenotypic transformation does not require specific growth factors, but more generally reflects the ability of cells to react to the total concentration of various growth factors present in the assay.

A general problem inherent to all studies on the role of polypeptide growth factors in phenotypic transformation, is the fact that these experiments are generally carried out in serum-containing media. Serum by itself contains a variety of polypeptide growth factors, in particular platelet-derived growth factor (PDGF; Heldin and Westermark, 1984), insulin-like growth factors, TGFβ (Childs et al., 1982; Assoian et al., 1983; Rapp et al., 1983) and low concentrations of an EGF-like activity (Assoian et al., 1984a). The above problem is illustrated by the observation that when assayed in plasma-containing media instead of serum-containing media, anchorage-independent proliferation of NRK cells requires PDGF in addition to EGF and TGFβ (Assoian et al., 1984a). In this study we describe the use of growth factor-defined assay conditions for studying the role of polypeptide growth factors in phenotypic transformation of NRK cells, thus avoiding the contributions of growth factors from serum in the assay. In addition these conditions permit studies on the mechanism whereby TGFβ can induce its growth modulating effects in NRK cells.

Growth factor-defined culture conditions of NRK cells

Polypeptide growth factors generally require intact disulfide bridges for activity. We have used this observation by showing that treatment of fetal calf serum (FCS) with dithiothreitol followed by iodoacetamide, results in a serum preparation, designated SH-FCS, which lacks mitogenic activity completely (Van Zoelen et al., 1985). In addition these serum preparations have no detectable TGFβ activity, as demonstrated by the fact that residual soft agar colony formation was not observed in the presence of EGF (Van Zoelen et al., 1985). This so-called growth factor-inactivated serum still contains albumin, phospholipids, protease inhibitors and cell attachment factors, which are all factors that promote in vitro culturing of cell lines. When NRK cells are cultured in media supplemented with SH-FCS only, they become quiescent in the G_0-phase of the cell cycle. In the additional presence of insulin and EGF, however, cells proliferate rapidly, at a similar rate as in media supplemented with untreated FCS (Van Zoelen et al., 1985). This demonstrates that polypeptide growth factors are essential for proliferation of NRK cells. In addition it shows that by using growth factor-inactivated serum as a supplement, NRK cells can be cultured under growth factor-defined conditions, thus avoiding any contribution of growth factors from serum.

Various authors have shown that NRK cells can also be cultured under complete serum-free conditions. Bradshaw and Dubes (1983) have shown that NRK cells can be cultured in a medium supplemented with EGF, insulin, fibronectin and retinoic acid (RA). Cells can only be cultured for a very limited number of passages in this medium, and therefore this approach is not useful for studies on the induction of anchorage-independent proliferation. Still it is remarkable that RA, which has strong growth inhibitory and differentiation inducing effects on a variety of cell lines (reviewed by Schroder et al., 1983) can act as a strong growth stimulating factor in this cell line (Bradshaw and Dubes, 1983). Rizzino (1984) has developed a serum-free medium for NRK cells, supplemented with transferrin, laminin, insulin,

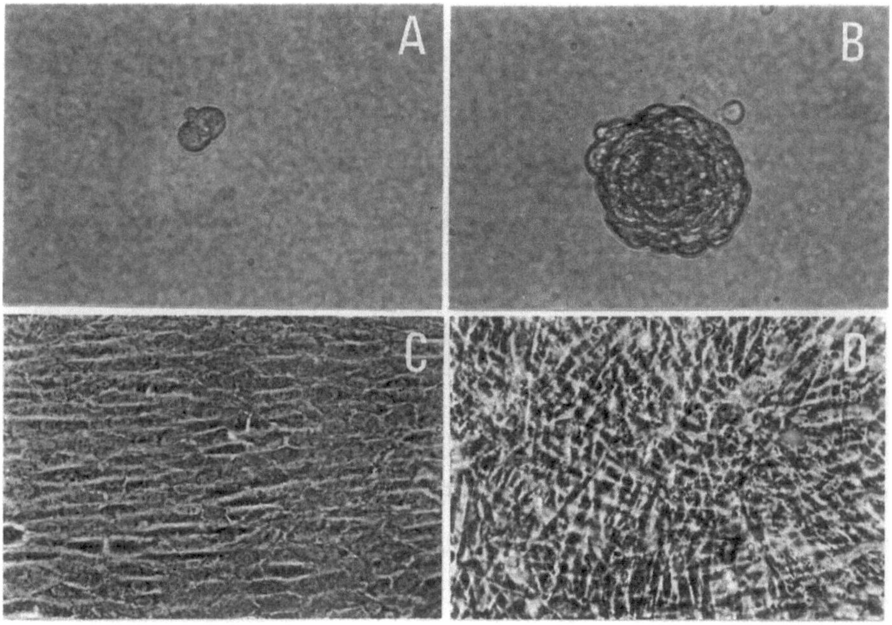

Fig. 1. Phenotypic transformation of NRK cells induced by EGF and TGFβ. NRK cells were cultured in medium supplemented with growth factor-inactivated serum, 5μg/ml insulin and 25ng/ml EGF (Van Zoelen et al., 1985). For soft agar growth experiments, $1x10^4$ trypsinized cells were inoculated into the agar in SH-FCS containing medium supplemented with 5μg/ml insulin and 2ng/ml EGF (a) or with 5μg/ml insulin, 2ng/ml EGF and 1ng/ml TGFβ (b). Colonies were scored after 14 days of inoculation. For morphological studies, monolayer cells were grown to confluency in the above medium, and subsequently cultured in SH-FCS contaning medium supplemented with 5μg/ml insulin and 2ng/ml EGF (c) or with 5μg/ml insulin, 2ng/ml EGF and 1ng/ml TGFβ (d). Cultures were photographed after 72 hours of incubation.

EGF, fibroblast growth factor (FGF) and high-density lipoprotein, which can also be used in soft agar growth experiments. As will be shown below, the growth factor requirements for induction of anchorage-independent proliferation are quite similar in this serum-free medium and in the growth factor-inactivated serum-containing media described in this study. The approach using SH-FCS-containing media has the advantage that cells can be routinely subcultured, similar as in regular serum-containing media. In addition we have shown that SH-FCS-containing media can be used in a general strategy for studying the growth factor requirements for proliferation of various tumor cell lines (Van Zoelen et al., 1987; Timmers et al., 1988; Van der Burg et al., 1988). Figure 1 illustrates that in SH-FCS-containing media polypeptide growth factors are able to induce phenotypic transformation of NRK cells. When cultured in the presence of EGF alone, NRK cells are unable to grow in semi-solid media such as soft agar (fig.1A), while in the additional presence of TGFβ they form progressively growing colonies in the agar, strongly resembling transformed cells (fig.1B). When cultured on a substrate, NRK cells grown to confluency in the presence of EGF form a uniform cellular monolayer, and stop dividing due to so-called contact inhibition (fig.1C). In the additional presence of TGFβ, however, cells acquire an altered morphology with an elongated shape, accompanied by loss of density-dependent inhibition of growth, resulting in the formation of cellular multilayers (fig.1D).

<u>Anchorage-independent proliferation of NRK cells</u>

Figure 2 shows the effects of various combinations of growth factors on the induction of NRK soft agar growth (fig.2A) in relation to their mitogenic effects on NRK cells (fig.2B). All experiments were carried out in SH-FCS-containing media in the presence of insulin (Van Zoelen et al., 1986). The growth factors tested in this study were EGF, TGFβ, PDGF and RA. In spite of an earlier report that RA may block phenotypic transformation of NRK cells (Todaro et al., 1978), later studies have shown that in serum-containing media RA may enhance the effects of TGFβ in inducing NRK soft agar growth (Jetten, 1983; Roberts et al., 1984). Figure 2B shows that by themselves EGF and PDGF are strongly mitogenic for quiescent NRK cells. In contrast, TGFβ and RA alone have hardly any growth stimulating activity. Figure 2A shows that none of the four factors tested in this study can induce soft agar growth of NRK cells by itself, and that under all conditions combinations of

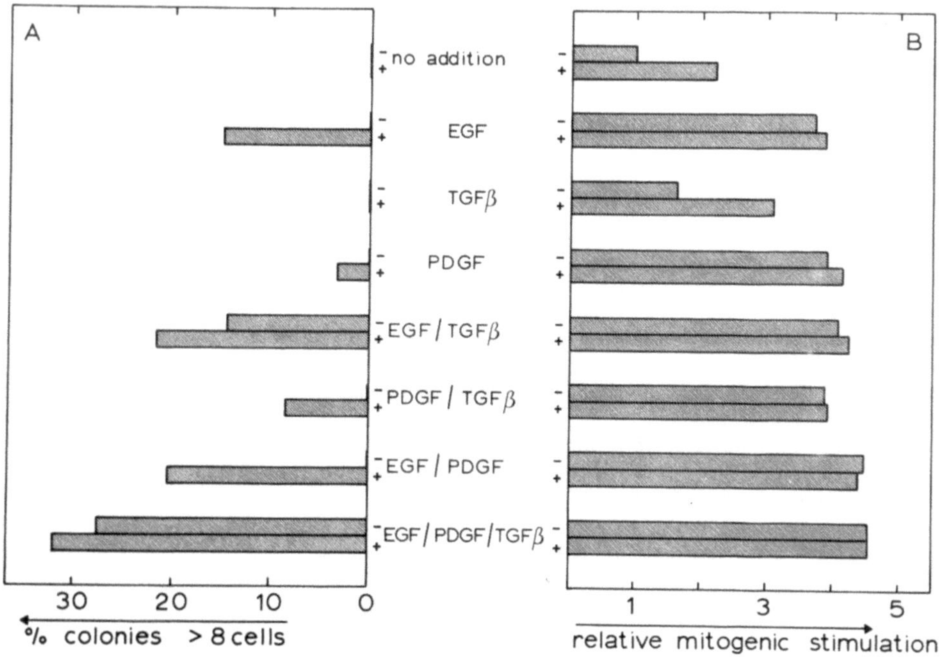

Fig. 2. Induction of NRK soft agar growth (A) and mitogenic activity on NRK cells (B) exerted by EGF (2ng/ml), PDGF (10ng/ml), and TGFβ (4ng/ml) in the absence (−) and presence (+) of retinoic acid (50ng/ml). All assays were performed in SH-FCS containing medium in the presence of 5μg/ml insulin, as described by Van Zoelen et al. (1986). For the mitogenic activity assay (B), cells were made quiescent for 72 hours in SH-FCS medium, and the growth factors to be tested added in the presence of 5μg/ml insulin. Stimulation is expressed relative to that induced by 5μg/ml insulin. Soft agar growth experiments (A) were carried out as in the legend of fig.1.

factors are required to induce this process. Remarkably, under these growth factor-defined conditions RA is nearly as potent as TGFβ in mediating the transforming activity of EGF. This observation, combined with the lack of direct mitogenic activity of these two factors, suggests that RA and TGFβ have similar modes of action on this cell line. In addition these two factors share the ability to inhibit proliferation of a variety of other cell lines (Schroder et al., 1983; Sporn et al., 1987), indicating that TGFβ and RA belong to a family of so-called modulating growth factors.

The data of fig.2A further show that none of the four factors tested is essential for NRK soft agar growth, since any combination of three out of these four factors was found to be sufficient to induce this process. The picture emerging from these data indicates that anchorage-independent proliferation of NRK cells requires the combination of a mitogenic polypeptide growth factor (PDGF or EGF) and a modulating agent, either TGFβ or RA (Van Zoelen et al., 1986). Later studies using various combinations of EGF and PDGF have shown that combinations of these two factors may also induce phenotypic transformation, that is without a requirement for a modulating agent (Van Zoelen et al., 1988). It was concluded from these data that anchorage-independent proliferation of NRK cells can be induced by various combinations of growth factors, and that the factors originally identified to induce this process (EGF + TGFβ) are not essential. These conclusions basically agree with the observations of Rizzino et al. (1986), that under serum-free conditions NRK cells can be induced to grow in soft agar by combinations of mitogenic polypeptide growth factors in the absence of TGFβ.

Density-dependent inhibition of growth of NRK cells

Besides morphological alterations and induction of anchorage-independent proliferation, loss of density-dependent inhibition of growth is an important characteristic of phenotypic transformation. The growth factor requirements for loss of density-dependent inhibition of growth of NRK cells can easily be studied under complete serum-free conditions (Van Zoelen et al., 1988). NRK cells are grown to confluency in serum-containing media, and then transferred for three days to serum-free medium in the absence of polypeptide growth factors. Growth factors to be tested are then added (time zero), and the incorporation of ^3H-thymidine, indicative for the presence of cells in the S-phase of the cell cycle, is measured over 2 hour time intervals. This approach is illustrated in fig.3. Upon additon of EGF,

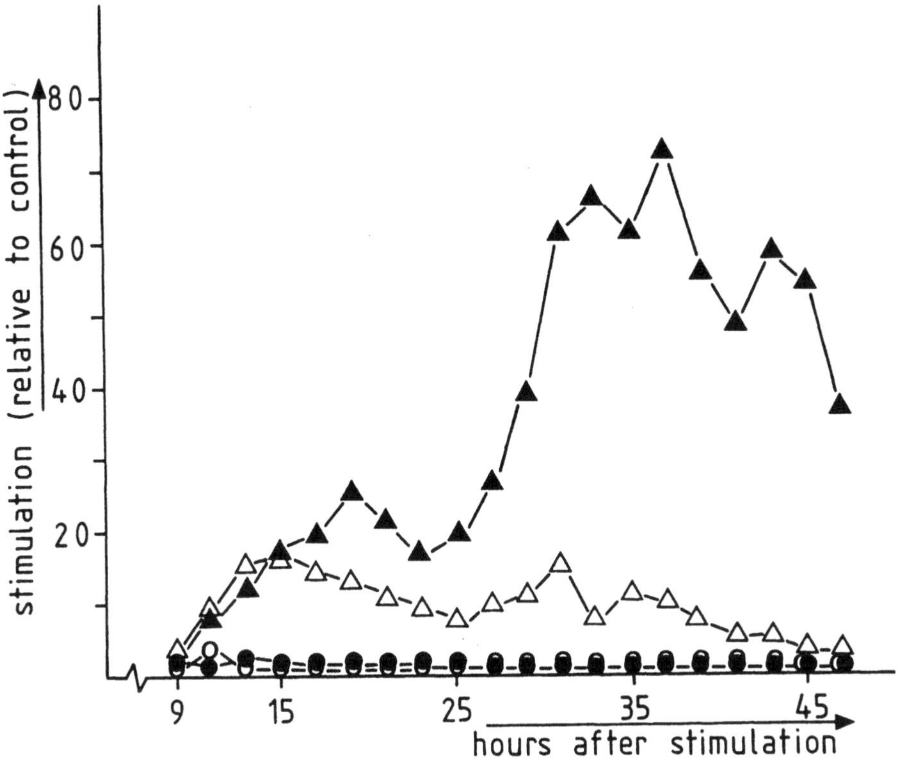

Fig. 3. Loss of density-dependent inhibition of growth of NRK cells induced by EGF and TGFβ. NRK cells were grown to confluency in serum-containing medium, and subsequently for 72 hours in serum-free medium, as described by Van Zoelen et al. (1988). Growth factors to be tested were then added (t=0) and the rate of ^3H-thymidine uptake (0.5μCi/well) was determined during 2-hour pulses; (○——○) 5μg/ml insulin; (●——●) 1ng/ml TGFβ + 5μg/ml insulin; (△——△) 5ng/ml EGF + 5μg/ml insulin; (▲——▲) 5ng/ml EGF + 1 ng/ml TGFβ + 5μg/ml insulin.

cells enter the cell cycle and traverse the S-phase between 9 and 25 hours after stimulation. Subsequently DNA synthesis decreases continuously, because cells are unable to proliferate due to density-dependent inhibition of growth. Obviously cells can undergo one round of DNA duplication under these experimental conditions, when stimulated by EGF alone. Treatment of the cells with TGFβ alone does not result in mitogenic stimulation of the cells. In the presence of a combination of EGF and TGFβ, however, cells progress through the first cell cycle, as observed in the case of EGF alone, but now ^{3}H-thymidine incorporation increases again after completion of the first cell cycle, indicating that under these conditions loss of density-dependent inhibition of growth has taken place. We have shown that this enhanced ^{3}H-thymidine incorporation is paralleled by an increase in cell number and total DNA content (Van Zoelen et al., 1988).

Using this approach, loss of density-dependent inhibition of growth can be quantified by measuring ^{3}H-thymidine uptake in the cells between 43 and 45 hours after stimulation by growth factors. Low levels of incorporation of the label are indicative of contact inhibition, high levels of incorporation denote loss of density-dependency of proliferation. Figure 4 shows this parameter for various combinations of growth factors. As in the previous experiments in soft agar, combinations of EGF, PDGF, TGFβ and RA were tested. This figure shows that combinations of growth factors are required to induce this process, but as in the soft agar growth experiments, none of the growth factors tested was essential for the process. A comparison with the requirements for soft agar growth shows that the growth factor requirements for both aspects of phenotypic transformation are very similar, and thus that essentially similar results are obtained under serum-free conditions and in SH-FCS containing media. One can conclude from these experiments that phenotypic transformation of NRK cells does not require specific polypeptide growth factors, but rather reflects a response of these cells to multiple growth factors. In this respect the behaviour of NRK cells strongly parallels that of a number of other non-tumorigenic indicator cell lines (Kaplan and Ozanne, 1983). It is remarkable, however, that TGFβ and RA, which by themselves are not mitogenic for NRK cells, have such a strong transformation mediating effect on this cell line.

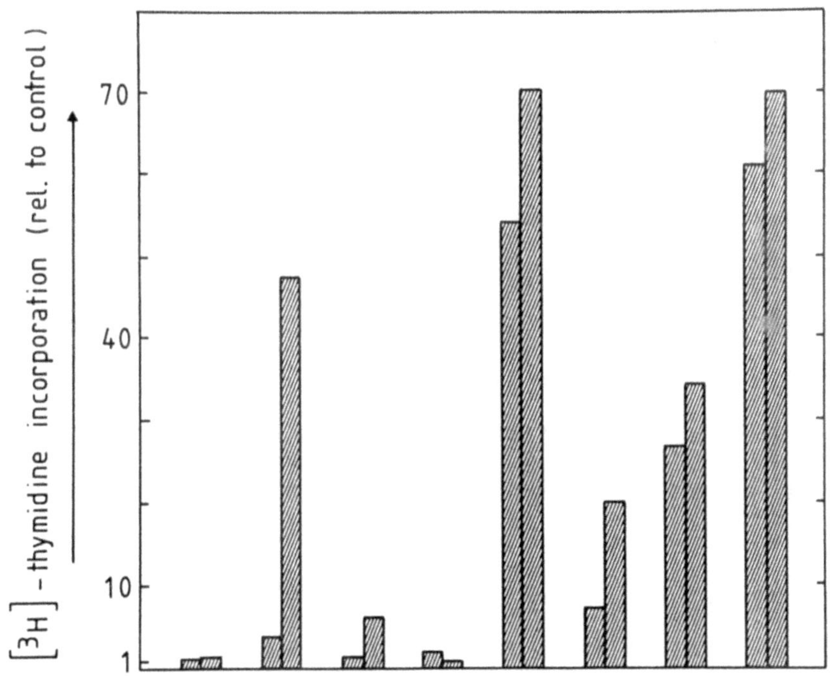

EGF	- -	+ +	- -	- -	+ +	- -	+ +	+ +
TGFβ	- -	- -	+ +	- -	+ +	+ +	- -	+ +
PDGF	- -	- -	- -	+ +	- -	+ +	+ +	+ +
RA	- +	- +	- +	- +	- +	- +	- +	- +
%S.A.G.	- -	- 15	- -	- 3	15 22	- 9	1 20	27 32

Fig. 4. Effect of polypeptide growth factors on loss of density-dependent inhibition of growth of NRK cells. [^{3}H]-thymidine uptake (0.5μCi/well) was determined between 43 and 45 hours after mitogenic stimulation of confluent serum-free cultures. Growth factors were tested in the additional presence of 5μg/ml insulin at the following concentrations: 5ng/ml EGF, 1ng/ml TGFβ, 5ng/ml PDGF and 50ng/ml RA. As a comparison, the percentage of soft agar growth (S.A.G.) colonies induced by these factors under SH-FCS containing assay conditions is indicated (see fig.2.).

Mode of action of TGFβ and retinoic acid

TGFβ exerts its action by binding to a set of specific plasma membrane receptors (Cheifetz et al., 1987). In contrast to a variety of other growth factors, TGFβ does not induce tyrosine phosphokinase activity in its receptor, neither does it induce rapid degradation of inositol-containing phospholipids (Libby et al., 1986). Biological effects induced by TGFβ all require protein synthesis, and the current hypothesis implies that TGFβ exerts its action by inducing genes for other growth regulating molecules (reviewed by Sporn et al., 1987). In this respect it has been shown that TGFβ induces genes for extracellular matrix components (fibronectin, collagens), interferon-β and protease inhibitors. Importantly it has been shown that TGFβ induces activation of the c-sis oncogene, which is paralleled by production of PDGF by these cells (Leof et al., 1986). The consequences of these observations will be discussed below. Retinoic acid has recently been shown to act by binding to a specific nuclear receptor (Petkovich et al., 1987; Giguere et al., 1987). To what extent similar genes are induced by RA as by TGFβ will require further investigations.

Shipley et al. (1985) have shown that TGFβ exerts its mitogenic activity on AKR-2B cells after a prolonged prereplicative time interval. This so-called delayed induction of mitogenesis is most likely caused by the fact that TGFβ first has to induce the cells to produce growth stimulating hormones, in particular PDGF (Leof et al., 1986), which can subsequently stimulate the cells to proliferate. In this model the growth stimulating activity of TGFβ would be an indirect one, by activating genes for other growth factors. In contrast to AKR-2B cells, NRK cells cannot be stimulated by TGFβ alone (see fig.3). In this cell line TGFβ can exert its action only in the presence of other growth factors, in particular EGF. The data of fig.4 show that phenotypic transformation of NRK cells can be induced by a combination of two growth stimulating hormones, in particular EGF and PDGF, but also by EGF in combination with modulating agents such as TGFβ and RA, which by themselves have no growth stimulating activity. This could be explained in the above model if these modulating agents in the presence of EGF were able to induce NRK cells to produce PDGF.

In order to test this hypothesis, confluent NRK cultures were put in serum-free medium and subsequently treated with EGF in combination with insulin. After 45 hours cells had undergone one round of division, and incorporation of ^{3}H-thymidine was reduced again to background levels (see fig.3). Subsequently (time zero) cells were incubated with either TGFβ, RA or PDGF, and ^{3}H-thymidine incorporation was

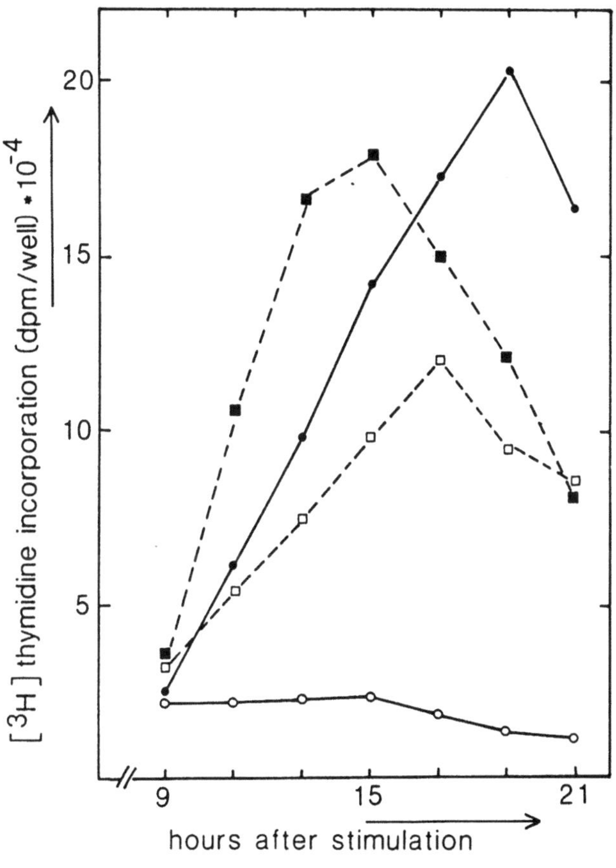

Fig. 5. Time course of loss of density-dependent inhibition of growth of EGF-pretreated NRK cells, induced by PDGF, TGFβ or RA. Confluent cultures were incubated for 72 hours on serum-free medium, as described by Van Zoelen et al. (1988). EGF (5ng/ml) and insulin (5μg/ml) were then added, and cells incubated for another 45 hours until they had reached quiescence again. Subsequently, additional growth factors were added (○——○:5ng/ml EGF; ■– –■: 5ng/ml PDGF; ●——● : 2ng/ml TGFβ; □ – – □ : 50ng/ml RA), and ^{3}H-thymidine incorporation (0.5μCi/well) was determined during 2-hour time pulses.

measured as a function of time. Cells are density-inhibited for growth under these conditions, and therefore induction of ^{3}H-thymidine incorporation is a direct measure for phenotypic transformation. Figure 5 shows that in the additional presence of PDGF, cells enter the S-phase 9 hours after addition of this growth factor, reaching maximum ^{3}H-thymidine incorporation 15 hours after addition of the mitogen. In the presence of TGFβ, however, maximum incorporation of ^{3}H-thymidine occurs only 19 hours after addition of the factor, although higher levels of incorporation are reached than in the case of PDGF. An intermediate situation is observed in the case of RA. These data show that induction of phenotypic transformation by TGFβ and RA in NRK cells occurs after a prereplicative delay, when compared to induction by PDGF. This suggests that in NRK cells TGFβ and RA may indeed act by inducing genes for other growth stimulating factors, similarly as observed in AKR-2B cells. The nature of this induced growth factor will require further investigation.

The exact role of EGF in modulating the effects of TGFβ and RA in NRK cells is still unclear. It has been shown that both RA (Roberts et al., 1984) and TGFβ (Assoian et al., 1984b; Assoian, 1985) enhance the number of EGF receptors in this cell line, or alter the receptor affinity for EGF binding (Massagué, 1985a). To what extent these alterations play a role in the phenotypic transformation process is unclear. Others have shown that TGFβ may act as a so-called competence factor for growth stimulation (Goustin et al., 1987), that TGFβ has insulin-like activity (Brown and Holley, 1987), or just requires insulin-like growth factors for optimal activity (Massagué et al., 1985b). It is well established that TGFβ induces genes for various extracellular matrix components, including fibronectin and collagen (Ignotz and Massagué, 1986). These authors have shown that fibronectin, added externally to NRK cells in the presence of EGF, can induce anchorage-independent proliferation of this cell line. Moreover, synthetic peptides blocking the cellular attachment of fibronectin were found to inhibit TGFβ-induced phenotypic transformation of NRK cells (Ignotz and Massagué, 1986). Care should be taken in interpretation of experiments using fibronectin, since in our experience and that of others (Fava and McClure, 1987) fibronectin preparations may contain considerable amounts of TGFβ. Still the induction of extracellular matrix components seems an essential aspect of TGFβ activity. Other attempts have been made to block TGFβ activity by specific inhibitors such as oxidized polyamines (Frolik et al., 1984), dexamethasone (Roberts et al., 1984), dimethylformamide (Levine and Crandall, 1987) and suramin (Coffey et al., 1987), but in our experience these agents block growth factor action in general, and are not very specific for TGFβ.

In conclusion, it can be stated that two models presently prevail for explaining the role of TGFβ in phenotypic transformation of NRK cells. In the first model the induction of extracellular matrix components by TGFβ is emphasized. It can well be imagined that under anchorage-independent conditions an extracellular matrix covering the cell surface is required for intercellular attachment, and therefore for colony formation. In the second model the induction of growth factor production by cells due to TGFβ activity is emphasized. It can indeed be expected that density-dependent inhibition of growth can be overcome by addition of additional growth factors (PDGF) or modulating agents which induce the cell to produce such growth stimulating factors in an autocrine fashion. The data presented in this study show, however, that the growth factor requirements for both aspects of phenotypic transformation, namely anchorage-independent proliferation and loss of density-dependent inhibition of growth, are nearly identical. Therefore, the same cellular mechanisms will most likely underlie both aspects of phenotypic transformation. The present results therefore urge for additional models to understand the role of polypeptide growth factors in phenotypic transformation of non-transformed indicator cells.

Acknowlegdements

I am grateful to Ms. T.M.J. Ward-van Oostwaard for technical assistance.

References

Anzano MA, Roberts AB, Meyers CA, Komoriya A, Lamb LC, Smith JM, Sporn MB (1982) Synergistic interaction of two classes of transforming growth factors from murine sarcoma cells. Cancer Res. 42:4776–4778

Assoian RK, Komoriya A, Meyers CA, Miller DM, Sporn MB (1983) Transforming growth factor-β in human platelets. Identification of a major storage site purification and characterization. J Biol Chem 258:7155–7160

Assoian RK, Grotendorst GR, Miller DM, Sporn MB (1984a) Cellular transformation by coordinated action of three peptide growth factors from human platelets. Nature 309:804–806

Assoian RK, Frolik CA, Roberts AB, Miller DM, Sporn MB (1984b) Transforming growth factor-β controls receptor levels for epidermal growth factor in NRK cells. Cell 36:35–41

Assoian RK (1985) Biphasic effects of type β transforming growth factor on epidermal growth factor receptors in NRK fibroblasts. J Biol Chem 260:9613–9617

Bradshaw GL, Dubes GR (1983) Supplementary factors required for serum-free culture of rat kidney cells of line NRK-49F. In vitro 19:735–742

Brown KD, Holley RW (1987) Insulin-like synergistic stimulation of DNA synthesis in Swiss 3T3 cells by the BSC-1 cell-derived growth inhibitor related to transforming growth factor type β. Proc Natl Acad Sci USA 84:3743–3747

Cheifetz S, Weatherbee JA, Tsang MLS, Anderson JK, Mole JE, Lucas R, Massagué J (1987) The transforming growth factor-β system a complex pattern of cross-reactive ligands and receptors. Cell 48:409–415

Childs CB, Proper JA, Tucker RF, Moses HL (1982) Serum contains a platelet-derived transforming growth factor. Proc Natl Acad Sci USA 79:5312–5316

Cinofe MA, Fidler IJ (1980) Correlation of patterns of anchorage-independent growth with in vivo behaviour of cells from a murine fibrosarcoma. Proc Natl Acad Sci USA 77:1039–1043

Coffey RJ, Leof EB, Shipley GD, Moses HL (1987) Suramin inhibition of growth factor receptor binding and mitogenicity in AKR-2B cells. J Cell Physiol 132:143–148

DeLarco JE, Todaro GJ (1978) Growth factors from murine sarcoma virus-transformed cells. Proc Natl Acad Sci USA 75:4001–4005

Derynck R (1986) Transforming growth factor-α: structure and biological activities. J Cell Biochem 32:293–304

Fava RA, McClure DB (1987) Fibronectin-associated transforming growth factor. J Cell Physiol 131:184–189

Frolik CA, Roller PP, Cone JL, Dart LL, Smith DM, Sporn MB (1984) Inhibition of transforming growth factor-induced cell growth in soft agar by oxidized polyamines. Arch Biochem Biophys 230:93–102

Giguere V, Ong ES, Segui P, Evans RM (1987) Identification of a receptor for the morphogen retinoic acid. Nature 330:624–629

Goustin AS, Leof EB, Shipley GD, Moses HL (1986) Growth factors and cancer. Cancer Res 46:1015–1029

Goustin AS, Nuttall GA, Leof EB, Ranganathan G, Moses HL (1987) Transforming growth factor type β can act as a potent competence factor for AKR-2B cells. Exp Cell Res 172:293–303

Heldin CH, Westermark B (1984) Growth factors: mechanism of action and relation to oncogenes. Cell 37:9–20

Ignotz RA, Massagué J (1986) Transforming growth factor-β stimulates the expression of fibronectin and collagen and their incorporation into the extracellular matrix. J Biol Chem 261:4337–4345

Jetten AM (1983) Action of retinoids on the anchorage-independent growth of normal rat kidney fibroblasts induced by 12-0-tetradecanoyl phorbol-13-acetate or sarcoma growth factor. Cancer Res 43:68–72

Kaplan PL, Anderson M, Ozanne B (1982) Transforming growth factor(s) production enables cells to grow in the absence of serum: an autocrine system. Proc Natl Acad Sci USA 79:485–489

Kaplan PL, Ozanne B (1983) Cellular responsiveness to growth factors correlates with a cell's ability to express the transformed phenotype. Cell 33:931–938.

Leof EB, Proper JA, Goustin AS, Shipley GD, DiCorleto PE, Moses HL (1986) Induction of c–sis mRNA and activity similar to platelet–derived growth factor by transforming growth factor β: a proposed model for indirect mitogenesis involving autocrine activity. Proc Natl Acad Sci USA 83:2453–2457

Levine AE, Crandall CA (1987) Effects of N, N dimethyl formamide and retinoic acid on transforming growth factor–β induced mitogenesis in AKR–2B mouse embryo fibroblasts. Cancer Res 47:4278–4282

Libby J, Martinez R, Weber MJ (1986) Tyrosine phosphorylation in cells treated with transforming growth factor–β. J Cell Physiol 129:159–166

Massagué J (1985a) Transforming growth factor–β modulates the high–affinity receptors for epidermal growth factor and transforming growth factor–α. J Cell Biol 100:1508–1514

Massagué J, Kelly B, Mottola C (1985b) Stimulation by insulin–like growth factors is required for cellular transformation by type β transforming growth factor. J Biol Chem 260:4551–4554

Moses HL, Tucker RF, Leof EB, Coffey RJ, Halper J, Shipley GD (1985) Type–β transforming growth factor is a growth stimulator and a growth inhibitor. In: "Growth factor and Transformation". Cancer cells 3:65–71. Cold Spring Harbor Laboratory

Ozanne B, Wheeler T, Kaplan PL (1982) Cells transformed by RNA and DNA tumor viruses produce transforming growth factors. Feder Proceed 41:3004–3007

Petkovich M, Brand NJ, Krust A, Chambon P (1987) A human retinoic acid receptor which belongs to the family of nuclear receptors. Nature 330:444–450

Rapp UR, Gunnell M, Marquardt H (1983) Normal mouse serum contains peptides with induce fibroblasts to grow in soft agar. J Cell Biochem 21:29–38

Rizzino A (1984) Behavior of transforming growth factors in serum–free media: an improved assay for transforming growth factors. In vitro 20:815–822

Rizzino A, Ruff E, Rizzino H (1986) Induction and modulation of anchorage–independent growth by platelet–derived growth factor fibroblast growth factor and transforming growth factor–β. Cancer Res 46:2816–2820

Roberts AB, Anzano MA, Lamb LC, Smith JM, Sporn MB (1984) Antagonistic actions of retinoic acid and dexamethasone on anchorage–independent growth and epidermal growth factor binding of normal rat kidney cells. Cancer Res 44:1635–1641

Schroder EW, Rapaport E, Black PH (1983) Retinoids and cell proliferation. Cancer Surveys 2:223–241

Shin SI, Freedman VH, Risser R, Pollack R (1975) Tumorigenicity of virus–transformed cells in nude mice is correlated specifically with anchorage–independent growth in vitro. Proc Natl Acad Sci USA 72:4435–4439

Shipley GD, Tucker RF, Moses HL (1985) Type β transforming growth factor/growth inhibitor stimulates entry of monolayer cultures of AKR–2B cells into S phase after a prolonged prereplicative interval. Proc Natl Acad Sci USA 82: 4147–4151

Sporn MB, Roberts AB, Wakefield LM, de Crombrugghe B (1987) Some recent advances in the chemistry and biology of transforming growth factor–beta. J Cell Biol 105:1039–1045

Timmers HTM, Van Zoelen EJJ, Bos JL, Van der Eb AJ (1988) Cells transformed by adenovirus type 12 but not by type 5 are dependent on insulin or insulin–like growth factor I for their proliferation. J Biol Chem 263:1329–1335

Todaro GJ, De Larco JE, Sporn MB (1987) Retinoids block phenotypic cell transformation produced by sarcoma growth factor. Nature 276:272–274

Van der Burg B, Rutteman GR, Blankenstein MA, De Laat SW, Van Zoelen EJJ (1988) Mitogenic stimulation of human breast cancer cells in a growth factor-defined medium: synergistic action of insulin and estrogen. J Cell Physiol 134:101–108

Van Zoelen EJJ, Van Oostwaard TMJ, Van der Saag PT, De Laat SW (1985) Phenotypic transformation of normal rat kidney cells in a growth factor-defined medium: Induction by a neuroblastoma-derived transforming growth factor independently of the EGF receptor. J Cell Physiol 123:151–160

Van Zoelen EJJ, Van Oostwaard TMJ, De Laat SW (1986) Transforming growth factor-β and retinoic acid modulate phenotypic transformation of normal rat kidney cells induced by epidermal growth factor and platelet-derived growth factor. J Biol Chem 261:5003–5009

Van Zoelen EJJ, Van Rooijen MA, Van Oostwaard TMJ, De Laat SW (1987) Production of transforming growth factors by simian sarcoma virus-transformed cells. Cancer Res 47:1582–1587

Van Zoelen EJJ, Van Oostwaard TMJ, De Laat SW (1988) The role of polypeptide growth factors in phenotypic transformation of normal rat kidney cells. J Biol Chem 263:64–68

Functions of the erbA and erbB Oncogenes in Avian Erythroblatosis

Björn Vennström, Hartmut Beug, Douglas Forrest, Ann Johnsson, Khash Khazaie, Alberto Munoz, Jan Sap, Axel Ullrich*, and Martin Zenke

Differentiation Programme
European Biology Laboratory
D-6900 Heidelberg, F.R.G.

INTRODUCTION

The avian erythroblastosis virus (AEV) induces an acute erythroid leukemia and sarcomas in virus-infected chickens. AEV strain ES4 was isolated already during the thirties by Engelbreth-Holm by serial passage of a weak leukemia inducing virus (RAV1) in chicks. During the mid-seventies T. Graf demonstrated that a biologically cloned AEV causes both types of diseases, and transforms erythroblasts as well as fibroblasts in vitro.

AEV has acquired two host cell-derived oncogenes, v-erbA and v-erbB (Vennström and Bishop, 1982). As a consequence, viral genes normally needed for replication have been deleted and the virus requires a helper virus (Rous associated virus, RAV) for replication. The v-erbA gene, expressed from a genome sized mRNA, is fused to part of the gag gene, and the corresponding $P75^{gag-v-erbA}$ protein has recently been shown to be localized in the nucleus (Sap et al. 1986). In contrast, v-erbB is translated from a subgenome mRNA into a 66-74 kd integral plasma membrane glycoprotein (Beug and Hayman, 1984, Privalsky et al. 1984).

Several groups had previously shown that the nuclear receptors for steroid receptors bear homology to the v-erbA gene, and subsequent parallel work by my own and R. Evans' group demonstrated that the cellular gene, c-erbA, is a nuclear

* Genetech Inc. South San Francisco, California, U.S.A.

NATO ASI Series, Vol. H26
Cell to Cell Signals in Mammalian Development
Edited by S. W. de Laat et al.

receptor for thyroid hormones (Sap et al. 1986, Weinberger et al. 1986). In addition, we have shown that $P75^{gag-v-erbA}$ binds to DNA but lacks hormone binding capacity (Sap et al. 1986). This deficiency in ligand binding is due to a least two mutations in the domain of $P75^{gag-erbA}$ homologous to the hormone binding domain of steroid receptors (Munoz et al. 1988). Finally, in contrast to the situation with steroid receptors, several genes for the thyroid receptor seem to exist both in the mammalian and avian genomes (Weinberger et al. 1986, Jansson et al. 1983, and BV, unpublished).

Thyroid hormones (triiodonthyronine, T3; and thyroxine, T4) have a broad spectrum of effects. They affect metabolism in many types of cells by enhancing cell metabolism, induce growth hormone production in the pituitary, and also repress prolactin and thyroid stimulating hormones. In the brain T3 was required for proper dendrite formation and myelination. In addition, T3 elevates the levels of Na^+/K^+ATPase activity in brain cells.

The v-erbB gene was identified as representing part of the 160 kd plasma membrane receptor for epidermal growth factor (EGFr; Downward et al. 1984, Ullrich et al. 1984, Lax et al. 1988). V-erbB^{ES4} lacks all except 70 amino acids of the ligand binding domain, and also 73 amino acids of the C-terminus (Choi et al. 1986, Scotting et al, 1987). In this domain the EGFr has three tyrosines which become phosphorylated in an autophosphorylation reaction as a consequence of ligand binding (Gullick et al. 1985). It is thought that this domain functions in regulating the tyrosine kinase activity of the catalytic region of the receptor, and J. Schlessinger and collaborators have indeed shown that the v-erbB protein possesses a constitutive tyrosine kinase activity (Lax et al. 1985), suggesting that the oncogenic activity of v-erbB is due to a constitutive mitogenic signalling. In addition, it is possible that the v-erbB protein has a different substrate specificity as compared to the normal receptor, resulting from the deletions and the many point mutations it has accumulated as well.

The results by Frykberg et al. and Sealy et al. (1983) established that v-erbB is the major transforming gene of AEV, and that v-erbA alone lacks transforming activity. In addition, Yamamoto et al. (1983) demonstrated that a leukemogenic and sarcomagenic virus, AEV-H, carried v-erbB as its oncogene, and Fung et al. (1983) showed that erythroleukemias induced by RAV-1 contained c-erbB genes activated by proviral insertion. Our own and subsequent results furthermore showed that v-erbA enhances the erythroblast transforming activity of v-erbB by blocking the residual, spontaneous, differentiation exhibited by such cells, and by enabling them to grow in standard tissue culture media (Beug et al, Damm et al. 1986).

The question of whether or not the cooperativity between v-erbA and v-erbB was due to specific and direct interaction between the two corresponding proteins, or to action along distinct pathways was clarified by investigating the activity of v-erbA in erythroblasts transformed by other oncogenes (Kahn et al. 1985). The fact that v-erbA blocked differentiation and enabled growth in standard tissue culture media of also v-src, v-sea and v-Ha-ras transformed erythroblasts demonstrated, that the cooperativity was largely independent of the type of primary oncogene used, and that v-erbA was likely to act upon pathways separate from those of the primary oncogene.

The work described above laid the ground for further study of the activities of v-erbA and v-erbB in erythroblasts. Subsequent work mainly by Beug and collaborators has resulted in the following understanding of the functions of the erb oncogenes in transformation of avian cells.

The effects of v-erbB in haemopoetic cells and fibroblasts appear to be distinct. In the erythroblasts v-erbB (as well as the tyrosine kinase related oncogenes and v-Ha-ras) has two major effects: it induces self-renewal of the cells, and makes them independent of the erythroid growth factor erythropoietin (EPO) for both growth and differentiation (Beug et al. 1985). One hypothesis is that the high expression of the mitogenic v-erbB protein induces the self-renewal, perhaps by acting on substrates not normally modified in differentiating erythroblasts, as these cells do not normally contain EGF receptors (Beug et al. 1986). In contrast, the conferment of EPO independence for differentiation suggests that the v-erbB protein has activities which overlap with those of the EPO receptor.

In chicken fibroblasts, the major effect of v-erbB seems to be its mitogenic activity. V-erbB expressing cells cannot be arrested in G_0, and express neither EGF nor PDGF receptors (A.J. and BV, unpublished). Furthermore, the cells exhibit several parameters of transformation, such as loss of contact inhibition, transformed morphology, and increased hexose transport. Expression of plasminogen activator activity, and disarrangement of actin and fibronectin networks are also increased, but not to levels observed in e.g. v-src transformed cells (Frykberg et al.1983). The data suggest that these effects of v-erbB, when combined, are responsible for its transforming and tumorigenic effect on fibroblastic cells.

In chicken fibroblasts v-erbA elicits only minor effects. When acting alone, an increased cellular lifespan and a reduced serum requirement have been observed (Gandrillon et al. 1987). When acting in concert with v-erbB an increased effect on several transformation parameters was observed (Frykberg et al. 1983). Furthermore, v-erbA enhances the transforming activity of weakly transforming v-erbB genes (Jansson et al. 1987), and promotes tumor formation by v-erbB^{ES4} in ovo (Gandrillon et al. 1987). Since v-erbA is a homologue of a thyroid hormone receptor it is likely that it affects transcription of one or several genes important for erythroid cell differentiation. These genes might be other than those recognized by the c-erbA protein, since P75$^{gag-erbA}$ contains two amino acid changes in its putative DNA binding region. At any rate, when bound to a presumptive regulatory element, P75$^{gag-erbA}$ could either up- or down regulate the transcription of the corresponding gene. The effect is furthermore likely to be constitutive, as the protein does not bind ligand (Sap et al. 1986). If downregulation is the mechanism by which $P75^{gag-erbA}$ acts, it seems likely that the expression of the affected erythroid specific gene(s) would be required for proper erythroid differentiation. Conversely, if v–erbA upregulates transcription, the expression would inhibit differentiation.

We report here that a new thyroid hormone receptor–related gene has a tissue specific expression pattern distinct form that of the previously described c–erbA gene. Furthermore, the v–erbA protein acts as repressor of expression of the gene for band 3, the major anion transporter in red blood cells, whereas chimeric v–/c–erbA proteins that bind T3, can act as inducers of expression for band 3.

Finally, we demonstrate that the human EGFr can replace v–erbB in most aspects of transformation in chicken cells, and that introduction of mutations enhances these transforming effects. However, the viral erbB gene is still more potent in transformation of fibroblasts and erythroblasts.

Results and Discussion

Isolation and characterization of new c–erbA related genes

Four cDNAs encoding thyroid hormone receptors from different species have been characterized to date. Based on their amino acid homologies they have been classified into two groups, alpha and beta. In general, they are highly homologous (Fig. 1), with the human β receptor being the most divergent in overall amino acid homology. Of particular interest is however that the human α receptor differs entirely from all the others in domain F, as does the human β receptor in domain B. Moreover, the receptors appear to differ in their patterns of expression in the different species. The rat α receptor has been described to be expressed as a 2.6 kb mRNA mainly in brain tissues, but also at a lower level in heart and kidney (Thompson et al. 1987). In contrast, the chicken α receptor is ubiquitously expressed (two mRNAs, 4.5 and 3.0 kb), with elevated levels (~10x) found only in

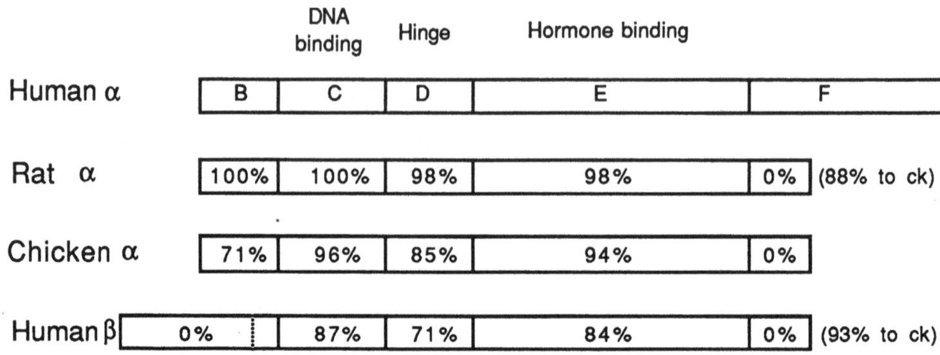

Figure 1. Comparison of thyroid hormone receptors. The percent homology refers to the human α receptor unless stated otherwise. The human α receptor was described by Pfahl et al. (1987); the rat α by Thompson et al. (1986); the chicken α by Sap et al. (1986); and the human β by Weinberger et al. (1986).

embryonic blood cells (Hentzen et al. 1987, D.F. and B.V., unpublished). Expression of the human α receptor has not been studied extensively, but it appears to be expressed (5 kb mRNA) at low levels in may tissues (Weinberger et al. 1986). Recent results (D.F. and B.V., unpublished) showed that a chicken c-erbA related gene is expressed in allantoic membrane, kidney and brain tissues as a 7 kb mRNA. Partial DNA sequence analysis of a genomic fragment has indicated that it encodes one exon from the ligand binding domain, and that it is more homologous to the β than to the α type of receptors.

The patterns of expression of the α and β receptors consequently seem to differ in different species. Since thyroid hormone receptors have important roles in cell metabolism as well as differentiation, e.g. in brain and tadpole development, we intend to study the nature of the chicken "β" receptor, and further compare its ligand binding properties, expression, and role in cell physiology.

Comparison of the effects of the viral and cellular erbA genes on erythroid differentiation

A general problem in studying T3 regulated transcription is that the only gene known to be regulated directly is the rat growth hormone gene. In contrast, the human gene is not responsive in the rat pituitary cells (GH_1 cells); moreover, introduction of the human β or the chicken α receptor into receptor-negative line of GH_1 cells does not lead to induction of growth hormone production (H. Samuels, personal communication), and neither does the v-erbA protein, postulated to act as repressor of transcription, reduce growth hormone mRNA synthesis in normal GH_1 cells (A.M. and B.V., unpublished).

In a search for target genes for v-erbA activity, Zenke et al. (1988) discovered that specifically the expression of band 3, the major anion transporter of red cells, is severely inhibited in v-erbA containing cells. It is therefore possible that the stringent growth conditions of v-erbB transformed erythroblasts (which mainly is a strict requirement for a medium pH between 7.5 and 7.7; Damm et al. 1986) is due to an uncontrolled and therefore abnormal ion exchange activity in the immature erythroblasts, thus resulting in toxic intracellular ion conditions. One major effect of v-erbA would then be an inhibition of band 3 expression, thereby circumventing the toxicity imposed by the ion transporter. Indeed, our experiments showed that inhibition by the use of specific drugs of band 3 activity in v-erbB transformed erythroblasts allowed their growth in media with normally toxic ion conditions. Although band 3 transcription is turned off in v-erbA containing erythroblasts, it was not possible to show unambiguously that this occurred directly at the transcriptional level due to the hormone unresponsiveness of the v-erbA protein, i.e. it was not possible to modulate its activity (Zenke et al. 1988). However, we were recently able to restore ligand binding to a gag-erbA protein by exchanging the mutant carboxy-terminal region of v-erbA with that of c-erbA (Munoz et al. 1988). Insertion of this gene, and other similar constructions that lack the gag domain, into avian retrovirus vectors that express the chimeric genes as well as a ts erythroblast-transforming gene allowed the treatment of primary, transformed erythroblasts with T3 in ts shift induced differentiation experiments. This enabled us to monitor whether or not induction of hormone dependent band 3 transcription takes place during erythroblast differentiation in the presence of cycloheximide, i.e. if T3 induction is a direct response to the receptor. The experiments showed that a hormone-binding gag-erbA protein induces band 3 transcription in T3 treated erythroblasts, although repression was observed in the absence of hormone.

The mechanism by which v-erbA blocks differentiation remains an enigma. It is possible that expression of other, as yet unidentified, genes is affected by $P75^{gag-erbA}$, and that these hypothetical genes have a controlling function in erythropoiesis. To determine the role of band 3 in this process we intend to introduce its cDNA into erythroblasts expressing v-erbA as well. If the cells resemble tsAEV transformed erythroblasts we can conclude that the suppression of band 3 expression by v-erbA plays little role in erythroid cell development; however, if the cells have the phenotype of v-erbB transformed erythroblasts the results would suggest an important role for band 3.

Alternatively, the lack of band 3 expression in the $P75^{gag-erbA}$ containing cells could be the cause of the differentiation block. Band 3 is not only an ion transporter, but also an anchor for the erythrocyte cytoskeleton, as well as for several enzymes. The possibility therefore exists that a lack of band 3 would have profound effects on the intracellular organization of components, the proper localization of which would be required for differentiation.

Delineation of normal and transforming functions of overexpressed erbB genes

Amplified erbB/EGFr genes have been detected in human glioblastomas and epidermoid tumors. In addition, high expression of EGF receptors have been found in human tumor cell lines carrying diploid erbB/EGFr genes. Both these and the amplified genes appear to be unmutated, and it is unclear how they contribute to neoplasia. In contrast, the v-erbB genes of the AEV strains ES4 and H are very efficient in transforming primary chicken fibroblasts and erythroblasts. We have therefore investigated the transforming properties of the human EGFr in avian cells.

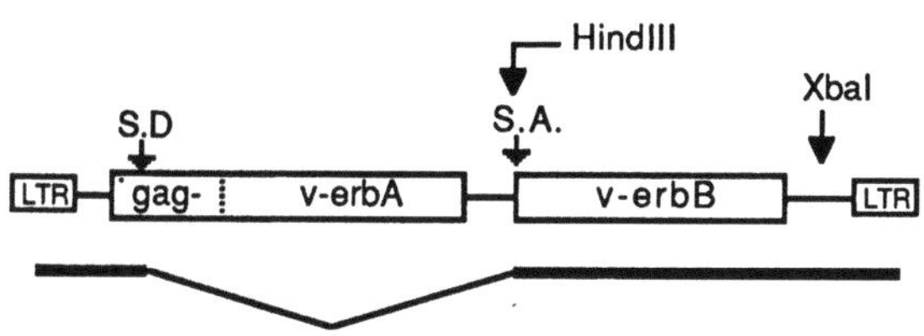

Figure 2. Structure of the AEV genome. Insertion of the human EGFR was done by excising v-erbB with HindIII and Apal, followed by insertion of cDNAs encoding the receptor molecules shown in Fig. 3. The heavy line indicates the structure of the sub-genomic mRNA encoding erbB sequences.

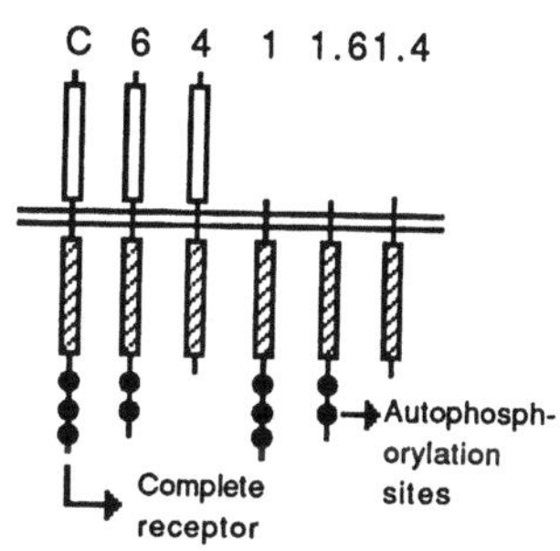

Figure 3. Structure of EGFr cDNAs replacing v-erbB in AEV constructs.

Accordingly, the v-erbB gene was deleted from AEV, and replaced with the complete or mutated human receptor cDNA (Fig.2 and 3). In addition, to monitor the contribution of v-erbA to transformation constructs with or without active erbA genes were generated.

Erythoblast transformation was obtained with all the human EGFr genes but constructs 4 and 1.4. This was achieved without addition of EGF to the culture medium in case of viruses also encoding v-erbA. Without v-erbA transformation was achieved with C and 6 only when ligand was added to the culture medium. Subsequently, we showed that erythroblasts transformed by C still required EPO for spontaneous differentiation, whereas cells transformed by construct 6 or v-erbB readily differentiated in the absence of EPO. This suggests that unlike v-erbB^{ES4}, the human EGFr does not replace the activity of the EPO receptor, possibly due to a more stringent control of mitogenic signal transmission.

Fibroblast transformation experiments showed that the human genes transform avian cells poorly in the absence of EGF in the culture medium, with increasing expression of transformation parameters as the EGFr molecule becomes more mutated as in constructs 1.6 and 1.4. EGF induces however a phenotypic transformation to cells expressing EGF receptors containing a ligand binding domain, as in C, 6 and 4, despite the finding that the latter cells express no high affinity binding sites.

The receptor proteins activated by ligand elicited stronger effects than those lacking binding domains, demonstrating that truncation of the ligand binding domain causes only a weak constitutive activation of the receptors. However, truncation of C-terminal regions also amplified the transforming and/or mitogenic effects. It is possible that full transformation activity, independent of ligand, requires the presence of not only N- and C-terminal deletions but also point mutations as in the v-erbB gene.

LITERATURE CITED

Beug H, Kahn P, Vennström B, Hayman MJ and Graf T (1985) How do retroviral oncogenes induce transformation in avian erythroid cells? Proc R Soc Lond B226:121-126

Beug H and Hayman MJ (1984) Temperature-sensitive mutants of avian erythroblastosis virus: surface expression of the erbB product correlates with transformation. Cell 36:963-972

Beug H, Hayman M, Raines M-B, Kung H-J and Vennström B (1986) RAV-1 induced erythroleukemic cells exhibit a weakly transformed phenotype in vitro, and release c-erbB containing retroviruses unable to transform fibroblasts. J Virol 57:1127-1138

Choi O, Trainor C, Graf T, Beug H and Engel JD (1986) A single amino acid substitution in v-erbB confers a thermolabile phenotype to ts167 avian erythroblastosis virus-transformed erythroid cells. Mol Cell Biol 6: 1751-1759

Damm K, Beug H, Graf T and Vennström B (1986) Cooperativity between the oncogenes of avian erythroblastosis virus (AEV): a mutant v-erbB gene defective in erythroblast transformation can be fully complemented by a highly active v-erbA gene. EMBO J 6:375-382

Downward J, Yarden Y, Mayes E, Scrace G, Totty N, Stockwell P, Ullrich A, Schlessinger J and Waterfield MD (1984) Close similarity of epidermal growth factor receptor and v-erbB oncogene protein sequences. Nature 307:521-527

Frykberg L, Palmieri S, Beug H, Graf T, Hayman MJ and Vennström B (1983) Transforming capacities of avian erythroblastosis virus mutants deleted in the erbA or erbB oncogenes. Cell 32:227-238

Fung YKT, Lewis WG, Kung HJ and Crittenden LB (1983) Activation of the cellular oncogene c-erbB by LTR insertion: molecular basis for induction of erythroblastosis by avian leukosis virus. Cell 33:357-368

Gandrillon O, Jurdic P, Benchaibi M, Xiao J-H, Ghysdael J and Samarut J (1987) Expression of the v-erbA oncogene in chicken embryo fibroblasts stimulates their proliferation in vitro and enhances tumor growth in vivo. Cell 49:687-697

Gullick W, Downward J, Waterfield M (1985) Antibodies to the autophosphorylation site of the epidermal growth factors receptor protein-kinase as probes of structure and function. EMBO J 4:2869-2877

Hentzen D, Renucci A, Le Guellec D, Benchaibi M, Jurdic P, Gandrillon O and Samarut J (1987) The chicken c-erbA proto-oncogene is preferentially expressed in erythrocytic cells during late stages of differentiation. Mol Cell Biol 7:2416-2424

Hughes S and Kosik E (1984) Mutagenesis of the region between env and src of RSV for the purpose of constructing helper-independent vectors. Virology 136:89-99

Jansson M, Beug H, Graf T, Gray C and Vennström B (1987) The viral erbA gene restores weakly transforming activities of mutant v-erbB genes in transformation of chicken fibroblasts and erythroblasts. Oncogene 1:167-173

Kahn P, Frykberg L, Brady C, Stanley IJ, Beug H, Vennström B and Graf T (1986a) V-erbA cooperates with sarcoma oncogenes in leukaemic cell transformation. Cell 45:349-356

Lax I, Kris R, Sasson I, Ullrich A, Hayman MJ, Beug H and Schlessinger J (1985) Activation of c-erbB in avian leukosis virus induced erythroblastosis leads to the expression of a truncated EGF-receptor kinase. EMBO J 4:3179-3182

Lax I, Johnsson A, Sap J, Ullrich A, Vennström B, Givol D and Schlessinger A (1988) Mol Cell Biol. submitted

Munoz A, Zenke M, Gehring U, Sap J, Beug H and Vennström B (1987) Characterization of the hormone-binding domain of the chicken c-erbA/thyroid hormone receptor protein. EMBO J in press

Privalski ML and Bishop JM Z (1984) Subcellular localization of the v-erbB protein, the product of a transforming gene of avian erythroblastosis virus. Virology 135:356-368

Sap J, Munoz A, Damm K, Ghysdael J, Leutz A, Beug H and Vennström B (1986) The c-erbA protein is a high affinity receptor for thyroid hormone. Nature 324:635-640

Scotting P, Vennström B, Jansen M, Graf T, Beug H and Hayman M (1987) Common site of mutation in the erbB gene of avian erythroblastosis virus mutants that are temperature sensitive for transformation. Oncogene Reserach 1:265-278

Sealy L, Privalsky ML, Moscovici G, Moscovici C and Bishop JM (1983) Site-specific mutagenesis of avian erythroblastosis virus: erbB is required for oncogenicity. Virology 130:155-178

Thompson C, Weinberger C, Lebo R, Evans R (1987) Identification of a novel thyroid hormone receptor in the mammalian nervous system. Science 237:1610–1614

Ullrich A, Coussens L, Hayflick JS, Dull TJ, Gray A, Tam AW, Lee J, Yarden Y, Libermann TA, Schlessinger J, Downward J, Mayes ELV, Whittle N, Waterfield MD and Seeburg PH (1984) Human epidermal growth factor receptor cDNA sequence and aberrant expression of the amplified gene in A431 epidermoid carcinoma cells. Nature 309:418–425

Vennström B and Bishop JM (1982) Isolation and characterization of chicken DNA homologous to the two putative oncogenes of avian erythroblastosis virus. Cell 28:135–143

Weinberger C, Thompson C, Ong E, Gruol D and Evans R (1986) The c–erbA gene endoces a thyroid hormone receptor. Nature 324:641–646

Zenke M, Kahn P, Disela C, Vennström B, Leutz A, Hayman M, Choi HR, Yew N, Engel J and Beug H. Arrest of erythroid differentiation by the v–erbA oncogene: Negative control of anion transporter gene expression. Cell, in press.

C-*FOS* AND POLYOMA MIDDLE T ONCOGENE EXPRESSION IN TRANSGENIC MICE AND EMBRYONAL STEM CELL CHIMAERAS

Erwin F. Wagner, R. Lindsay Williams and Ulrich Rüther

European Molecular Biology Laboratory (EMBL)
Meyerhofstraße 1
6900 Heidelberg
F.R.G.

It is widely accepted that oncogene products play an important role in the genetic control of cellular differentiation and development. Many efforts are directed towards understanding the function of the proto-oncogenes (c-*onc*, the cellular homologues of retroviral transforming genes, v-*onc*), which are often expressed in embryos and adult tissues in a remarkable tissue-, cell type- and stage specific form (for review see Kahn and Graf, 1986; Adamson, 1987). The role of c-*onc* genes in growth control is supported by the notion that the products of several oncogenes are either growth factors or growth-factor receptors (Bishop, 1987). A large number of important studies on oncogenes were performed in cell culture systems; however, they give only limited insights into oncogene function in the context of the whole organism. Our studies focus on investigating the role oncogenes have in mouse development as well as in the differentiation of multipotent stem cells. We are using transgenic mice and embryonal stem (ES) cells as powerful systems to test the possible role of oncogenes *in vivo* (Wagner and Stewart, 1986). Here we report our results on c-*fos* and polyoma middle T oncogene expression in transgenic mice and ES cell chimaeras.

I. c-*fos* expression in transgenic mice

We have chosen to study the function of the c-*fos* proto-oncogene in normal development, differentiation and growth control. The v-*fos* oncogene has been detected in two mouse sarcoma viruses (MSVs), the

NATO ASI Series, Vol. H26
Cell to Cell Signals in Mammalian Development
Edited by S. W. de Laat et al.

FBJ-MSV and the FBR-MSV. Although the two viruses contain structurally different forms of the *fos* gene, they both induce osteosarcoma-like tumours in newborn mice (Müller, 1986). The reason for this tumour type specificity is not known. The function of the c-*fos* gene has been the subject of intense investigation. Present evidence suggests that the c-*fos* gene product exerts different functions in complex biological processes such as differentiation and growth control (Verma, 1986). To study the importance of c-*fos* in development and to elucidate the mechanism(s) responsible for the specificity of *fos*-induced tumorigenesis, we have generated transgenic mice expressing deregulated c-*fos* genes in a variety of tissues.

MT-c-*fos* mice: Initially we introduced the murine c-*fos* gene under the control of the glucocorticoid- and heavy-metal-inducible human metallothionein promoter (MT) into fertilized mouse eggs. In seven independent mouse lines exogenous c-*fos* expression was undetectable. When transgenic mice were injected with cycloheximide, a protein synthesis inhibitor known to "superinduce" c-*fos* expression *in vitro*, presumably by preventing c-*fos* mRNA degradation, exogenous c-*fos* transcripts were found in most tissues analyzed (Rüther et al., 1987). This result suggests that the exogenous gene is transcribed in several tissues and that accumulation of c-*fos* mRNA - at least in some tissues - is prevented by post-transcriptional regulatory mechanisms. We therefore next used a DNA construct, which lacked the 3' mRNA destabilising sequence and replaced it with a long terminal repeat (LTR) from the FBJ-MSV virus.

MT-c-*fos*LTR mice: High levels of exogenous c-*fos* expression were detected in two independent mouse lines, in particular in pancreas, kidney, brain, heart, bone and muscle; induction with $CdCl_2$ increased RNA levels about 10-fold in these tissues. No constitutive or inducible c-*fos* RNA expression was detectable in the liver, spleen, thymus and bone marrow in mice of either strain. The c-*fos* expressing transgenic mice developed characteristic swellings on the leg bones as early as 2-3 weeks after birth (Rüther et al., 1987). A pathohistological examination revealed marked disturbances of bone remodelling, which is characterised by bone marrow fibrosis and greatly enhanced formation of new bone. The specific bone lesions did not markedly increase in size beyond the 4th week after birth; however, in older animals bone tumours, mainly chondrosarcomas, were observed (15% mostly males older than 9 months). Despite expression in other tissues, bone cells appeared to be particularly susceptable to *fos* induced disturbances in these animals. This target cell specificity is reminiscent of the osteosarcoma-like tumours induced by the v-*fos* transducing retroviruses FBJ-MSV and FBR-MSV.

To widen our studies, we next exchanged the promoter region from MT to MHC H2-K^b, which should lead to a broader expression pattern in transgenic mice.

H2-c-*fos*LTR mice: As expected, the exchange of the promoter region gave a) stable exogenous c-*fos* expression in almost all tissues and b) a much more pronounced phenotype with bone lesions occurring on almost all bones in the body (Figure 1). Only two founder animals were obtained and both failed to produce transgenic offspring. The histological examination revealed extensive bone alterations with numerous preneoplastic lesions particularly in the head, chest and pelvis area. These mice further support the previous findings and allow us to correlate the expression of c-*fos* with the observed phenotype specific for bone tissue.

Finally, we wanted to find out if the exchange of the promoter region can confer stability of c-*fos* expression in mice carrying H2-K^b c-*fos* sequences without the alteration on the 3' non-translated region of c-*fos*.

H2-c-*fos* mice: Five independent transgenic mouse lines were generated, which expressed the exogenous c-*fos* gene predominantly in hemopoietic organs such as spleen and thymus (Rüther et al., 1988). These mice have enlarged spleens and hyperplastic thymuses with an increased number of thymic epithelial cells. The exogenous c-*fos* expression specifically affects T-cell development in the thymus, thereby increasing the fraction of mature thymocytes. Results obtained with bone marrow radiation chimaeras suggest that the altered distribution of T-cell subsets is not a direct effect of c-*fos* expression within the T-cell lineage. It appears that c-*fos* specifically stimulates the proliferation of thymic epithelial cells, and may thus affect indirectly T-cell development. No changes in the

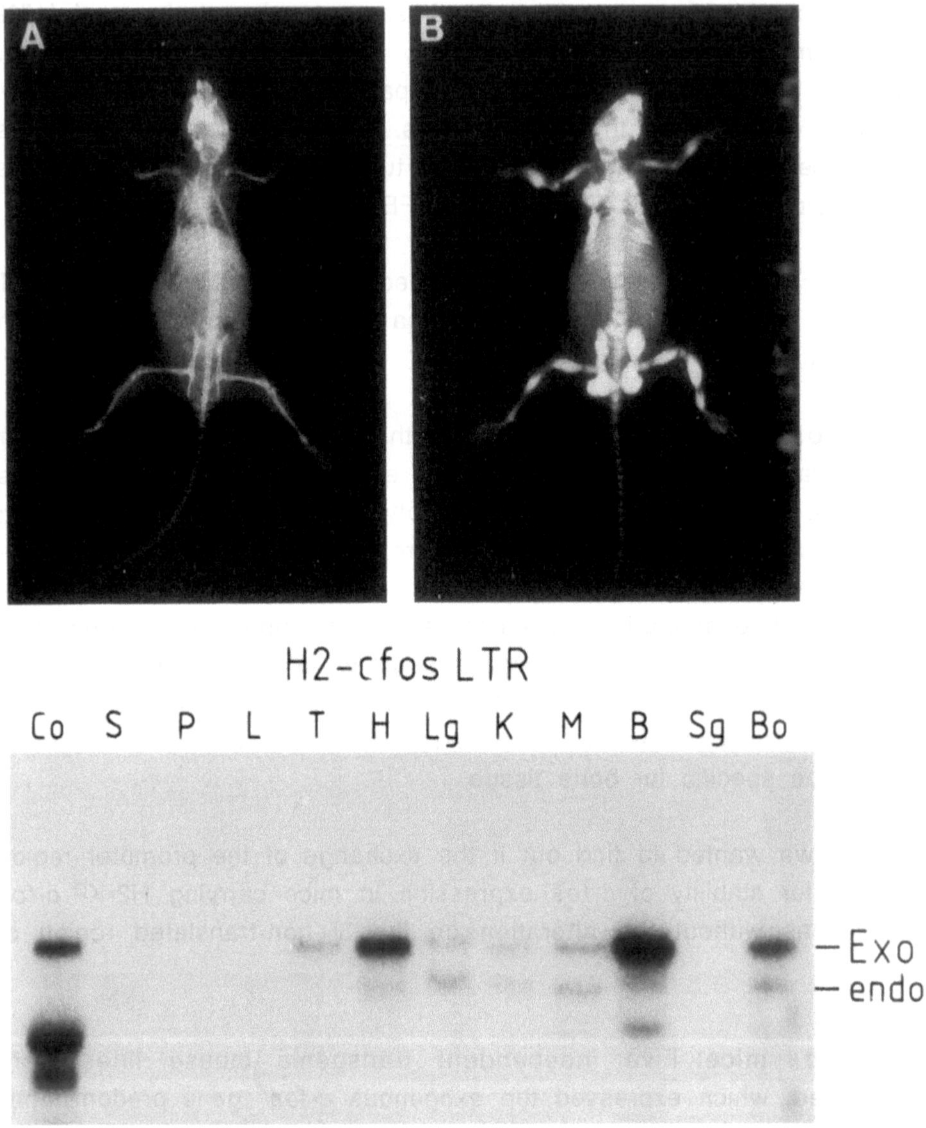

Figure 1: X-ray and expression analysis of a H2-c-*fos* LTR transgenic mouse. A. control mouse; B. transgenic mouse with drastic bone alterations. The same mouse was analyzed by Northern blotting, shown in the lower part of the figure. Exo, exogenous; endo, endogenous c-*fos* transcripts. Letters on top of the figure are: Co, control cell line; S, spleen; P, pancreas; L, liver; T, thymus; H, heart; Lg, lung; K, kidney; M, muscle; B, brain; Sg, salivary gland; Bo, bone.

proportion of hemopoietic cell lineages are seen in the spleen and these mice do not develop lymphoid malignancies. T- and B-cell function, however, is impaired and H2-c-*fos* mice are immune deficient (Rüther et al., 1988).

In summary, we have identified two distinct cell populations which are sensitive to c-*fos* action. In both instances *fos* appears to stimulate the proliferation of the target cells and future studies will investigate the mechanism by which c-*fos* expression perturbs the normal growth control pathways in these cells.

II. Polyoma middle T oncogene expression in transgenic mice and in ES cell chimaeras

The aim of these studies is to determine the role of the c-*src* gene in growth control by modulation of the pp60$^{c\text{-}src}$ tyrosine kinase activity during embryogenesis and in the adult animal (for review see Wagner and Müller, 1986). These experiments should identify stages of development or cell types where tyrosine kinase activity is crucial for growth control. One method for altering the activity of pp60$^{c\text{-}src}$ is to express the middle T antigen of polyoma virus which has been shown to complex with and modulate its activity *in vitro* (Courtneidge and Smith, 1983). More recently middle T antigen has also been shown to complex with another member of the *src* tyrosine kinase family pp62$^{c\text{-}yes}$ (Kornbluth et al., 1987). Thus the *in vivo* expression of middle T may also help in the identification and characterisation of the roles of other members of the *src* family.

Middle T expression in transgenic mice: Initial studies to investigate the consequence of middle T expression *in vivo* were carried out by DNA microinjection of the middle T gene under the control of the constitutive HSV thymidine kinase (TK) promoter and four transgenic lines were established. Mice from one of the lines developed abdominal tumours, diagnosed as bilateral adrenal corticocarcinomas, 3-8 weeks after birth. This indicates that the middle T antigen which is expressed in these tumours can act as a potent oncogene in these cells (Williams, R.L., Wagner, E.F. and

Courtneidge, S.A. in preparation). The middle T antigen was not expressed in any of the other tissues except for the brain (as early as mid-gestation) in two transgenic lines. These mice also developed cavernous haemangiomas in various tissues with latency periods of 2 weeks to 9 months most likely due to spontaneous activation of the TK promoter in the endothelial cells. Similar results have been observed by Bautch *et al.* (1987) when mice with a middle T transgene expressed from its own promoter developed haemangiomas with varying latency periods. These mice also express the middle T antigen in the testis with no observable phenotype (Bautch et al. 1987). Taken together these results demonstrate that certain tissues such as endothelial and adrenal cortex are sensitive to the expression of middle T antigen whilst cell types in the brain and testis apparently tolerate middle T expression with no observable phenotype.

Middle T expression via retroviral vectors in ES cell chimaeras: Retrovirus vectors were also used for the introduction of the middle T gene into mice since these vectors provide a flexible and efficient method for the introduction of new genetic material into a wide variety of cell types *in vitro* and *in vivo* (Wagner and Stewart, 1986). The retrovirus vector constructed (N-TKmT) expresses the selectable neomycin *(neo)* resistance gene from the Moloney Murine Leukemia virus long terminal repeat (LTR) and the middle T gene from an internal TK promoter (Williams et al., 1988). Initial experiments to test the action of middle T *in vivo* were carried out by injection of virus into newborn and adult mice. All the mice analysed developed haemangiomas in a variety of tissues including the lung, spleen, liver, fat pads, intestine and muscles which again indicated that middle T was acting as a potent oncogene on endothelial cells.

In order to assure middle T expression during early stages of development, ES cells were used as a powerful alternative gene transfer system. ES cells provide an advantage over other methods of *in vivo* gene transfer since they can be cultured and manipulated *in vitro* and then returned into the embryonic environment by injection into blastocysts (for review see Robertson, 1987). In the resulting chimaeric embryos they can contribute normally to all somatic tissues including the germ-line allowing the investigation of the effect of middle T on normal development.

The N-TKmT vector was introduced into ES cells by infection and *neo* resistant clones were selected with G418. To produce chimaeric embryos, ES cell clones expressing middle T were injected into blastocysts which were then transferred into pseudopregnant animals. The chimaeric embryos developed normally until mid-gestation when they were arrested and displayed the presence of multiple haemangiomas on the yolk sacs (Williams *et al.*, 1988). Embryos with levels of chimaerism of greater than 40% did not show any normal blood vessels while embryos with less than 40% chimaerism displayed varying degrees of vessel development interrupted by the presence of haemangiomas. Mice derived from one ES cell clone, which produced extremely low chimaeras, were not arrested during development and survived until 2-6 weeks after birth when they also developed multiple haemangiomas. The number of haemangiomas and the time of their onset being again dependent on the degree of chimaerism.

Histological analysis of the haemangiomas on yolk sacs of chimaeric embryos indicated that the tumours consisted of a thin sheath of endothelial cells enclosing primitive blood cells such as nucleated erythrocytes (Figure 2). The amount of tumour tissue present in the haemangiomas in infected animals or chimaeric embryos made molecular analysis difficult. Therefore cell lines were established from these tumours by selection for G418 resistant cells. The cell lines were identified as endothelial cells by the presence of the endothelial cell marker von Willebrand factor. The cells appeared to be immortal as they have been maintained in culture for over 14 months, however, after approximately 6 - 8 months the cells no longer package von Willebrand factor into discrete granules but display a general background fluorescence for this antigen.

Injection of the middle T expressing endothelial cells into mice resulted in the appearance of cavernous haemangiomas within 3 to 5 days. The extreme rapidity of the tumour induction even with cells maintained in culture for over 14 months suggested that there may be a host component to the tumour. Therefore, the endothelial cells were injected into a non-syngeneic mouse strain to analyse the origin of the tumour. Surprisingly the tumours did not reveal any marker contribution from the injected cells implying that they were inducing host endothelial cell proliferation. Further investigations were carried out by injecting cells treated with mitomycin C to prevent

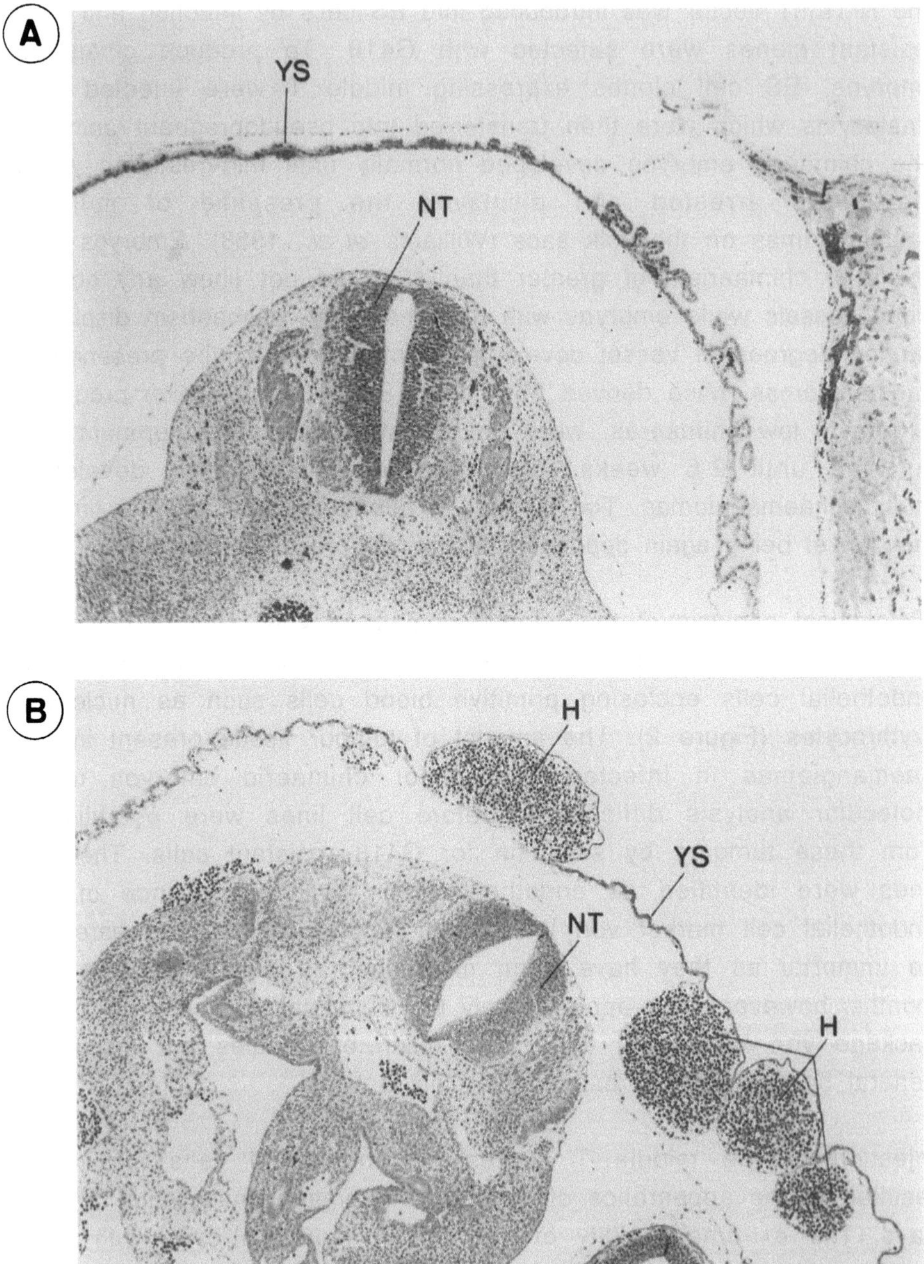

Figure 2: Histological analysis of a normal (A) and a middle T expressing chimaeric embryo at midgestation (day 12). In B four blood filled haemangiomas (H) along the yolk sac (YS) are clearly visible. NT, neural tube.

cell division. These cells transiently induced haemangiomas further suggesting that the endothelial cells are inducing host derived tumours. The mechanism by which these tumours are induced is currently under investigation.

These studies demonstrate that expression of middle T oncogene is sufficient to disrupt the growth control of two distinct cell populations (endothelial and adrenal cortex cells). Future experiments will determine whether the tyrosine kinase encoded by the *src* gene is a key component in the endothelial tumor induction. In combination with mutant forms of *src* and middle T genes we aim to dissect the molecular function of tyrosine kinases in cellular differentiation and development.

Acknowledgements: We would like to acknowledge the valuable contribution of our colleagues Sara Courtneidge, Christa Garber and Dymitri Komitowski in this work and we thank Sibylle Reichling for typing the manuscript.

References:

Adamson ED (1987) Oncogenes in development. Development 99: 449-471

Bautch VL, Toda S, Hassell JA, Hanahan D (1987) Endothelial cell tumors develop in transgenic mice carrying polyoma virus middle T oncogene. Cell 51: 529-538

Bishop JM (1987) The molecular genetics of cancer. Science 245: 305-311

Courtneidge SA, Smith AE (1983) Polyoma virus transforming protein associates with the product of the c-*src* cellular gene. Nature 303: 435-439

Kahn P, Graf T (eds) (1986) Oncogenes and growth control. Springer Verlag Berlin Heidelberg New York

Kornbluth S, Sudol M, Hanafusa H (1987) Association of the polyoma virus middle T antigen with c-*yes* protein. Nature 325, 171-173

Müller R (1986) Cellular and viral *fos* genes: structure, regulation of expression and biological properties of their encoded products. Biochim et Biophys Acta 823: 207-225

Robertson EJ (ed) (1987) Teratocarcinomas and embryonic stem cells: A practical approach. Oxford: IRL Press

Rüther U, Garber C, Komitowski D, Müller R, Wagner EF (1987) Deregulated *c-fos* expression interfers with normal bone development in transgenic mice. Nature 325: 412-416

Rüther U, Müller W, Sumida T, Tokuhisa T, Rajewsky K, Wagner EF (1988) *c-fos* expression interfers with thymus development in transgenic mice. Cell in press

Verma IM (1986) Proto-oncogene fos: a multi-faceted gene. Trends Genet 2: 93-96

Wagner EF, Müller R (1986) A role for proto-oncogenes in differentiation? In Oncogenes and Growth Control. P. Kahn, T. Graf (eds) Springer Verlag Heidelberg 18-26

Wagner EF, Stewart CL (1986) Integration and expression of genes introduced into mouse embryos. In Approaches to mammalian embryonic development. J Rossant, RA Peterson (eds) University Press, Cambridge 509-549

Williams RL, Courtneidge SA, Wagner EF (1988) Embryonic lethalities and endothelial tumors in chimaeric mice expressing polyoma virus middle T oncogene. Cell 52: 121-131

Expression of Proto-oncogene _int-1_ is Restricted to Specific Regions of the Developing Mouse Neural Tube

David G. Wilkinson, Juliet A. Bailes & Andrew P. McMahon
Laboratory of Developmental Biochemistry
National Institute for Medical Research
The Ridgeway
Mill Hill, London NW7 1AA, UK.

SUMMARY

The proto-oncogene _int-1_ is implicated in mouse mammary tumour formation and encodes a protein of unknown function that is probably secreted. _int-1_ has a highly restricted pattern of normal expression, limited to the mature testis and a transient expression during mouse development. To gain clues as to the normal function of _int-1_ we have examined the embryonic sites of _int-1_ RNA accumulation by _in situ_ hybridization. _int-1_ expression is restricted to a subset of cells in the developing central nervous system. In the spinal cord, these cells comprise the non-neuronal roof plate. Possible functions of _int-1_ in the organization of the mouse neural tube are presented in light of the recent finding that the _Drosophila_ homologue of _int-1_ is _wingless_, a segment polarity gene.

INTRODUCTION

Integration of mouse mammary tumour virus in the proximity of certain cellular genes has been implicated in the formation of mammary tumours (Nusse, 1986). One such region flanks a gene, _int-1_, which is not detectably expressed in the normal mammary gland but is activated by viral promoter elements in tumours (Nusse and Varmus, 1982). Since viral integration always occurs outside of the protein coding sequences, it appears that abberent expression of a normal _int-1_ protein may be involved in mammary tumour formation (Ooyen and Nusse, 1984). Indeed, _int-1_ has been shown to transform certain mammary epithelial cell lines (Brown et al., 1986; Rijsewijk et al., 1987a). _int-1_ encodes a 37 KD cysteine-rich, glycosylated polypeptide with a long hydrophobic leader, some of which appears to be cleaved off to generate the mature protein (Fung et al., 1985; Brown et al., 1987). Circumstantial evidence suggests that _int-1_ protein is on the secretory pathway (Papkoff et al., 1987; McMahon, unpublished observations).

Normal expression of _int-1_ has been demonstrated in post-meiotic spermatids and during mouse development (Jakobovits et al., 1986;

NATO ASI Series, Vol. H26
Cell to Cell Signals in Mammalian Development
Edited by S. W. de Laat et al.

Shackleford and Varmus, 1987). As a step towards understanding the role of _int-1_ in the development of the mouse, we have utilized _in situ_ hybridization to define the temporal and spatial pattern of _int-1_ RNA accumulation (Wilkinson et al., 1987). We find that _int-1_ expression is restricted to a specific group of cells in the developing neural tube. These results are discussed in light of the identity of _Drosophila_ _int-1_ with _wingless_, a gene involved in pattern formation in the _Drosophila_ embryo (Rijsewijk et al., 1987b).

RESULTS

int-1 transcripts are first detected at 8.5 days of development in the head folds comprising the anterior end of the neural plate (Fig. 1, 1-2) which is destined to form the brain. The spinal cord is formed from more posterior regions of the neural plate. Transverse sections reveal _int-1_ to be expressed throughout the anterior neural plate (Fig. 1, 3-4), but restricted to its lateral edges in more posterior regions (Fig. 1, 5-6). These lateral edges subsequently meet and fuse to form the dorsal wall of the neural tube, and indeed this was found to be the site of _int-1_ expression in the early spinal cord (Fig. 1, 7-8).

Analysis of serial transverse sections of the 10.5 day embryo revealed a more complex pattern of _int-1_ expression in the brain. To visualize this more clearly we used computer-aided reconstructions of the sites of _int-1_ expression in the neural tube of 10.5-14.5 day embryos (see Wilkinson et al., 1987). A similar pattern of _int-1_ expression was seen at both of these stages. The major site of expression was in the dorsal wall of the spinal cord and specific regions of the brain. In addition, _int-1_ transcripts were detected in certain ventral regions of the brain and in a loop of cells located at the midbrain-hindbrain junction (Wilkinson et al., 1987). Although this spatial pattern remains essentially constant between 10.5-14.5 days of development, dramatic morphological changes are occurring in the neural tube. The newly closed spinal cord consists of a one-cell-thick epithelium. As described above, _int-1_ is expressed in cells of the dorsal wall, termed the roof plate, and this is shown in more detail in Fig. 2, 1-2. Subsequently, the proliferation and migration of cells leads to a thickening of the spinal cord and a diminution of its central canal. Cells derived from the lateral walls differentiate into neuronal and glial cells arranged in a complex, bilaterally symmetric pattern, while the roof plate generates only glial cells. int-1 continues to be expressed in cells located at the dorsal midline during these morphogenetic events. At 14.5 days, int-1 transcripts are detected in a group of dorsal cells that are triangular in cross-section (Fig. 2, 3-4). Two days later, the int-1-expressing cells form a narrow strip extending from the dorsal surface towards the vestigial canal of the spinal cord (Fig. 2, 5-6).

DISCUSSION

int-1 is expressed in glial cells located at the dorsal midline during the period that the complex organisation of the spinal cord becomes established. Our ideas as to the possible role of int-1 are guided by two observations. Firstly, int-1 protein may be secreted (Papkoff et al., 1987; McMahon, unpublished observations) and, secondly, Drosophila int-1 is wingless, a segment polarity gene (Rijsewijk et al., 1987b). Both the non-cell autonomous phenotype of wingless mutants (Morata and Lawrence, 1977) and the narrow band of wingless RNA expression compared with the broader segmental domain deleted in mutants (Baker, 1987) suggest that wingless protein is secreted and acts on surrounding cells to establish normal pattern within each segment. In view of the substantial primary sequence homology between Drosophila wingless/int-1 and mouse int-1, it

---

Figure 1. Expression of int-1 in the neural plate and early neural tube. In situ hybridization was performed with antisense int-1 probe against 6μM mouse embryo sections. Following autoradiography and staining with toluidine blue, photomicrographs were taken under brightfield (left-hand panels) and darkfield (right-hand panels) illumination. 1,2: longitudinal section of 8.5 day embryo; 3-6: transverse sections through increasingly more posterior regions of the anterior neural plate of 8.5 day embryo; 7,8: transverse section through spinal cord of 10.5 day embryo. Note that due to curvature of the embryo, the neural tissue is transected twice in the transverse sections. Arrows indicate sites of int-1 expression. NP, neural plate; SC, spinal cord; HF, head folds. Fig.1. Copyright by Cell Press.

Figure 2. int-1 expression during morphogenesis of the spinal cord. In situ hybridization was performed against transverse sections through the spinal cord of embryos at various stages of development. Brightfield (left-hand panels) and darkfield (right-hand panels) photomicrographs are presented. 1,2: 10.5 days; 3,4: 14.5 days; 5,6: 16.5 days. Note that these photomicrographs are taken at different magnifications as there is a considerable increase in size of the spinal cord throughout this period. Arrows indicate sites of int-1 expression. Fig.2. 2.3, 2.4 Copyright by Cell Press.

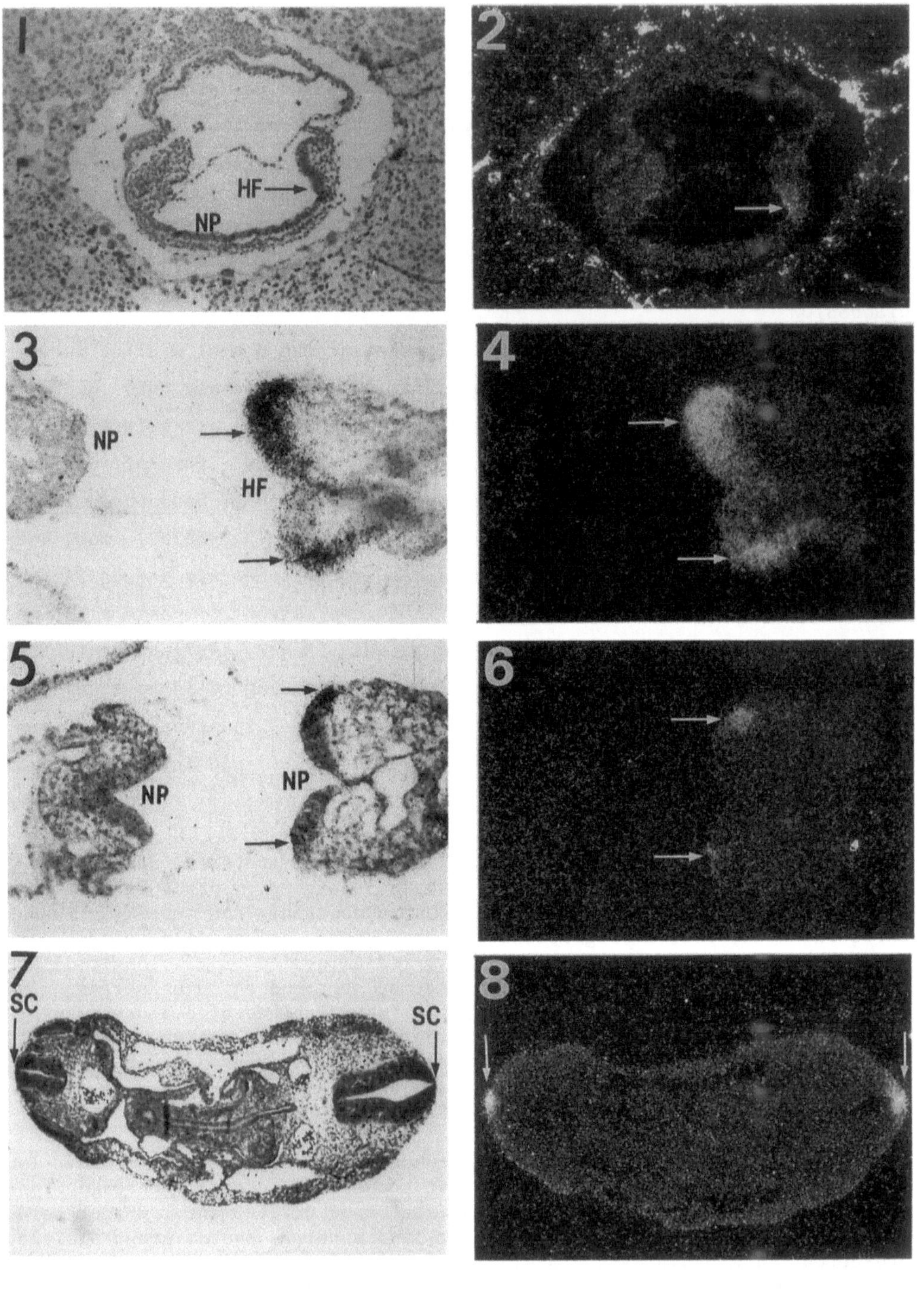
1
HF
NP
2
3
NP
HF
4
5
NP
NP
6
7
SC
SC
8

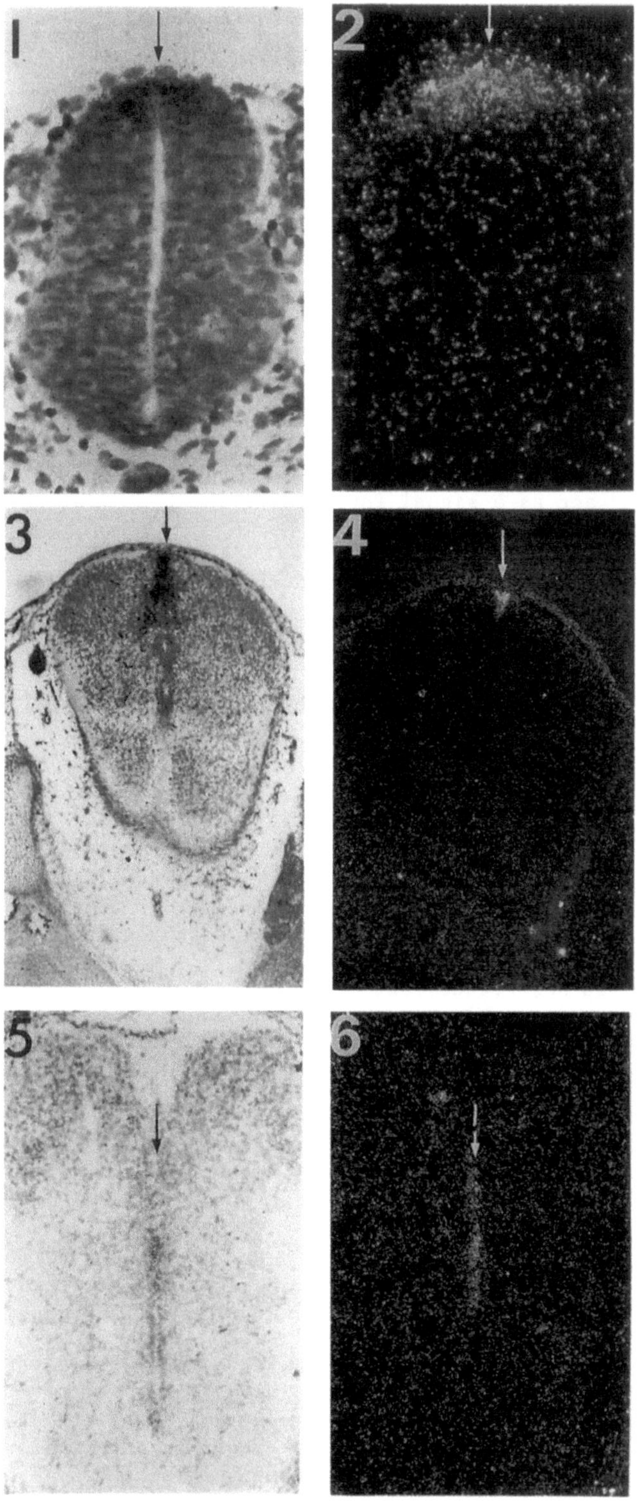
1
2
3
4
5
6

seems likely that the latter may also act as an intercellular signal. What process might such a signal be involved in?

The glial cells that express int-1 are believed to have a structural role in spinal cord morphogenesis, linking the receding central canal to the dorsal surface to form the dorsal median septum. In addition, these glial cells appear to form channels and it has been suggested that these are involved in the guidance of the axons of sensory neurons along the spinal cord (Altman and Bayer, 1984). These sensory neuron axons are grouped into "dorsal columns" that lie each side of, but do not cross, the dorsal midline. There is experimental evidence that the pathway of certain spinal cord neurons is guided by localized cues in lower vertebrates (Katz et al., 1980; Katz and Lasek, 1981). It seems possible that int-1 may act as a signal involved in the organization of the dorsal columns. Predictions of this idea are that these sensory neurons are responsive to int-1 protein and that an absence of int-1 will result in their abnormal migration. We are currently testing these predictions. Clearly, a better understanding of the role of int-1 in the mouse neural tube awaits thorough characterization of the biological effects of int-1 protein and identification of its putative receptor.

REFERENCES

Altman J, Bayer SA (1984) The development of the rat spinal cord. Springer-Verlag Berlin Heidelberg New York Tokyo

Baker NE (1987) Molecular cloning of sequences from wingless, a segment polarity gene in Drosophila: the spatial distribution of a transcript in embryos. EMBO J 6:1765-1773

Brown AMC, Wildin RS, Prendergast TJ, Varmus HE (1986) A retrovirus vector expressing the putative mammary oncogene int-1 causes partial transformation of a mammalian epithelial cell line. Cell 46:1001-1009

Brown AMC, Papkoff J, Fung YKT, Shackleford GM, Varmus HE (1987) Identification of protein products encoded by the proto-oncogene int-1. Mol Cell Biol 7:3971-3977

Fung YKT, Shackleford GM, Brown AMC, Sanders GS, Varmus HE (1985) Nucleotide sequence and expression in vitro of cDNA derived from mRNA of int-1, a provirally activated mouse mammary oncogene. Mol Cell Biol 5:3337-3344

Jakobovits A, Shackleford GM, Varmus HE, Martin GR (1986) Two proto-oncogenes implicated in mammary carcinogenesis, int-1 and int-2, are independently regulated during mouse development. Proc Natl Acad Sci. USA 83:7806-7810

Katz MJ, Lasek RJ (1981) Substrate pathways demonstrated by transplanted Mauthner axons. J Comp Neurol 195:627-641

Katz MJ, Lasek RJ, Nauta HJW (1980) Ontogeny of substrate pathways and the origin of the neural circuit pattern. Neuroscience 5:821-833

Morata G, Lawrence PA (1977) The development of wingless, a homeotic mutation of Drosophila. Develop Biol 56:227-240

Nusse R (1986) The activation of cellular oncogenes by retroviral insertion. Trends Genet 2:244-247

Nusse R, Varmus HE (1982) Many tumors induced by the mouse mammary tumour virus contain a provirus integrated in the same region of the host genome. Cell 31:99-209

Papkoff J, Brown AMC, Varmus HE (1987) The int-1 proto-oncogene products are glycoproteins that appear to enter the secretory pathway. Mol Cell Biol 7:3978-3984

Rijsewijk FAM, van Deemter L, Wagenaar E, Sonnenberg A, Nusse R (1987a). Transfection of the int-1 mammary oncogene in cuboidal RAC mammary cell line results in morphological transformation and tumorigenicity. EMBO J 6:127-131

Rijsewijk FAM, Schuermann M, Wagenaar E, Darren P, Weigel D, Nusse R (1987b). The *Drosophila* homolog of the mouse mammary oncogene int-1 is identical to the segment polarity gene *wingless*. Cell 50:649-657

Shackleford GM, Varmus HE (1987). Expression of the proto-oncogene int-1 is restricted to postmeiotic male germ cells and the neural tube of mid-gestational embryos. Cell 50:89-95

van Ooyen A, Nusse R (1984). Structure and nucleotide sequence of the putative mammary oncogene int-1; proviral insertions leave the protein-encoding domain intact. Cell 39:233-240

Wilkinson DG, Bailes JA, McMahon AP (1987). Expression of the proto-oncogene int-1 is restricted to specific neural cells in the developing mouse embryo. Cell 50:79-88

INDEX

NATO ASI Series H

NATO ASI Series H

Vol. 21: **Cellular and Molecular Basis of Synaptic Transmission.**
Edited by H. Zimmermann. 547 pages. 1988.

Vol. 22: **Neural Development and Regeneration. Cellular and Molecular Aspects.**
Edited by A. Gorio, J. R. Perez-Polo, J. de Vellis, and B. Haber. 711 pages. 1988.

Vol. 23: **The Semiotics of Cellular Communication in the Immune System.**
Edited by E. E. Sercarz, F. Celada, N. A. Mitchison, and T. Tada. 326 pages. 1988.

Vol. 24: **Bacteria, Complement and the Phagocytic Cell.**
Edited by F. C. Cabello und C. Pruzzo. 372 pages. 1988.

Vol. 25: **Nicotinic Acetylcholine Receptors in the Nervous System.**
Edited by F. Clementi, C. Gotti, and E. Sher. 424 pages. 1988.

Vol. 26: **Cell to Cell Signals in Mammalian Development.**
Edited by S. W. de Laat, J. G. Bluemink, and C. L. Mummery. 322 pages. 1989.

MIX
Papier aus verantwortungsvollen Quellen
Paper from responsible sources
FSC® C105338

If you have any concerns about our products,
you can contact us on
ProductSafety@springernature.com

In case Publisher is established outside the EU,
the EU authorized representative is:
Springer Nature Customer Service Center GmbH
Europaplatz 3, 69115 Heidelberg, Germany

Printed by Libri Plureos GmbH
in Hamburg, Germany